Girish Nath Jha (Ed.)

Sanskrit Computational Linguistics

4th International Symposium
New Delhi, India, December 10-12, 2010
Proceedings

Springer

Series Editors

Randy Goebel, University of Alberta, Edmonton, Canada
Jörg Siekmann, University of Saarland, Saarbrücken, Germany
Wolfgang Wahlster, DFKI and University of Saarland, Saarbrücken, Germany

Volume Editor

Girish Nath Jha
Special Center for Sanskrit Studies
Jawaharlal Nehru University
New Delhi 110067
India
E-mail: girishjha@gmail.com

Library of Congress Control Number: 2010939861

CR Subject Classification (1998): I.2.7, F.4.2, I.2, F.4.1, J.5, I.2.7

LNCS Sublibrary: SL 7 – Artificial Intelligence

ISSN 0302-9743

ISBN-10 3-642-17527-9 Springer Berlin Heidelberg New York
ISBN-13 978-3-642-17527-5 Springer Berlin Heidelberg New York

springer.com

© Springer-Verlag Berlin Heidelberg 2010
Printed in Germany

Typesetting: Camera-ready by author, data conversion by Scientific Publishing Services, Chennai, India
Printed on acid-free paper 06/3180

Preface

It is with great pleasure that I present the selected papers from the 4th International Sanskrit Computational Linguistics Symposium (4i-SCLS) to you. The event is being hosted by the Jawaharlal Nehru University, the premier research University of India during (December 10–12, 2010) at the Special Center for Sanskrit Studies. The first symposium was organized at INRIA, France, by Gérard Huet in 2007, the second at Brown University, USA, by Peter Scharf in 2008, and the third was organized at the University of Hyderabad by Amba Kulkarni in January 2009. The Sanskrit computational linguistics community is relatively young, and the foundation for this kind of formal meeting to exchange ideas between Sanskritists, linguists and computer scientists was given by Prof. Huet and Prof. Amba Kulkarni. My hearty thanks to both of them for bringing about this unification of scholars under one umbrella.

The 4i-SCLS saw excellent response from the scholars. We received more than 31 papers, which were examined by our Program Committee members to shortlist 18 papers for publication presented in this volume. The papers can be categorized under the following broad areas:

1. Phonology and speech technology
2. Morphology and shallow parsing
3. Syntax, semantics and parsing
4. Lexical resources, annotation and search
5. Machine translation and ambiguity resolution
6. Computer simulation of Aṣṭādhyāyī

Some of the notable misses were the speech corpora annotation, image processing techniques like OCR, and also the papers written in Sanskrit. Efforts will be made to ensure wider participation by scholars in future events.

Under category 1, the following papers were selected:

Wiebke Peterson and Silke Hamann (Unversity of Duesseldorf) in their paper "On the generalizability of Pāṇini's Pratyāhāra Technique to Other Languages" have experimented the Pāṇini's *pratyāhāra* technique to German with good results. That there are universal principles in Pāṇini's grammar has been a well-known fact among linguists, but actually applying them to languages other than Sanskrit has not been done a great deal. The present paper is therefore very important in this context. One of the more significant conclusions arrived at by the authors is that they have found this technique to be better than the feature-based techniques used by modern phonology. The paper "Building a Prototype Text to Speech for Sanskrit" by Baiju Mahananda, Raju C.M.S., Ramalinga Reddy Patil, Narayana Jha, Shrinivasa Varakhedi, and Kishore Prahallad presents a prototype TTS system for Sanskrit. The paper is significant because the group is the first to develop a prototype using a simplified phone set and the well-known *Festvox* engine.

The following three papers were presented under category 2:

"Rule Interaction, Blocking and Derivation in Pāṇini" by Rama Nath Sharma (University of Hawaii) is based on an earlier paper by the author, and is the title of the keynote talk to be delivered in the inaugural session. It is a detailed account of rule applications, interpretations and interplay, blocking in the complex process of obtaining a syntactic form from stems. Prof. Sharma's mastery of the structure and processes in the Aṣṭādhyāyī can be seen in the description of the blocking parameters and the derivation procedures. Peter Scharf (Brown University) in his paper titled "Rule-Blocking and Forward-Looking Conditions in the Computational Modeling of Pāṇinian Derivation" demonstrates the mechanism of rule blocking and looking ahead in Pāṇini through XML schema and regular expressions compiled into Perl code. The paper, besides discussing this in the context of derivations and other preprocessing implementations, gives an option to those Pāṇinian enthusiasts who can learn a simple tag like schema (not a programming language yet) and hope to convert Pāṇini's rules into XML. Then a routine done by Hyman can convert all that into a Perl program. The paper titled "Sanskrit Compound Processor" by Anil Kimar, Vipul Mittal and Amba Kulkarni (University of Hyderabad) presents a compound processor for Sanskrit which does segmentations and type identification based on manually annotated data. Sanskrit derivational morphology has not been worked on a great deal due to various reasons. The paper is significant in this context and also because it will help in identifying complex words not found in the dictionary.

Under category 3, the following papers were selected:

"Designing a constraint-based parser for Sanskrit" by Amba Kulkarni, Sheetal Pokar and Devendra Shukl (University of Hyderabad) is an attempt toward parsing simple one-verb sentences of Sanskrit prose with proper annotation and morphological information. The *śābdabodha* and *mīmāṃsā* parameters are brought in to reduce ambiguities and far-fetched possibilities. As Sanskrit is a highly inflected language with relatively free word-order, morphological analyses are generally assigned a greater role. However, the importance of parsers cannot be underestimated when one considers processing constituents higher than words—sentence and discourse. In the paper "Generative Graph Grammar of Neo-Vaiśeṣika Formal Ontology (NVFO)", Rajesh Tavva and Navjyoti Singh (IIIT Hyderabad) describe a formal model of *Vaiśeṣika* ontology for computing *śābdabodha* of sentences (among other things). The approach described is foundational and bottom–up and the proposed grammar based on graphs can correctly distinguish well-formed graphs from others. The proposed framework is generative and can help model ontologies effectively without the known disadvantages of the top–down approach. The paper "Headedness and Modification in Nyāya Morphosyntactic Analysis: Towards a Bracket-Parsing Model" by Malhar Kulkarni, Anuja Ajotikar, Tanuja Ajotikar, Dipesh Katira, Chinmay

Dharurkar and Chaitali Dangarikar (IIT Bombay) present an alternative model for sentence analysis using the *prakāratā* and *viśeṣyatā* concepts of the *Navya-Nyāya* school of philosophy. The paper is a comprehensive account of the alternative parsing model which the authors have tried on simple and complex sentences as well as discourses with anaphors and ellipsis.

Under category 4, the following six papers were presented:

In the paper "Citation Matching in Sanskrit Corpora Using Local Alignment", the authors Abhinandan S. Prasad and Shrisha Rao from IIIT Bangalore experiment with a citation matching technique used in bioinformatics for Sanskrit corpora. Of particular interest is their method of approximate matching of Mahābhārata citations in *Mahābhārata-tātparyanirṇaya* using the Smith-Waterman-Gotoh algorithm which is generally used in scientific documents. Diwakar Mani in "RDBMS-Based Lexical Resource for Indian Heritage: The Case of Mahābhārata (MB)" presents his research work done at Jawaharlal Nehru University. The indexing system developed as part of this research is done in Java/JSP and MS SQL server with a Unicode input/output mechanism. The system is live at http://sanskrit.jnu.ac.in/mb. The author has used BORI's authoritative version of the MB text and has created a database system which allows three kinds of searches – direct search, alphabetical search and the search by structure of MB. The search results are listed as index of words, details of which can be obtained by following the link. As an extension, not typical of indices, the system also links the results with other lexical resources available on the internet through simple http connections. The significance of the work rests in the fact that in future, interlinked database resources of heritage texts can be developed which will make classical indological research more attractive. In another paper in this category—"Evaluating Tagsets for Sanskrit" Madhav Gopal, Diwakar Mishra and Priyanka Singh from Jawaharlal Nehru University compare existing tagsets for Sanskrit: JNU Sanskrit tagset (JPOS), Sanskrit consortium tagset (CPOS), MSRI-Sanskrit tagset (IL-POST), IIIT Hyderabad tagset (ILMT POS) and CIIL Mysore tagset for their LDCIL project (LDCPOS). The main goal behind this enterprise is to check the suitability of existing tagsets for Sanskrit. Indian language computing groups have created numerous tagsets and therefore it is very important that these are evaluated based on standard procedures. In this context, the paper by Oliver Hellwig titled "Performance of a Lexical and POS Tagger for Sanskrit" is very relevant. The paper reports testing the performance of *SanskritTagger* on more than 34 Sanskrit texts and does estimations of the error rates in automatic processing of Sanskrit. They also note the problems like segmentation, homonymy, complexities of Sanskrit morphology and syntax which affect the accuracy of automatic processing. "Knowledge Structure in Amarakośa" by Sivaja S. Nair and Amba Kulkarni presents a structural description of Amarakośa bringing out complex relationships that lexical items may have across *kāṇḍas* and *vargas*. The paper seems to suggest how a wordnet based on Amarakośa structure can be evolved for Indian languages. In "Gloss in Sanskrit Wordnet", Malhar Kulkarni,

Iravati Kukarni, Chaitali Dangarikar and Pushpak Bhattacharya (IIT Mumbai) highlight the role of glosses in the wordnet and MRDs. Their paper also describes the experience in building Sanskrit wordnet which is based on the Hindi wordnet. The authors also report on the various wordnet activities on Indian languages based on the Hindi wordnet done at IIT Mumbai.

Under category 5, the following papers were presented:

"Vibhakti Divergence Between Sanskrit and Hindi" by Preeti Shukla, Devanand Shukl and Amba Kulkarni (University of Hyderabad) focuses on a specific kind of variance between a pair of languages under MT. Mapping Sanskrit (which is a synthetic language) *vibhaktis* to Hindi (which is post-positional in nature) is not easy due to many-to-one as well as one-to-many cases, and also because of many new features emerging in Hindi in the course of evolution. The authors examine this problem at the level of divergence and discuss each case in detail. In "Anaphora Resolution Algorithm for Sanskrit", Pravin Pralayankar and Sobha Lalitha Devi (AU-KBC Research Centre, Anna University) have continued some of the recent work in this area with a more definite mechanism for resolving anaphora in Sanskrit. They have also implemented a system and have given performance evaluation based on 200 anaphoric sentences. The work on Sanskrit syntax and discourse has been minimal, and therefore the paper becomes very important for future research and development in this direction. The paper by Brendan S. Gillon (McGill University)—"Linguistic Investigations into Ellipsis in Classical Sanskrit – provides a detailed account of the distribution of Ellipsis in Sanskrit, its classification and linguistic analysis. Since not many scholars have ventured in this area of Sanskrit linguistics, this work is certainly going to be a fundamental document for researchers working this area.

Under category 6, the following papers were presented:

"Asiddhatva Principle in Computational Model of Aṣṭādhyāyī" by Sridhar Subbanna (Rashtriya Sanskrit Vidyapeetha, Tirupati) and Srinivasa Varakhedi (Sanskrit Academy, Osmania University, Hyderabad) has evolved from a previous work by the same authors on the conflict resolution techniques in Aṣṭādhyāyī. In this paper they have tried to map the *asiddhatva* principle to the concept of 'filter' so that all the cases of *asiddha* operations can be accounted for. In "Modeling Aṣṭādhyāyī—An Approach Based on the Methodology of Ancillary Disciplines (Vedāṅga)", the author (Anand Mishra, University of Heidelberg) extends his previous works (as presented in the previous symposia at Paris and Hyderabad) and has adopted a new perspective. Since Pāṇini's grammar is situated in the tradition of Vedāṅga (four of which deal with language), the organization of the text may have certain inheritances from the tradition. The author has exploited this fact in terms of the content, structure and operations of methodological enquiry of Vedāṅga and Aṣṭādhyāyī and has put together a generalized model with broader coverage.

This volume would not have been possible without the active support of my Ph.D. students Diwakar Mishra (from the Special Center for Sanskrit Studies) and Ritesh Kumar (from the Center for Linguistics). Their LaTex skills and unending enthusiasm were critical in finalizing this volume.

I also thank all the members of the Program Committee who took pains to review the papers in time despite their hectic schedules. My special thanks to the members of the Steering Committee, Local Advisory Committee and Organizing Committee for facilitating this event. I also thank the sponsors (Dept. of I.T., C.I.I.L. Mysore, Microsoft Research India and others who may confirm support later) whose generous support was key to organizing this symposium. Finally, a special mention of Springer for accepting to publish this volume.

I hope the readers will like the volume and it will lead to the popularization of computational linguistics amongst Sanskrit scholars and students.

November 2010 Girish Nath Jha

Conference Organization

Conference Chair

Girish Nath Jha	Jawaharlal Nehru University, New Delhi, India

Steering Committee

Brendan S. Gillon	McGill University, Montreal, Quebec, Canada
Gerard Huet	INRIA, Rocquencourt, Paris, France
Girish Nath Jha	Jawaharlal Nehru University, New Delhi, India
Amba P. Kulkarni	University of Hyderabad, India
Malhar A. Kulkarni	I.I.T. Mumbai, India
Peter M. Scharf	Brown University, USA

Program Committee

Pushpak Bhattacharya	I.I.T. Mumbai
Brendan S. Gillon	McGill University, Montreal, Quebec, Canada
Oliver Hellwig	University of Berlin, Germany
Gerard Huet	INRIA, Rocquencourt, Paris, France
Girish Nath Jha	Jawaharlal Nehru University, New Delhi, India
Amba P. Kulkarni	University of Hyderabad, India
Malhar A. Kulkarni	I.I.T. Mumbai, India
Shoba L.	Anna University (AU-KBC), Chennai, India
K.V. Ramkrishnamacharyulu	Rashtriya Sanskrit Vidyapeeth, Tirupati, India
Peter M. Scharf	Brown Universiry, USA
Lalit Tripathi	Rashtriya Sanskrit Sansthan, Allahabad, India
Srinivasa Varakhedi	Sanskrit Academy, Osmania University, Hyderabad, India

Local Organizing Committee

Santosh Kumar Shukla	Jawaharlal Nehru University, India
Ramnath Jha	Jawaharlal Nehru University, India
Hariram Mishra	Jawaharlal Nehru University, India
Rajnish Mishra	Jawaharlal Nehru University, India

Local Advisory Committee

Sankar Basu	Jawaharlal Nehru University, India
Devendra Chaube	Jawaharlal Nehru University, India
Shashiprabha Kumar	Jawaharlal Nehru University, India
D. K. Lobiyal	Jawaharlal Nehru University, India
Varyam Singh	Jawaharlal Nehru University, India

Table of Contents

Rule Interaction, Blocking and Derivation in Pāṇini*

Rama Nath Sharma

University of Hawaii at Manoa
Honolulu, HI, U.S.A.
rama@hawaii.edu

1 *Sāmānya, Viśeṣa and Śeṣa*

Pāṇini's grammar is a class of rules formulated based on generalization abstracted from usage, so that the vast oceans of words could be properly understood. This class of rules will consist of general (*utsarga*) and their related particulars (*viśeṣa*). A general rule, since it is to be formulated with certain generalizations made about its scope of application, must yield to its related particulars which would necessarily require delineation of their own particular scope of application. A particular rule is thus formulated with particular properties relative to generalized properties. A general rule is supposed to pervade its scope of application in its entirety. It is in this sense that it is called *vyāpaka* (pervader). Since a particular rule is formulated with particular properties relative to the general, the scope of application of a particular must then be extracted from within the scope of its general counterpart. A related particular is called pervaded (*vyāpya*), since its scope of application is to be carved out from within the general scope of its corresponding *utsarga*, the pervader (*vyāpaka*). Rules whose application cannot be captured within the related class of general and particular have been classed as residual (*śeṣa*). A residual would thus fall outside of the applicational scope of its general and particular counterparts. For, it refers to a proposal which is yet to be made, close to its context (*upayuktād anyaḥ śeṣaḥ*).

Rules of Pāṇini's grammar have been formulated with brevity though certainly not at the expense of clarity. Consideration of brevity demands that rule be formulated with terms and symbols. Condition of explicitness requires proper interpretation of rules, so that they could reach their desired context of application. Two rules may share a single context of application (*sāvakāśa*), thereby presenting the difficulty of selecting concurrent (*yugapad*) application, application in turn (*paryāya*), application of one by blocking the application of the other (*bādhaka*), or application of one over the other based on it being left with no context of application (*niravakāśa*). For a rule should apply only once

* Based on Laguage, Grammar and Linguistics my paper published in the History of Indian Science, Philosophy and Culture in Indian Civilization (2010, Vol. VI.4: pp. 134-250).

G.N. Jha (Ed.): Sanskrit Computational Linguistics, LNCS 6465, pp. 1–20, 2010.

in realizing its desired goal of application (*lakṣye lakṣaṇaṃ sakṛd eva pravartate*). A rule which is formulated against a provision already made available to a rule, blocks the application of the other, obligatorily (*nitya; yena nāprāpte yo vidhir ārabhyate sa tasya bādhako bhavati*). A prior (*pūrva*) rule may block the application of a rule which is subsequent (*para*) in the order of enumeration. A rule whose condition of application is internal (*antaraṅga*) to the other conditioned externally (*bahiraṅga*), blocks the application of the other. Relative power of application of these rules is determined in this order: *pūrva* 'prior', *para* 'subsequent', *nitya* 'obligatory', *antaraṅga* 'externally conditioned' and *apavāda* 'exception, particular'.

Rules of grammar are likened to rain clouds which rain over land and water indiscriminately (*parjanyaval lakṣaṇāni bhavanti*). This indiscriminate raining of clouds over land and water with no consideration whatsoever of its beneficent, or malevolent, results, is not acceptable in the world of grammar. A *parjanya* enjoys a great deal of freedom when it comes to accomplishing its desired goal of raining (*varṣaṇa*). A rule of grammar, on the other hand, does not enjoy as much freedom when it comes to its desired application. It has to be carefully restrained or its scope of application must be delineated with care, so that it can accomplish its desired (*iṣṭa*) goal.

Proper delineation of context of rule application depends on interpretation of rules. Since rules are interpreted as a sentence, and also since rules also share single sentence relationship with rules placed at different places in grammar, interpretation of rules become very difficult. This implies that interpretation of rules is two-fold, (i) where rules which share single sentence relationship are placed together in one place, and (ii) where rules which share single sentence relationship are placed at different places.

It is generally agreed that proper understanding of a rule is possible only when one considers the exposition of the learned (PŚ1: *vyākhyānato viśeṣapratipattir na hi sandehād alakṣaṇaṃ*). It is in this sense that Pāṇinīyas accept the notion of *ekavākyatā* 'single sentence-ness' for interpretation of a rule. A rule in its physical context can be interpreted as a single sentence rather easily. To interpret a rule as a single sentence in its functional context is not so easy. For, that rule could not be properly understood unless the single sentence interpretation of its physical context is brought close to its functional context of application. Now consider two interpretive conventions (PŚ2-3) which both are complementary to each other, and which scholars consider operative in the Aṣṭādhyāyī:

1. *yathoddeśaṃ saṃjñā-paribhāṣham*
2. *kāryakālaṃ saṃjñā-paribhāṣham*

These two views about interpretation of technical names (*saṃjñā*), and interpretive conventions (*paribhāṣā*), are called *yathoddeśa* 'not going beyond the place where taught' and *kāryakāla* 'taught at the place where operative.' A student may like to understand a rule, for example 1.1.2 *aden guṇaḥ* , right at the place where it is taught in the grammar. Some other student may like to understand this rule where, for example, rule 6.1.87 *ād guṇaḥ* is taught with the use of the term *guṇa*. He understands 1.1.2 *adeṅ guṇaḥ* as a rule which assigns the name *guṇa* to *aT* and *eṄ*. He also learns that *eṄ*, per 1.1.71 *ādir antyena sahetā* with reference to inventory of sounds listed by the *Śiva-sūtra*, is an abbreviated symbol used with the denotation of *e* and *o*. An understanding of 1.1.70 *taparas tatkālasya* further explains that a specification made with *-t* (which shows here with the uppercase *T*) denotes duration (*kāla*) of 'that which is used with *T*.' He thus understands that *guṇa* is a name (*saṃjñā*) assigned to vowels a, e, and o. Rules 1.1.3 *iko guṇavṛddhī* , 1.1.49 *ṣaṣṭhī sthāneyogā* and 1.1.67 *tasmād ity uttarasya*, similarly teach him that 'a specification made with the genitive (*ṣaṣṭhī*) means 'in place of' (that which is specified with the genitive),' and 'a specification made with the ablative (*pañcamī*) means 'after' (that which is specified with the ablative).' In addition, a replacement ordered with express mention of the terms *guṇa* and *vṛiddhi* must come in place of a vowel denoted by the abbreviated symbol *iK* (1.1.3 *iko guṇavṛddhī*). It is at the time of understanding the function (*kārya*), and thereby application of rule 6.1.87 *ād guṇaḥ* , that his *saṃskāra* 'memory impressions' of understanding 1.1.2 *adeṅ guṇaḥ* , 1.1.3 *iko guṇavṛddhī* , 1.1.70 *taparas tatkālasya*, 1.1.71 *ādir antyena sahetā* and 1.1.67 *tasmād ity uttarasya* must be brought close to the context of 6.1.87 *ād guṇaḥ* , for its proper interpretation and application. The ablative of *āT* and the term *guṇa*, must serve as a mark (*liṅga*) for reconstruction of the full context of this rule by memory so that its application could be accomplished. Note that this rule is put in the domain (*adhikāra*) of 6.1.72 *saṃhitāyām*, where *saṃhitāyām* 7/1, *aci* 7/1 (6.1.77 *iko yaṇ aci*) and *ekaḥ* 1/1 *pūrvaparayoḥ* 6/1 (6.1.84 *ekaḥ pūrvaparayoḥ*) must be carried over to its context. The locative (*saptamī*) of *aci* 7/1 and the genitive dual (*ṣaṣṭhī dvivacana*) of *pūrvaparayoḥ 6/2* must also serve as marks for reconstructing memory impressions of rules 1.1.3 *iko guṇavṛddhī* , 1.1.49 *ṣaṣṭhī sthāneyogā* and 1.1.67 *tasmād ity uttarasya*. This, in turn, facilitates his full understanding of rule 6.1.87. I now quote single sentence interpretation of this rule from the Kāśikāvṛtti:

> *avarṇāt paro yo'c aci ca pūrvo yo avarṇaḥ tayoḥ pūrvaparayoḥ avarṇācoḥ sthāne eko guṇādeśo bhavati*
> 'a single replacement termed *guṇa* comes in place of both, the (*guṇa*) vowel which follows *a*, and the (*guṇa* vowel) a which precedes it, within the scope of *saṃhitā* 'close proximity between sounds'

Notice that this preceding is a single sentence interpretation of rule 6.1. 87 *ād guṇaḥ* , presented in the physical context of *saṃhitā* (6.1.72 *saṃhitāyām*). But this interpretation cannot be considered *yathoddeśa* 'place where *guṇa* is taught' interpretation. For, it involves bringing memory impressions of rules placed outside the domain of rule 6.1.87 *ād guṇaḥ*. It must then be viewed as the *kāryakāla* at time when the rule is functional interpretation whereby rules necessary for proper interpretation of a rule are brought close to its context from the outside of its domain. It is to be noted that no derivation is possible without taking help from the *kāryakāla* view. What triggers this interpretation in the derivational process? It is triggered by the mark of definitional terms, and interpretive conventions (*paribhāṣā*). We see that *guṇa* is the mark of the definitional term *guṇa* in rule 6.1.87 *ād guṇaḥ* . We also find *pañcamī* 'ablative', *ṣaṣṭhī* 'genitive' and *saptamī* 'locative' in words which are brought close to the context of rule 6.1.87 *ād guṇaḥ* , and trigger the reconstruction of *pañcamī*, *ṣaṣṭhī*, and *saptamī* from domains of rules placed outside this domain. This reconstruction is triggered by definitional and interpretational terms, and is accomplished by scanning rules where they have been used. This reconstruction entails bringing memory impressions of rules such as 1.1.2 *aden̄ guṇaḥ* , 1.1.3 *iko guṇavṛddhī* , 1..49 *ṣaṣṭhī sthāneyogā*, 1.1.67 *tasmād ity uttarasya* and *tasminn iti nirdiṣṭe pūrvasya*, to facilitate proper interpretation of rule 6.1.87 *ād guṇaḥ* . I have shown in the derivational section how derivations cannot be carried out without reference to antecedents of definitional and interpretational terms. I have used the word Referential Index (RI) for reconstruction of rule contexts triggered by marks of terms on hand.

2 General Blocking Considerations

Rule-interaction has been studied in the literature from the point of view of possibility of rule application (*prāpti-sambhava*). Once this possibility of application is ascertained in a context, we look at the context and give some serious consideration towards establishing the blocked-blocker relationship (*bādha-cintā*). Rules whose possibility of application is ascertained in a given context are called *sāvakāśa* (with valid scope of application). If two rules A and B become applicable in a given context Z, a concurrent application of both rules is impossible. There are two possibilities:

1. Apply rules A and B in turn (*paryāya*), or
2. Apply only one rule by blocking the application of the other
3. A rule (*kakṣaṇa*) applies only once (*sakṛd*) to reach its desired goal (*iṣṭa*)

Here are some generally established blocking considerations:

1. B blocks the application of A if B is a particular (*viśeṣa*) related to its general (*utsarga*) counterpart A;
2. The *ākaḍārīya* proposal of *vipratiṣedha* 'conflict among rules of equal strength' whereby B blocks the application of A if B is subsequent (*para*) in the order of enumeration (1.4.1 *ā kaḍārād ekā saṃjñā* and 1.4.2 *vipratiṣedhe paraṃ kāryam*).
3. The *ābhīya* proposal of rule suspension (*asiddhatva*) of 6.4.22 *asiddhavad atrābhāt*.
4. The *tripādī* proposal of rule (8.2.1 *pūrvatrāsiddham*) suspension (*asiddhatva*).
5. B blocks the application of A if B may be rendered without any scope of application (*niravakāśa*).
6. B blocks the application of A if B is obligatory (*nitya*).
7. B blocks A if B is internally conditioned (*antaraṅga*), as against A which is externally conditioned (*bahiraṅga*).

8. B blocks A if B is placed higher in relative hierarchy of rules in interaction. Consider the following interpretive convention of relative blocking from the *Paribhāṣenduśekhara* of Nāgeśa: *pūrvaparanityāntaraṅgāpavādānāmuttarottaraṃ balīyaḥ* 'prior (*pūrva*), subsequent (*para*), obligatory (*nitya*), internally conditioned (*antaraṅga*) and exception (*apavāda*), are considered more powerful in this order.'

These blocking considerations express closely relate to the *utsargāpavāda* dichotomy of (1) general (*sāmānya*), particular (*viśeṣa*) and residual (*śeṣa*). It is in this sense that Pāṇini's grammar, the Aṣṭādhyāyī, is considered a set of ordered rules capable of deriving the infinity of utterances of the Sanskrit language.

3 Derivational System of the *Aṣṭādhyāyī*

The goal of grammar is to derive correct words (*śabda-niṣpatti*) of the language. The tradition uses *śabda* 'word' in the general sense of an utterance which, given its basic purpose of serving as means of communication, can be referred to as a sentence. A word in its technical sense is accepted as fully derived, a pada which ends in two sets of affixes, namely *sUP* and *tiṄ* (1.4.14 *suptiṅantaṃ padam*). This yields two *pada* types, *subanta* 'that which ends in a *sUP*, and *tiṅanta* 'that which ends in a *tiṄ*.' Recall that this grammar imagines constituency of words in bases (*prakṛti*) and affixes (*pratyaya*), and as a result of operations (*kārya*) carried out with application of rules on input strings, yields a fully derived word. There are two types of bases, namely *prātipadika* 'nominal stem' and *dhātu* 'verb root':

1.2.45 - *arthavad adhātur apratyayaḥ prātipadikam* 'a non-root and non-affix word-form (*śabda-rūpa*) which carries a meaning (*arthavad*) is termed *prātipadika* nominal stem.'
1.2.46 - *kṛttaddhitasamāsāś ca* 'a word form which ends in affixes termed *kṛt* (3.1.93 *kṛd atiṅ*) and *taddhita* (4.1.76 *taddhitāḥ*), or which is termed *samāsa* (2.1.3 *prāk-kaḍārāt samāsaḥ*), is also termed *prātipadika* nominal stem.'
1.3.1 - *bhūvādayo dhātavaḥ* 'word-forms which are listed in groups headed by *bhū* 'to be, become,' and its likes, are termed *dhātu.*
3.1.33 - *sanādyantā dhātavaḥ* 'word-forms which end affixes saN, etc., are also termed *dhātu.*'

Note that nominal stems (*prātipadika*), and verb roots (*dhātu*) will be here considered as base-input (*prakṛti*). Affixes which are introduced after base- inputs are classified into three groups of (i) *ṅyāP* 'those which are formed with a common *Ṅī* (*ṄīP* /*ṄīṢ* /*ṄīN*) and *āP* (*CāP* /*ṬāP* /*ḌāP*), (ii) Vibhakti : 'nominal inflectional endings (*sUP*)' and 'verbal inflectional endings (*tiṄ*),' (iii) *kṛt* (3.1.93 *kṛd atiṅ*) and *taddhita* (4.1.76 *taddhitāḥ*). Now consider the following rules:

3.1.7 - *dhātoḥ karmaṇaḥ samānakartṛkād icchāyāṃ vā* (*san*) 'affix *saN* is, optionally, introduced after a verb root used with the denotation of object of *iṣ* 'to desire, wish,' provided its agent (*kartṛ*) is the same as the agent of *iṣ*.'

For example, *kartum icchati* ⟶ *cikīrṣati*, where *cikīrṣa* 'to wish to do' is a root derived with affix *saN*, introduced after the verbal root *ḌUkṛÑ* 'to do.' This derived base input can then gain access to the domain of 3.1.91 *dhātoḥ* whereby *cikīrṣa*, with introduction of *LAṬ* ⟶ *tiP*, and *ŚaP* would yield *cikīrṣati*, a verbal *pada*.

3.1.8 *supa ātmanaḥ kyac* 'affix *KyaC* is, optionally, introduced after a *pada* which ends in a *sUP*, and is used with the denotation of an object wished for one's own (*ātmanaḥ*).'

For example, *ātmanaḥ putraṃ icchati* ⟶ *putrīyati*, where *putrīya* 'to wish a son of one's own' is a root derived with affix *saN* introduced after *putra* + *am*, a *pada* ending in *sUP*.

3.1.91 - *dhātoḥ* 'after a verb root'
3.2.123 - *vartamāne laṭ* 'affix *LAṬ* is introduced after a verb root when action is denoted at the current time'
3.4.77 - *lasya* 'in place of that which is formed with a *LA*'
3.4.78 - *tiptasjhi-sipthastha-mivbasmas-tātāñjhathāsāthāṃ-dhvamiḍvahimahiṅ* 'the affixes *tiP, tas, jhi*, etc.' For example, *pac* + *LAṬ* ⟶ *tiP* ⟶ *pac* + *ŚaP* + *tiP* = *pacati*, a verbal *pada* which ends in a *tiṄ*.

4.1.1 - *ṅyāp-prātipadikāt* 'an affix is introduced after that which ends in an affix, formed with *Ṅī* and *āP* (*ṬāP/CāP/ḌāP; ṄīP/ṄīṢ/ṄīN*), or after that which is termed a *prātipadika* (nominal stem)'
4.1.2 - *svaujasmauṭchaṣṭābhyāṃbhisṅebhyāṃbhyasṅasibhyāṃbhyasṅasosāṃ-ṅyossup* 'an affix denoted by *sUP* is introduced after that which ends in an affix formed with *Ṅī* and *āP*, or after that which is termed a *prātipadika*'

Table 1. The *tiṄ* affixes

	ekavacana 'singular'	*dvivacana* 'dual'	*bahuvacana* 'plural'	
		parasmaipada 'active'		
prathama	*tiP*	*tas*	*jhi*	3rd person
madhyama	*siP*	*thas*	*tha*	2nd person
uttama	*miP*	*vas*	*mas*	1st person
		ātmanepada 'middle'		
prathama	*ta*	*ātām*	*jha*	3rd person
madhyama	*thās*	*āthām*	*dhvam*	2nd person
uttama	*iṬ*	*vahi*	*mahiṄ*	1st person

Table 2. The *sUP* affixes

	ekavacana 'singular'	*dvivacana* 'dual'	*bahuvacana* 'plural'	
prathamā	*sU*	*au*	*Jas*	'nominative'
dvitīyā	*am*	*auṬ*	*Śas*	'accusative'
tṛtīyā	*Ṭā*	*bhyām*	*bhis*	'instrumental'
caturthī	*Ṅe*	*bhyām*	*bhyas*	'dative'
pañcamī	*ṄasI*	*bhyām*	*bhyas*	'ablative'
ṣaṣṭhī	*Ṅas*	*os*	*ām*	'genitive'
saptamī	*Ṅi*	*os*	*suP*	'locative'

4.1.3 *striyām* 'an affix is introduced after a nominal stem when feminine is denoted'
4.1.4 *ajādyataḥ ṭāp* 'affix *ṬāP* is introduced after a nominal stem...
4.1.5 *ṛnnebhyo ṅīp* 'affix *ṄīP* is introduced after a nominal stem extracted from the group headed by *aja* 'goat,' or one which ends in -a'
4.1.76 *taddhitāḥ* 'affixes termed *taddhita...*'

A form which ends in the feminine suffixes *ṬāP, ṄīP* etc., is not assigned the name *prātipadika.* It, however, gains access to the domain of 4.1.1 *ṅyāp-prātipadikāt* as a base-input again since its suffix is marked with *ṅyāp*. This time it must opt for application of 4.1.2 *svaujasmauṭ...*, whereby, with introduction of *sUP*, it subsequently yields a *pada.*

4.1.82 *samarthānāṃ prathamād vā* 'a *taddhita* affix, namely *aṆ* (read with
4.1.83 *prāg dīvyato'aṇ*) is introduced after the first among syntactically related nominal *pada.*'

Consider *upagu* + *Ṅas*, a nominal *pada* which, with introduction of the *taddhita* affix *aṆ*, yields *aupagava*, a nominal stem. This nominal stem must now get access to the domain of 4.1.1 *ṅyāp-prātipadikāt*, whereby, with introduction of *sUP* (4.1.2 *svaujasmauṭ...*) it yields *aupagavaḥ* 'male descendant of *upagu*' , a nominal *pada.* Note however, that a base-input ending in a *taddhita* affix may gain access to the domain of 4.1.3 *striyām* for yielding a nominal base ending in a feminine affix. The output of this application will then go for access to the domain of 4.1.1 *ṅyāp-prātipadikāt* for application of 4.1.2 *svaujasmauṭ...* This clearly establishes the cyclic nature of these domain accesses.

It is clear from the definitions of nominal stems and roots that they each have two sets of forms, simple and derived. A nominal *pada* which ends in a *sUP* can also serve as a base input for introduction of *taddhita* affixes under the provision of rule 4.1.82 *samarthānāṃ prathamād vā.* This will still yield a nominal stem, and with the introduction of a *sUP* would yield a nominal *pada.* A *kṛt* affix can be introduced after a verbal base under the co-occurrence condition of a nominal *pada*, whereby a form which ends in a *kṛt* affix is termed a nominal stem. For example, *kumbhakāra* 'pot-maker' which is a derived nominal stem. This can access the domain of 4.1.1 *ṅyāp-prātipadikāt*, where with the introduction of *sU* of 4.1.2 *svaujasmauṭ*... it yields *kumbhakāraḥ*, a nominal *pada*. A *pada* which ends in a *sUP* can also be combined with another, also ending in a *sUP*, to yield a compound (2.1.3 *saha supā; samāsa*), again termed a nominal stem. Affixes *saN*, etc., can be introduced after a base-input termed *dhātu*, whereby a form which ends in them is again termed *dhātu*, a derived verb root. An affix of this class, for example *KyaC*, can also be introduced after a nominal *pada*, for example *putra + am*, under some co-occurrence condition to yield *putrīya*, a verb root. This verb root can then yield a verbal *pada*, for example *putrīyati* 'he wishes a son of his own.' Finally, affixes formed with the two shared elements *Ṅī* and *āP* are introduced after a nominal stem to yield yet another complex base-input. Note that this set of six base-inputs which denote feminine are not classed as a nominal stem, or verb root. They are, characterized as ending in affixes formed with *Ṅī* and *āP* instead.

This controlled description of a fully derived word by way of bases (*prakṛti*), affixes (*pratyaya*) and operations (*kārya*) may give the impression to many that the Aṣṭādhyāyī is a morphological grammar, even more so because a *pada* is its final output. It is true that Pāṇini accepts *pada* as the final output of his grammar. But his *pada* ends in a *sUP*, or in a *tiṄ* (1.4.14 *suptiṅantaṃ padam*). These *sUP* and *tiṄ* affixes which come as terminal elements in a *pada* are introduced after bases which carry meaning. The *sUP* and *tiṄ* affixes themselves express meanings both grammatical and notional. The meaning of a base is always notional. Grammatical and notional meanings are expressed by affixes, including *sUP* and *tiṄ* . For example, consider *nara + sU + odana + am* ⟶ *naraḥ odanam*, and *pac + LAṬ* ⟶ *pac + ŚaP+tiP = pac + a + ti = pacati = naraḥ odanaṃ pacati* 'the man cooks rice,' where affix *LAṬ* is introduced after *pac*, a verbal base-input. Rule *laḥ karmaṇi ca bhāve cākarmakebhyaḥ* states that a *LA*-affix is introduced after a transitive (*sakarmaka*) verb root when *kartṛ* 'agent' and *karman* 'object' are denoted. This same *LAṬ* can also be introduced after an intransitive (*akarmaka*) verb root when *kartṛ* and *bhāva* 'root-sense' are denoted. The *-ti* of *pacati* is selected as a replacement of *LAṬ* with the choice of expressing *kartṛ* 'agent.'

This choice of expressing *kartṛ* with *-ti* has consequence for selecting the nominal inflectional ending *sUP* after *nara* and *odana* which happen to be the named *kartṛ* and *karman* of the sentence. Now consider rule 2.3.1 *anabhihite* which makes a restrictive provision for selection of *sUP*. This rule would allow the selection of a *sUP* only when the denotatum of *sUP* is not already expressed. The *dvitīyā ekavacana* 'accusative' ending *-am* of *sUP* which is selected for introduction after *odana* expresses the named object (*karman*) of *pac*. This is made possible because the *-ti* of *pacati* has expressed *kartṛ*, and the *karman* is not already expressed (2.3.1 *anabhihite*; 2.3.2 *karmaṇi dvitīyā*). The choice to express *kartṛ* by 2.3.18 *kartṛ-karaṇayos tṛtīyā* was not allowed in case of *nara*, the named agent of the sentence, because the *-ti* of *pacati* has already expressed it. It is for this reason that the *prathamā-ekavacana* 'nominative singular' ending of 2.3.46 *prātipadikārtha...* had to be introduced after *nara* to express nothing but the sense of the nominal stem (*prātipadikārtha*). If a choice is made to express the *karman* with the verbal *pada* at the time of selection of *tiṄ*, a replacement of *LAṬ*, rule 2.3.1 *anabhihite* would not allow introduction of *-am* after *odana* to express the *karman.* The nominative singular ending *sU* would be then introduced after *odana* to express its nominal stem notion. The *kartṛ* of the sentence would then be expressed by 2.3.18 *kartṛ-karaṇayos tṛtīyā* . The sentence would then be *nareṇa odanaṃ pacyate*, a passive counterpart of the active *naraḥ odanaṃ pacati.* This clearly shows that a *pada* in Pāṇini expresses grammatical and notional relations in a sentence. Furthermore, its derivation must adhere to certain selectional restrictions which bear upon the derivation of a sentence as a whole. I shall subsequently show how derivation of complex bases is related closely to expression of *kartṛ*, *karman* and *bhāva*, thereby yielding complex sentential strings. This expression of *kartṛ, karman*, and *bhāva* directly relates to derivation of simple and complex sentences, and the derivational paths the strings follow.

Pāṇini derives words (*pada*; 1.4.14 *suptiṅantaṃ padam*) by first extracting them from sentences, and then by analyzing their constituency in terms of bases and affixes. A formal string of base(s) and affix(es) is then processed through a network of rule applications to yield a fully derived *pada.* Since a *pada* necessarily carries an impression of grammatical and notional relations, and such relations are part of sentential meaning, a *pada* shares dependency relationship with a sentence. A *pada* cannot be fully derived without reference to its syntactic context. It thus becomes necessary for grammar to first present an abstract syntactic representation of a sentence. Since action (*kriyā*) forms the central denotatum of a Sanskrit sentence, its abstract syntactic representation is presented as an action-complex with participants, namely the *kārakas.* Pāṇini sets up six *kāraka* categories, namely *apādāna, sampradāna, karaṇa, adhikaraṇa* and *kartṛ*, in this order.

Note that agent (*kartṛ*) is a *kāraka* which is independent of all other *kārakas* in the sense that the action (*kriyā*) must necessarily have it as a participant. All other *kārakas* are named by the action depending on its own nature. It is not necessary for all six *kārakas* to participate in accomplishment of all actions. It is also not necessary that a speaker may not look at the role of a given *kāraka* as that of some other. Consider the following sentences where *apādāna* could not be brought as a participant *kāraka*:

1. *bhṛtyaḥ* vane *kāṣṭhaiḥ sthālyāṃ odanaṃ pacati kumbhakārāya* 'the servant cooks rice in a pot with wood in the forest for the pot-maker'
2. *bhṛtyena vane kāṣṭhaiḥ sthālyāṃ odanaṃ pacyate kumbhakārāya* 'rice is cooked in a pot with wood in the forest for the pot-maker by the servant'

These two sentences are related in the sense that (1) is an action, and the other is its passive counterpart. Since Pāṇini derives them with a common string, I shall present the abstract conceptual structure (CS) of the active, namely the first sentence. A CS would constitute an action-complex where < action> will be central. Each CS will obligatorily have at least one participant, namely agent < *kartṛ* >, who will bring other participating *kārakas* into action, if the speaker so desires.

Let us now return to the CS of sentences (1):

> CS1: *bhṛtyaḥ vane kāṣṭhaiḥ sthālyām odanam pacati* 'x engages in accomplishing the action named y (softening) intended for z at a place named r in m, a receptacle'

This sentence has five participants in its action complex, where since the agent is expressed with the verbal *pada pacati.* In fact *pacati*, by itself, can be accepted as a single *pada* sentence with its third person singular agent already expressed. I shall next show the derivation of *pacati* 'he / she / it cooks,' along with the derivation of other *padas* of the sentence (1).

Participant(s) < *kartṛ* >
Action <*viklitti* 'softening' > (3) *pacati* 'he cooks'

step #1
pac 'to cook' <*dhātu / sakarmaka*>
1.3.1 *bhūvādayo dhātavaḥ*

step #2
pac $\longrightarrow$ *pac + LAṬ*
3.1.91 *dhātoḥ*
{3.1.1 *pratyayaḥ*
3.1.2 *paraś ca,*
3.1.3 *ādyudāttaś ca,*
3.1.4 *anudāttau suppitau*} <*anudātta*>
#1.1.67 *tasmād ity uttarasya* <*pañcamī* >
3.2.123 *vartamāne laṭ* <*LAṬ* >
'affix *LAṬ* is introduced after a verb root when action is denoted at the current time'
3.4.69 *laḥ karmaṇi ca bhāve cākarmakebhyaḥ* <*kartṛ* >

> 'a *LA*-affix is introduced after a transitive (*sakarmaka*) verb root when *kartṛ* 'agent' and *karman* 'object' are denoted; it is introduced after an intransitive (*akarmaka*) verb root when *kartṛ* 'agent' and '*bhāva*' are denoted'

$\longrightarrow$ *pac + LAṬ*
<*dhātu / sakarmaka / pratyaya / anudātta /vartamāna / kartṛ/ LAṬ* >

step #3
pac + (*LAṬ* $\longrightarrow$ *tiP*) = *pac + tiP*
3.4.77 *lasya* 'in place of that which is formed with *LA (LAṬ)*'
#1.1.49 *ṣaṣṭhī sthāneyogā* <*ṣaṣṭhī* >
3.4.78 *tip-tas-jhi-sip-thas-tha-mip-vas-mas-ta-ātām-jha-thās-āthām- dhvam-iḍ-vahi-mahiṅ*'
$\longrightarrow$ *pac + tiP* <*tiṄ* >

tiṄ-selection
1.4.99 *laḥ parasmaipadam 'a LA-replacement is termed parasmaipada'*
<*tiṄ/ parasmaipada*>
1.4.100 *taṅānāv ātmanepadam* 'replacements of *LA* denoted by *taṄ*, and also *āna*, are termed *ātmanepada*'
<*tiṄ-ātmanepada*>
1.4.102 *tāny ekavacana-dvivacana-bahuvacanāny ekaśaḥ*
'elements of triads of *tiṄ* are termed *ekavacana, dvivacana* and *bahuvacana*, one after the other'
<*ekavacana*>
1.4.104 *vibhaktiś ca* <*vibhakti*>
'triads of *sUP*, and *tiṄ*, are termed *vibhakti*'
1.4.107 *tiṅas trīṇi trīṇi prathamamadhyamottamāḥ* 'each triad of *tiṄ* is termed *prathama, madhyama* and *uttama*'
<*prathama...*>
1.4.108 *śeṣe prathamaḥ* 'a *prathama* 'the first triplet of verbal ending' is used when the remainder, i.e., tad 'third personal pronominal,' whether explicitly stated or implicitly assumed, shares co-referential relation with it.'
step #4
<*prathama*>
pac + tiP ⟶ *pac + ŚaP + tiP* <*ŚaP*>
3.4.113 *tiṅśit sārvadhātukam* 'that which is a *tiṄ*, or is marked with *Ś* as an *it*, is termed *sārvadhātuaka*'
<*sārvadhātuka*>
3.1.67 *sārvadhātuke yak* '*yaK* is introduced after a verb root when an affix termed *sārvadhātuka* follows'
#1.1.66 *tasminn iti nirdiṣṭe pūrvasya*
<*sārvadhātuka*>
3.1.68 *kartari śap* 'affix *ŚaP* is introduced after a verb root when a *sārvadhātuka* with the denotation of *kartṛ* follows' <*sārvadhātuka*>
⟶ *tiṄ*-selection
⟶ *pac + (Ś ⟶ φ) a (P ⟶ φ) + ti (P ⟶ φ)*
step #5
1.4.14 *suptiṅantaṃ padam*
pac +a + ti ⟶ *pacati 'he /she / it cooks'* <*pada*> Referential Index
<*dhātu / sakarmaka / pratyaya / anudātta /vartamāna / kartṛ/ LAṬ/ tiṄ / prathama / ekavacana / pada*>

Let us now return to the derivation of (3) *pacati*, our base-input *pac* activates the grammatical device and is assigned the term *dhtu* by the rule 1.3.1 *bhūvādayo dhātavaḥ* of the Controlling Domain (CD; *adhyāya* one). It is then sent for scanning domain headings in the Obligatory Domain (OD; *adhyāya* three through five) of the grammar for possible rule application. It locates the domain of 3.1.91 *dhātoḥ* for possible rule application because *dhātoḥ* is the heading (*adhikāra*) of the domain and it also contains the term *dhātu* with which the base input is identified. Term assignment (Naming a term) thus guides a base-input in locating domain of possible rule application. Further scanning of this domain, especially in view of its CS marker of <*vartamāna*> 'current time,' facilitates application of rule 3.2.123 *vartamāne laṭ*. This rule must be interpreted with the obligatory *anuvṛtti* of 3.1.1 *pratyayaḥ*, 3.1.2 *paraś ca*, 3.1.3 *ādyudāttaś ca* and 3.1.4 *anudāttau suppitau*. For these form the rule-context of the larger domain of which 3.1.91 *dhātoḥ* is an interior domain. This is how 3.1.123 *vartamāne laṭ* yields the meaning, 'affix (*pratyaya*) *laṭ* is introduced after the transitive (*sakarmaka*) verb root *pac* when action is accomplished at the current time (*vartamāna*); it is also marked with *udātta* at the beginning (3.1.3 *ādyudāttaś ca*). Since rule 3.1.91 *dhātoḥ* is marked with *pañcamī* 'ablative,' rule 1.1.67 *tasmād ity uttarasya* becomes operative. This assures that affix *laṭ* is introduced not just after (3.1.2 *paraś ca*) *pac* , but 'immediately after' *pac*. Our string *pac* + *LAṬ* must now scan the domain looking for application of a rule guided by introduction of the term *LAṬ*. We find rule 3.4.69 *laḥ karmaṇi ca bhāve cākarmakebhyaḥ* whereby we learn that a *LA*-affix is introduced after an intransitive verb root when *kartṛ* and *karman* are to be denoted. Our string *pac* + *LAṬ*, where *pac* is marked with <*sakarmaka*> and *LAṬ* is marked with <*pratyaya*, initial *udātta* (3.1.1-3.13), + *kartṛ*, + *karman* (3.4.69) >, now moves to the application of rules 3.4.77 *lasya* and 3.4.78 *tip-tas-jhi-sip-thas-tha-mip-vas- mas...* where it is faced with the problem of selecting one *tiṄ*-element out of eighteen. Recourse must now be taken to reconstruction of referential index triggered by the terms *LAṬ* and *tiṄ* whereby rules 1.4.99, 1.4.100, 1.4.102, 1.4.104, 1.4.107 and 1.4.108 are brought close to application of rule 43.4..77 and 3.4.78. This is how we select *tiP*, an active (*parasmaipada*) third personal (*prathama*) singular (*ekavacana*) ending (*vibhakti*) with the denotation of *kartṛ* 'agent.' Two things must be noted here: (i) selection of *-ti* is made on the basis of choice expression of *kartṛ* 'agent' with reference to 3.4.69 *laḥ karmaṇi ca bhāve cākarmakebhyaḥ*; (ii) the *anudātta* accent assigned to *LAṬ* in view of 3.1.2 *ādyudāttaś ca* must be replaced with the *anudātta* of 3.1.4 *anudāttau suppitau*, based on P as an *it* of *tiP*. The following is the summary representation of *tiṄ*-selection:

1.4.99 *laḥ parasmaipadam* <*parasmaipada*>
1.4.100 *taṅānāv ātmanepadam* < *ātmanepada*>
1.4.101 *tiṅastrīṇitrīṇiprathamamadhya...* <*prathama*>
1.4.102 *tāny ekavacana-dvivacana...* <*ekavacana*>
1.4.104 *vibhaktiś ca* <*vibhakti*>
1.4.108 *śeṣe prathamaḥ* <*prathama*>
1.4.22 *dvyekayor dvivacanaikavacane* <*ekavacana*>

The *-ti* of our string *pac* + *LAṬ* $\longrightarrow$ *ti* (*P* $\longrightarrow \phi$) = *pac* + *ti* is now assigned the term *sārvadhātuka* by rule 3.4.113 *tiṅśit sārvadhātukam*. This triggers the reconstruction of the term <*sārvadhātuka*> which, via reverse scanning, leads to application of rule 3.1.68 *kartari ŚaP* read with 3.1.67 *sārvadhātuke yak*. This rule introduces affix *ŚaP* after the verb root *pac* under the right condition of a *sārvadhātuka* affix, namely *-ti*, used with the denotation of *kartṛ* 'agent.' Notice that rules 3.4.78 *tiptasjhisipthastha..* which introduces *-ti* is in the fourth quarter of *adhyāya* three. Rule 3.1.68 *kartari śap* which introduces *ŚaP* is placed in the first quarter of *adhyāya* three, even before 3.1.91 *dhātoḥ* within whose domain 3.2.123 *vartamāne laṭ* introduced affix *LAṬ*. The selection of *-ti*, via reconstruction of *LAṬ* and *tiṄ*, not only helps the selection of *-ti* but also facilitates application of 3.1.68 *kartari śap*, by way of *-ti* as right condition of application. This application is made possible by reconstruction of the term <*sārvadhātuka*>. Our string $pac + (\acute{S} \longrightarrow \phi)a(P \longrightarrow \phi) + ti = pa + a + ti$ can exit this domain, and with assignment of new term in the Controlling Domain (CD), must gain access to domains of further rule application. A summary representation of terms which facilitated this derivation can be made in the form of a string with rule numbers as follows:

<*dhātu / sakarmaka (1.3.1) / pratyaya (3.1.1) / anudātta (3.1.4) / vartamāna (3.2.123) /kartṛ (3.4.68) / LAṬ (3.2.123) / tiṄ (3.4.78) / parasmaipada (1.4.99) / prathama (1.4.101) / vibhakti (1.4.104) / ekavacana (1.4.12) / pada (1.4.14)*>

(4) *bhṛtyaḥ* 'servant'

Participant: <*kartṛ* >

Action *pac* 'to cook' <*viklitti/dhātu/sakarmaka /+ kartṛ* >

bhṛtyaḥ 'servant'

step #1

bhṛtya <*prātipadika*>

1.2.45 *arthavad adhātur apratyayaḥ prātipadikam* 'that which is meaningful (*arthavat*), non-root (*adhātuḥ*) and non-affix (*apratyayaḥ*) is termed *prātipadika*'

1.4.53 *svatantraḥ karttā*; '*kartṛ* is independent (*svatantra*)'

⟶ *bhṛtya* 'servant' <*prātipadika; kartṛ* >

step #2

bhṛtya ⟶ *bhṛtya + sU*

4.1.1 *ṅyāp-prātipadikāt* 'an affix occurs after that which is marked with *ṅyāp*, or else is a nominal stem'

4.1.2 *su-au-jas-am-auṭ-śas-ṭā-bhyām-bhis-ṅe-bhyām-bhyas-ṅasi-bhyām- bhyas-ṅas-os-ām-ṅi-os-sup* 'the nominal endings *sUP*'

⟶ *bhṛtya + sU*

#*sUP*-selection

2.3.1 *anabhihite* 'when not already expressed'

2.3.64 *prātipadikārthaliṅgaparimāṇavacanamātre prathamā* 'prathamā 'nominative' is used when sense of the nominal stem (*prātipadikārtha*), gender (*liṅga*) and number (*vacana*), alone, is to be expressed'

⟶*bhṛtya +sU* <*prātipadika/kartṛ/sUP/ sU/ ekavacana*>

1.4.22 *dvyekayor dvivacanaikavacane* '*dvivacana* 'dual' and *ekavacana* 'singular' occur when 'two-ness, duality' and 'one-ness, singularity' is to be denoted'

1.4.100 *...trīṇi trīṇi...* 'each triad...' <*ekavacana*>

1.4.102 *tāny ekavacana-dvivacana-bahuvacanāny ekaśaḥ* 'elements of triads of *tiṄ* are termed *ekavacana, dvivacana* and *bahuvacana*, one after the other'

<*ekavacana*>

1.4.103 *supaḥ* 'elements of triads of *sUP* are termed *ekavacana*, *dvivacana* and *bahuvacana*, one after the other' <*sUP*>

1.4.104 *vibhaktiś ca* 'triads of *sUP*, and *tiṄ*, are termed *vibhakti*'

step #3 <*vibhakti*>

bhṛtya + *s*(*U* ⟶ *ϕ*) = *bhṛtya* + *s*

1.4.13 *yasmāt pratyayavidhis tadādi...* <*aṅga*>

6.4.1 *aṅgasya* (no rule application in the *aṅga* domain)

1.4.14 *suptiṅantaṃ padam* <*pada*>

1.4.110 *virāmo' vasānam* <*avasāna*>

1.2.41 *apṛkta ekāl pratyayaḥ* 'an affix formed with a single sound segment is termed *apṛkta*' <*apṛkta*>

step #4

8.1.16 *padasya* 'of that which is a *pada*' <*pada*>

8.2.66 *sasajuṣo ruḥ* ⟶ *bhṛtya* + (*s*⟶*r* (*Ũ* ⟶ *ϕ*) ⟶ *bhṛtya* + *r*

8.3.15 *kharavasānayor visarjanīyaḥ* <*avasāna*>

⟶ *bhṛtya* + (*r*⟶ *ḥ*) = *bhṛtyaḥ* 'servant' <*pada*>

Referential Index

<*prātipadika* / *kartṛ*/ *vibhakti* / *prathamā*/ *sU* / *ekavacana* / *pada* / *avasāna* / *apṛkta*>

The derivation of *pada* (4) *bhṛtyaḥ* 'servant' begins with the base-input *bhṛtya* which is assigned the term *prātipadika* by rule 1.2.46 *arthavad adhātur apratyayaḥ...* of the Controlling Domain (CD). Our string *bhṛtya* <*prātipadika*> is now sent to the Obligatory Domain (OD) for locating an interior domain where possibility of rule application is indicated by the term *prātipadika*. The interior domain of 4.1.1 *ṅyāp-prātipadika* is selected because the rule is formed with the term *prātipadika* in it. The application of rule 4.1.2 *svaujasmauṭchaṣṭābhyām...* poses the problem of selecting one *sUP* affix out of twenty-one. Similar to *tiṄ*-selection of (1) *pacati*, recourse must be taken to reconstruct the referential index of *sUP*. The selection of the nominative (*prathamā*) singular (*ekavacana*) ending (*vibhakti*) will be made by bringing the following rules close to its context via reconstruction of referential index of *sUP*:

1.4.102 *tāny ekavacana-dvivacana-bahuvacanāny ekaśaḥ* 'elements of triads of *tiṄ* are termed *ekavacana, dvivacana* and *bahuvacana*, one after the other' <*ekavacana*>

1.4.103 *supaḥ* 'elements of triads of *sUP* are termed *ekavacana*, *dvivacana* and *bahuvacana*, one after the other'

1.4.104 *vibhaktiś ca* 'triads of *sUP*, and *tiṄ*, are termed *vibhakti*'

1.4.22 *dvyekayor dvivacanaikavacane* '*ekavacana* 'singular' and *dvivacana* 'dual' are used when singularity and duality is denoted'

2.3.1 *anabhihite* 'when not expressed otherwise'

2.3.46 *prātipadikārthaliṅgaparimāṇavacanamātre prathamā* '*prathamā* is introduced when nothing but the sense of the nominal stem (*prātipadikārtha*), gender (*liṅga*), measure (*parimāṇa*) and number (*vacana*) is denoted'

Note here that this selection of -*sU* is made under the restrictive provision of 2.3.1 *anabhihite.* This rule can allow the selection of a *sUP* if its denotaum is not already expressed otherwise. Since *pacati* has already expressed the named *kartṛ* of the sentence, *bhṛtya* must now be introduced with the nominative singular -*sU* to express nothing but its own sense (*prātipadikārtha*), per rule 2.3.46 *'prātipadikārthaliṅgaparimāṇavacanamātre prathamā.'* Our string *bhṛtya* +*s* $(U \longrightarrow \phi)$ is now sent to the Controlling domain where it is assigned the term *pada* (1.4.14 *suptiṅantaṃ padam*). This term then facilitates scanning of domain headings beyond *adhyāya* five (Obligatory Domain), and thereby access to the domain of rules headed by 8.1.16 *padasya.* Rule 8.2.66 *sasajuṣo ruḥ* then replaces the -*s* with *r* $(\tilde{U} \longrightarrow \phi)$, thereby yielding *bhṛtya + r.* This -*r* is then replaced with *visarjanīya* of rule 8.3.15 *kharavasānayor visarjanīyaḥ* . Note however that rule 8.3.15 turns the -*s* turned -*r* to *visarga* under the condition of -*s* termed *apṛkta* (1.2.41 *apṛkta ekāl pratyayaḥ*). We now have *bhṛtya* +(*r*⟶*ḥ*) = *bhṛtyaḥ*, a *pada.*

The derivational history of *bhṛtyaḥ* can be captured with the following string of terms of its referential index which guided the derivation:

<*prātipadika* (1.2.45) / *kartṛ* (1.4.53) / *sUP* (4.1.2) / *vibhakti* (1.4.104) / *prathamā* (2.3.46) / *sU* (4.1.2) /*ekavacana* (1.4.22) / *pada* (1.4.53) / *avasāna* (1.4.110) / *apṛkta* (1.2.41)>

(5) *vana* 'forest' <*prātipadika; adhikaraṇa*>
1.2.45 *arthavad adhātur apratyayaḥ prātipadikam*
4.1.1 *ṅyāp-prātipadikāt*
1.4.45 *ādhāro' dhikaraṇam*
#similar to step #2 of (4) *bhṛtyaḥ*
4.1.2 *svaujasmauṭ... # sUP-selection*
2.3.1 *anabhihite*
2.3.36 *saptamyadhikaraṇe ca*
⟶*vana + Ṅi* ⟶
6.1.87 *ād guṇaḥ (6.1.72 saṃhitāyām)*
(6) *kāṣṭha* 'wood' <*prātipadika; karaṇa*>
1.2.45 *arthavad adhātur apratyayaḥ prāipadikam*
4.1.1 *ṅyāp-prātipadikāt*
1.4.42 *sādhakatamam karaṇam*
#similar to step #2 of (4) *bhṛtyaḥ*
4.1.2 *svaujasmauṭ...*
2.3.1 *anabhihite*
2.3.18 *kartṛ-karaṇayos tṛtīyā*
⟶ *kāṣṭha + bhis*
1.4.13 *yasmāt pratyayavidhis tadādi pratyaye' ṅgam*
6.4.1 *aṅgasya kāṣṭha + bhis*
7.1.9 *ato bhis ais*
⟶ *kāṣṭha (bhis ⟶ ais)*
kāṣṭha + ai(s⟶ḥ)

kāṣṭhaiḥ 'woods,' a pada
(7) *odana* 'rice' *<prātipadika; karman>*
1.2.45 *arthavad adhātur apratyayaḥ prāipadikam*
4.1.1 *ṅyāp-prātipadikāt*
1.4.49 *kartur īpsitatamaṃ karma*
#similar to step #2 of (4) *bhṛtyaḥ*
4.1.2 *svaujasmauṭ...*
⟶*odana + am*
6.1.72 *saṃhitāyām*
6.1.106 *ami pūrvaḥ*
⟶*odan(a+a ⟶ a)m = odanam*
= *odanam 'rice,' a pada*
(8) *sthālī* *<prātipadika; adhikaraṇa*
1.4.45 *ādhāro' dhikaraṇam...*
(9) *kumbhakārāya*
1.2.45 arthavad *adhātur apratyayaḥ prāipadikam*
4.1.1*ṅyāp-prātipadikāt*
1.4.49 *karmaṇā yam abhipraiti sa sampradānam*
#similar to step #2 of (4) *bhṛtyah*
4.1.2 *svaujasmauṭ...*
2.3.1 *anabhihite*
2.3.13 *caturthī sampradāne*
⟶ *kumbhakāra + Ṅe*
1.4.13 *yasmāt pratyayavidhis tadādi pratyaye' ṅgam*
6.4.1 *aṅgasya*
7.1.13 *ṅer yaḥ*
⟶ *kumbhakāra + (Ṅe⟶ya)*
= *kubhakāra + ya*
7.3.102 *supi ca*
⟶ *kumbhakār (a ⟶ ā) + ya*
= *kumbhakārāya*
(10) *kumbhakāraḥ*
(a) *ḌUkṛÑ* ⟶ 1.3.1 *bhūvādayo dhātavaḥ* <*dhātu*>
⟶ 1.3.3 *halantyam*
⟶ 1.3.5 *ādirñiṭuḍavaḥ*
⟶ 1.3.9 *tasya lopaḥ* (*it*-deletion)
(ḌU ⟶ ϕ)kṛ(Ñ ⟶ ϕ) = kṛ
(b) *kṛ* ⟶ 3.1.91 *dhātoḥ*
1.1.62 *tasmād ity uttarasya*
3.1.1 *pratyayaḥ*
3.1.2 *paraś ca*
3.1.3 *ādyudāttaś ca* <*ādyudātta*>
3.2.1 *karmaṇy aṇ* <*karman*>

3.1.92 *tatropadama saptamīstham* <*upapada*> 'that which is specified in this domain of *dhātoḥ* with the *saptamī* 'locative' is termed *upapada* 'conjoined *pada*'

3.4.67 *kartari kṛt* <*kartṛ* >

kṛ + *a*(*Ṇ* $\longrightarrow \phi$) = *kṛ* + *a* (*it* -deletion)

1.4.49 *kartur īpsitatamaṃ karma* <*karman*>

2.3.1 *anabhihite*

2.3.65 *kartṛ-karmaṇoḥ kṛti* <*karman / kartṛ* >

$\longrightarrow$*kumbha* + *am*

$\longrightarrow$*kumbha* + *ām kṛ* + *a*

2.2.19 *upapadam atiṅe* <*upapada*>

2.1.3 *prākkaḍārāt samāsaḥ* <*samāsa*>

$\longrightarrow$*kumbha am kāra*

1.2.46 *kṛt-taddhitasamāsāś ca* <*prātipadika*>

2.4.71 *supo dhātuprātipadikayoḥ*

$\longrightarrow$*kumbha* + (*am* $\longrightarrow \phi$) + *kāra*

= *kumbhakāra* 'pot-maker'

Our base input for deriving *kumbhakāra* is verb root *kṛ* 'to do, make' which ,in turn, is assigned the term <*dhātu*>. This serves as a mark for guiding the string for access to the domain of 3.1.91 *dhātoḥ*. Rule 3.2.1 *karmaṇy aṇ* then applies to introduce affix *aṆ*. The locative singular (*saptamī-ekavacana*) of *karmaṇi* of this rule serves as a mark for bringing rule 3.1.92 *tatropapadaṃ saptamīstham* 'that which is specified with a locative (*saptamīstham*) in this domain of *dhātoḥ* is termed an *upapada* 'a conjoined *pada*.' If a choice is made to introduce affix *aṆ* after *kṛ*, a *pada* denoting *karman* must be brought close to this context of *kṛ*.

Selection of a nominal ending with the denotation of *karman* must be made in accordance with the condition of 2.3.1 *anabhihite* 'not already stated, otherwise.' We realize that the affix which is to be introduced, namely *aṆ*, is a *kṛt* (3.4.67 *kartari kṛt*) affix, and hence it would denote *kartṛ*. We may now select the genitive (*ṣaṣṭhī*) plural ending -*ām* of rule 2.3.36 *kartṛkarmaṇoḥ kṛti.* Note that 2.3.36 allows genitive to denote *kartṛ* or *karman*, when they are not already expressed otherwise, and when a *kṛt* (non-*tiṄ*; 3.1.93 *kṛd atiṅ*) affix follows in construction. This is what enables us to meet the condition of 2.3.1 *anabhihite*, and select genitive plural to express *karman.* The *kartṛ* is already expressed with *kāra*. The derivational string at this stage is: *kumbha* + *ām kṛ* + *aṆ*. It has the referential index of <*dhātu, upapada, pratyaya, kṛt, ady udātta*>. Rule 1.4.13 *yasmāt pratyayavidhis tadādi pratyaye' ṅgam* assigns the term *aṅga.* The string is sent to the domain of 6.4.1 *aṅgasya* where rule 7.2.115 *aco' ñṇiti* applies to replace *ṛ* of the *aṅga* with its *vṛddhi* counterpart *ār*. This application yields *kumbha* + *ām k*(*ṛ* $\longrightarrow$ *ār*) *a*(*Ṇ* $\longrightarrow \phi$) = *kumbha* + *am kāra*. The term *upapada* must now guide the derivation. Rule 2.2.19 *upapadam atiṅ* allows the formation of a compound (2.1.3 *samāsaḥ*) which yields *kumbha* + *am* + *kāra.* This string is then assigned the new term *prātipadika* 'nominal stem.' This leads to application of rule 2.4.71 *supo dhātuprātipadikayoḥ* whereby a *sUP*, here -*am of kumbha* + *am* + *kāra*, is subject to deletion by *LUK*. We now have *kumbha (am* $\longrightarrow$ ϕ*)kāra* = *kumbhakāra*, a derived nominal base (*prātipadika*). This completes the derivational history of fully derived words (*padas*) which form simple sentences.

The following is a flow-chart showing selectional restriction and agreement based of reconstruction of rule contexts by Referential Indices:

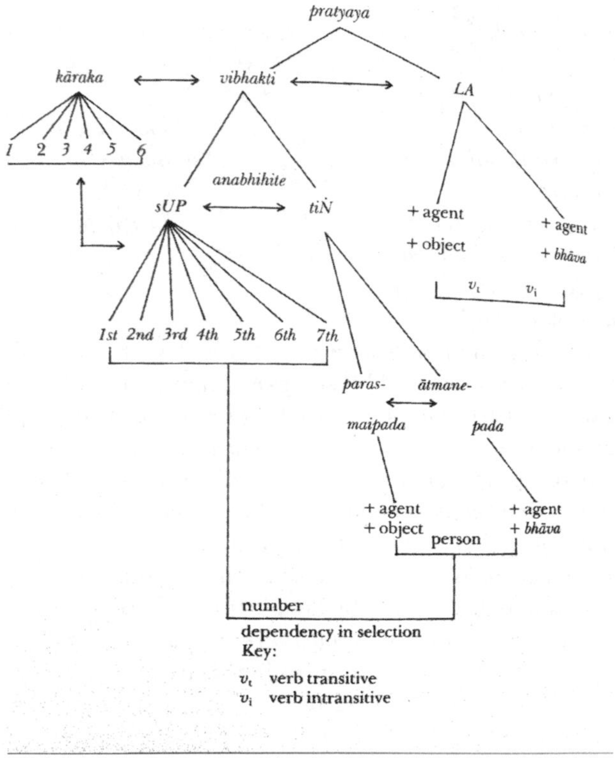

4 Derivational Mechanism

The derivational mechanism of the Aṣṭādhyāyī makes use of a network of bases and affixes to derive *padas*, with application of select operations. These operations are carried out, primarily at two levels of naming and expressing, following certain conventions.

5 Levels of Derivation

The derivational mechanism of the Aṣṭādhyāyī primarily operates on two levels of naming and expressing, with reference to the Action Complex (AC) of the Conceptual Structure (CS) of sentences:

(1) CS1:
'x accomplishes the action of making y at the current time'
bhṛtyaḥ ghaṭaṃ karoti 'servant makes a pot'
AC1: <*kartṛ* > Action: <*dhātu*>
Level 1: Naming: <term assignment>
bhṛtya <*kartṛ* > Action: *pac* <*dhātu*>
<*prātipadika* / *kartṛ* > <*dhātu* / *sakarmaka*>
1.2.45 *arthavad...* 1.3.1 *bhūvādayo...*
Level 2: <domain access / affix placement>
<*adhikāra* / *pratyaya*>
bhṛtya <*kartṛ*/ *prātipadika*> *pac* <*dhātu*/ *sakarmaka*>
1.2.45 *arthavad...* *3.1.7 dhātoḥ karmaṇaḥ...*
4.1.1 *ṅyāp-prāt...* *3.1.91 dhātoḥ*
4.1.2 *svaujas...*<*sUP* /*prātipadika*> 3.1.123 *vartamāne laṭ* <*LAṬ* >
4.1.3 *striyām* <*dhātu* / *laṭ* / *vartamāna*>
4.1.76 *taddhitāḥ* 3.4.77 *tasya*
4.1.82 *samarthānāḥ* 3.4.78 *tiptasjhi...* <*tiṄ*>

<rule Context>

3.1.1 *pratyayaḥ* 3.1.2 *paraś ca*
3.1.3 *ādyudāttaś ca* 3.1.4 *anudāttau suppitau*
<*sUP*-selection> <*tiṄ*-selection> <agreement>
1.4.99 *laḥ parasmaipadam* 1.4.100 *taṅānāv ātmanepadam*
1.4.102 *tāny ekavacana...* 1.4.103 *supaḥ*
1.4.104 *vibhaktiś ca* 1.4.105 *yuṣmady upapade...*
1.4.106 *prahāse ca manyopa...* 1.4.107 *asmady uttamaḥ*
1.4.108 *śeṣe prathamaḥ* 1.4.21 *bahuṣu bahuvacanam*
1.4.22 *dvyekayor dvivacanaikavacane*
6, Term Assignment (Naming; for exiting the Obligatory Domain)
1.4.109 *paraḥ sannikarṣaḥ saṃhitā* <6.1.72 *saṃhitāyām* >
1.4.13 *yasmāt pratyayavidhis...* <6.4.1 *aṅgasya*>
1.4.14 *suptiṅantaṃ padam* <8.1.16 *padasya*>
1.4.18 *yaci bham* <6.4.129 *bhasya*>
Other terms, depending on the base-input, and inputs yielded by individual applications, may be assigned to direct operations.
C. Convention

1. A base-input, i.e. *dhātu* and *prātipadika*, when made input to the controlling domain (CD; the first *adhyāya* of the grammar) activates this grammatical device with assignment of term (*saṃjñā-kārya*) to the base-input(s). For example, the assignment of the term <*dhātu*> to *pac* by 1.3.1 *bhūvādayo dhātavaḥ*, and of the term <*prātipadika*> to *bhṛtya* by 1.2.45 *arthavad adhātur...* This process can be called <Term Assignment>.
2. A base-input must gain access to the Obligatory Domain (OD; *adhyāya* three through five) of the grammar for scanning of domain headings (*adhikāra*) for possible rule application. A heading formed with the term assigned to the

base- input locates the domain for possible rule application. For example, *pac*, the base-input which is assigned the term <*dhātu*>, can opt for possible rule application in the domain of 3.1.5 *guptijkidbhyaḥ* san, 3.1.7 *dhātoḥ*..., and 3.1.91 *dhātoḥ*, if it meets required condition. Access to the domain of possible rule application is made fairly automatic by assignment of the term, this case, <*dhātu*>.

3. Access to, and application of a rule within a domain, must be made in consonance with syntactico-semantic specification of the CS, and terms currently assigned. This is the reason why *pac* was sent to the domain of 3.1.91 *dhātoḥ*.
4. Each time a definitional term and abbreviated symbol is introduced to the derivational string, recourse should be taken to scanning of the domain for possible explanation and application via reconstruction of referential index.
5. A referential index of definitional terms and symbol must be reconstructed by scanning rules beginning with the Controlling Domain (CD) to the domain which triggers scanning. Terms and symbols of the referential index alone guide a string for further location of domains and thereby rule application.
6. Operations directed by terms of the referential index must be performed in the order the terms and symbols appear on the index. Operations relative to term1 must be completed before an operation required by term2 is performed.
7. The string, after each application, is sent back to the (CD) for possible assignment of new term, and thereby termination of operation in that domain.
8. Result of each application must be sent to the Controlling Domain (CD) for assignment of term for scanning of domains of possible rule-application.

On the Generalizability of Pāṇini's Pratyāhāra-Technique to Other Languages

Wiebke Petersen and Silke Hamann

Heinrich-Heine-Universität Düsseldorf
wiebke.petersen@phil.uni-duesseldorf.de
hamann@phil.uni-duesseldorf.de

Abstract. Pāṇini defines the sound classes involved in grammatical rules by pratyāhāras, i.e., a two-letter code based on the order of the sounds in the Śivasūtras. In the present paper we demonstrate that Pāṇini's pratyāhāra method is generalizable to the description of the phonological systems of other languages by applying it to the sound classes and phonological alternations of German. Furthermore, we compare Pāṇini's pratyāhāra technique with the technique of describing phonological classes by phonological features, which is more common in Western phonology. It turns out that pratyāhāras perform better than features for the description of our sample of German phonological processes if one considers the quality criterion for class-description devices proposed by [10] which is based on the ratio of describable to actual classes.

Keywords: Panini, Sivasutras, phonological features, sound classes, formal concept analysis.

1 Introduction

The *Aṣṭādhyāyī* – Pāṇini's circa 2 500 years old grammar of Sanskrit – starts with the Śivasūtras, a list of the sounds of Sanskrit. Throughout his grammar, Pāṇini uses the so-called pratyāhāras – a two-letter code based on the order of the sounds in the Śivasūtras – to define the sound classes involved in grammatical rules. Since Pāṇini's time, the Śivasūtras have been studied intensively with focus on the following questions: Is the order of the sounds in the Śivasūtras determined by the rules in the *Aṣṭādhyāyī*? Why does one sound (the glottal [h]) occur twice in the list? Are the Śivasūtras minimal? And how did Pāṇini develop the Śivasūtras?

In the present article we make a first attempt to describe the phonological processes of another language with the help of a Śivasūtra-like list and pratyāhāras. For this purpose we chose German, another Indo-European language with a phonological system that differs considerably from Sanskrit. We propose a list of all German sounds in the style of the Śivasūtras that allows us to refer to the sound classes relevant for the description of German phonological processes in the form of pratyāhāras. This Śivasūtra-like list of sounds will be called *pratyāhāra sūtras* in the following. Our aim is to test what the formalization of

G.N. Jha (Ed.): Sanskrit Computational Linguistics, LNCS 6465, pp. 21–38, 2010.

pratyāhāra sūtras for another language entails, and how far such a description differs from the phonologically more common description of sound classes with phonological features.

The remainder of the paper is structured as follows. In section 2, we introduce two ways of referring to sound classes and phonological processes: the language-independent description of processes with phonological features and Pāṇini's approach with Śivasūtras and pratyāhāras that was especially designed for Sanskrit. Section 3 gives a short overview of the German sound system and the phonological processes that are the basis for our analysis of German. In the central section 4 we propose *pratyāhāra sūtras* for German. Finally, in section 5 we conclude with a brief discussion of our results.

2 Describing Sound Classes and Phonological Processes

Phonology is concerned with the sound system of a language and the alternations that these sounds undergo. In phonological theory, alternations between sounds are called phonological processes. In German, for instance, plural forms of nouns often involve a number of alternations compared to the singular forms, see the example in (1).

(1) *Hand* [hant] 'hand' - *Hände* [hɛndə] 'hands'

In (1), the vowel [a] in the singular alternates with the vowel [ɛ] in the plural, and the consonant [t] in the singular alternates with the consonant [d] in the plural. Generative phonologists describe such processes by assuming that only one of the alternants is the form that native speakers or listeners have stored in their mind, while the other alternating form is derived from it. The sound representations stored in the mind are usually called *underlying representations* and are denoted in slashes, and the sounds that the speakers actually produce or the listeners hear are termed *surface representations* and are denoted in square brackets. For the singular-plural alternation in (1), it is usually assumed that the /d/ is the underlying representation, and the [t] is derived from it by a process that is called *final devoicing* (see section 3.2.1 below).

Phonological processes can be described in terms of *phonological rules*. These rules are typically of the form "the underlying sound /A/ is realized as the surface sound [B] if it is preceded by sound C and/or followed by sound D". This can be formalized as

(2) $/A/ \rightarrow [B] \; / \; C\_\_ \; D$.

The variables A, B, C and D in (2) can stand for single sounds, but also for classes of sounds. Whole classes of sounds can thus undergo a phonological process, be the result of a phonological process or pose the context of such a process. Sound classes can be described in several ways. In the following two subsections, we shortly present the commonly employed phonological description with features (2.1) and the description by pratyāhāras introduced by Panini for Sanskrit (2.2).

2.1 Featural Descriptions

Instead of listing whole classes of sounds in phonological processes, phonological descriptions usually employ *phonological features*. The key assumption of this description technique is that every sound in a language can be described sufficiently by a set of binary features, making it distinct from all other sounds in the same language [7]. Phonological features have acoustic, articulatory or auditory definitions. For instance, the feature [+high] is defined as an articulation with a high tongue body. Accordingly, the feature [−high] also refers to the dimension of tongue position; it is defined as an articulation with a non-high tongue body. Sounds that share a feature and therefore a phonetic trait are called a *natural class*. The Spanish vowels [i u], for instance, are the only vowels in Spanish with the feature [+high] and therefore form a natural class in Spanish.

Unnatural classes of sounds are those that do not share a phonetic trait, and they usually do not occur as undergoer, result or context of a phonological process. A well-known example of an unnatural class is the context of the so-called *ruki*-rule in Sanskrit [[20], 61f.]. According to this process, the segments [r], [u], [k] and [i] cause retroflexion of the dental [s]. The four segments forming the context of this rule form no natural class but rather an arbitrary set of sounds, because they involve two types of vowels (the back vowel [u] and the front vowel [i]), and two types of consonants (a retroflex [r] and a velar [k]), which cannot be referred to by one or a few phonological features. To describe this class, all feature specifications of all four segments have to be given. A situation like this, where the context of a rule cannot be referred to by a single or a few features, is called a *disjunction* in the phonological literature [e.g.[8], 216]. The disjunct context of the *ruki*-rule probably diachronically emerged from the merger of several processes [see the discussion in [6]chapter 4.3.4].

'Binary phonological features' are a relatively modern concept; the first complete set of features has been proposed by [7]. However, the traditional Śikṣās- and Prātiśākhyas-literature, which even predates Pāṇini, already classifies sounds by phonetic criteria that can be interpreted as phonological features [cf. the elaborated varga system of the sparśas as described in [3,18]]. In his grammar, Pāṇini uses the varga system in addition to his pratyāhāra technique.

2.2 Pāṇini's Śivasūtras

Pāṇini's grammar of Sanskrit, the *Aṣṭādhyāyī*, is preceded by the 14 sūtras given in Figure 1, which are called Śivasūtras. Each single sūtra consists of a sequence of sounds which ends in a consonant. This last consonant of each sūtra is used meta-linguistically as a marker to indicate the end of a sūtra. In order to emphasize the technical nature of the end consonants, they are replaced by neutral marker elements M_i in Figure 1(III). Together the 14 Śivasūtras define a linear order on the sounds of Sanskrit. The order is such that more or less each class of sounds on which a phonological rule of Pāṇini's grammar operates forms

अइउण् ऋऌक् एओङ् ऐऔच् हयवरट् लण् ञमङणनम् झभञ् (I)
घढधष् जबगडदश् खफछठथचटतव् कपय् शषसर् हल्

a·i·uṇ ṛ·ḷk e·oṅ ai·auc hayavaraṭ laṇ ñamaṅaṇanam jhabhañ (II)
ghaḍhadhaṣ jabagaḍadaś khaphachaṭhathacaṭatav kapay śaṣasar hal

a i u M_1 ṛ ḷ M_2 e o M_3 ai au M_4 h y v r M_5 l M_6 ñ m ṅ ṇ n M_7 jh bh M_8 (III)
gh ḍh dh M_9 j b g ḍ d M_{10} kh ph ch ṭh th c ṭ t M_{11} k p M_{12} ś ṣ s M_{13} h M_{14}

Fig. 1. Pāṇini's Śivasūtras for Sanskrit (I: Devanāgarī script; II: Latin transcription; III: Analysis – the syllable-building vowels are left out and the meta-linguistically used consonants are replaced by neutral markers M_i)

an interval which ends immediately before a marker element.[1] As a result, Pāṇini could use a two letter code consisting of a sound and a marker called pratyāhāra in order to designate the sound classes in his grammar. A pratyāhāra denotes the continuous sequence of sounds in the interval between the sound and the marker (including the first sound, but non of the markers). E.g., the pair iM_2 in Figure 1 denotes the class [i, u, ṛ, ḷ].

Concerning the question of how Pāṇini developed the Śivasūtras, it is generally agreed upon that the order of the sounds in the Śivasūtras is primarily determined by the structural behavior of the sounds in the grammar rules and that the arrangement of the sounds is chosen such that brevity is maximized [cf.[17,12,1,9]]. [13] proves that Pāṇini's Śivasūtras are an optimal solution for the following task: Given the set of all phonological classes which are encoded as pratyāhāras in the *Aṣṭādhyāyī*, construct a list which is interrupted by markers such that each class can be denoted as a pratyāhāra. Choose the list where the fewest sounds are repeated and minimize its length. It follows from the proof that the duplication of the sound [h] in the Śivasūtras is not superfluous and that the number of markers and thereby the number of Śivasūtras cannot be reduced. In [14,15] it could be shown that there are nearly 12 000 000 alternative sound lists interrupted by markers which allow the formation of the required pratyāhāras and which are of the same length as the Śivasūtras.

3 The Phonemes and Phonological Processes of German

In order to describe the phonological processes of German with the technique of Pāṇini's pratyāhāras, we first have to establish the sound system of German and its phonological alternations. This is not a trivial task. While for Sanskrit it is

[1] As mentioned before, Pāṇini uses different description techniques in parallel. Hence, he states not every phonological rule in terms of pratyāhāras.

generally accepted that Pāṇini's grammar describes the phonological system of the language, no such undisputable description exists for German. Establishing the sound system of German involves decisions on whether certain sounds are considered to be mentally stored (underlyingly represented), or whether they are considered surface alternants that can be derived from another underlying form via a process. Underlying forms are usually meaning-distinguishing units with relatively unrestricted occurrences, and called *phonemes*, while those forms that are derived and have only a restricted context are called *allophones*. If a sound in a language is classified as an allophone, the phonological description of the language has to include a process to describe its derivation from an underlying phoneme. The velar nasal [ŋ], for instance, is considered by some phonologists [e.g.[19,21]] to be an allophone of the alveolar nasal phoneme/n/ in German because it only occurs after a vowel and before a syllable break, in the so-called *coda* position. Such decisions on the phoneme status of a sound are often made purely on theoretical grounds: [2], e.g., postulated that the number of phonemes should be as small as possible while the number of processes is unrestricted. The following description of German is based on [21] and [5] and the theoretical assumptions therein.

3.1 The Sounds of German

The consonants and vowels of German forming the basis of our analysis are given in Figure 2 and 3, respectively [based on [5], pp. 31, 62, 68].

In Figure 2, two consonants are given in brackets, namely the glottal plosive [ʔ] and the velar fricative [x]. These sounds are bracketed in Figure 2 because their occurrence is predictable from the context and can therefore be derived with a process: The glottal stop occurs before syllable-initial stressed vowels, and the velar fricative after low and back vowels, as described with the phonological processes in 3.2.3 and 3.2.5 below. All other sounds in Figure 2 can be considered phonemes of German.

The vowels of German are given in Figure 3. Again, the sounds given in brackets are considered to be allophones of underlying phonemes. These are the low vowel [ɐ], which is the realization of the German /ʁ/ in coda position and

	bilabial	labio-dental	alveolar	post-alveolar	palatal	velar	uvular	glottal
plosive	p b		t d			k g		(ʔ)
nasal	m		n			ŋ		
fricative		f v	s z	ʃ ʒ	ç	(x)	ʁ	h
affricate		pf	ts	tʃ dʒ				
approximant					j			
lateral			l					

Fig. 2. Consonants of German

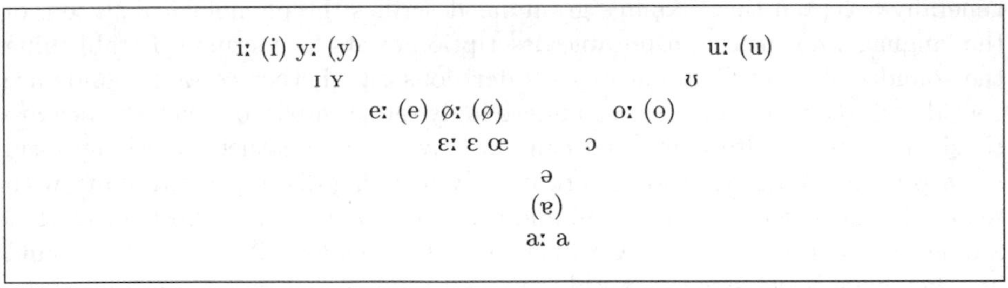

Fig. 3. Vowel triangle with vowels of German

therefore predictable, and the short tense vowels [i, y, u, e, ø, o]. These short tense vowels are allophones of the long tense /iː, yː, uː, eː, øː, oː/. The latter only occur in stressed position, while the former only occur in unstressed position, see the phonological process of vowel shortening in 3.2.4.

In addition to the vowels in Figure 3, German also has three *diphthongs*; these are vowels that change their quality during the articulation. The diphthongs of German are [ɔɪ̯, aɪ̯, aʊ̯].

3.2 Phonological Processes of German

The present description is restricted to processes that involve classes of sounds. These classes can be the undergoer, the result or the context of a phonological process. Processes where only single segments are involved are excluded because they are of no relevance for a description with Śivasūtras or phonological features. Such a process is for instance the vocalisation of the German /ʁ/ to [ɐ] in coda position [see e.g. [21] pp. 252ff. for a detailed description]. The following six processes meet this criterion and therefore seem to be relevant for our descriptions.

3.2.1 Final Devoicing

German, like many other languages, has a process of final devoicing that turns a class of voiced consonants into voiceless ones if they occur in word-final position [see e.g., [21], pp. 199ff.]. In the example in (1), the plural *Hände* [hɛndə] 'hands' is realized as [hant] with a final [t] in the singular.

The whole list of sounds that undergo German final devoicing are given in the formalization in (3).

(3) /b, d, g, v, z, ʒ/ → [p, t, k, f, s, ʃ] / __ word boundary

The group of sounds undergoing this process are traditionally described as *obstruents*, a term that refers to all plosives, fricatives, and affricates in a language. As we can see in Figure 3, German has more obstruent phonemes than the ones listed in rule (3). The input to the rule lacks all affricates and the fricatives [ç, ʁ, h]. While the voiceless phonemes [pf, ts, ç, h] cannot undergo final devoicing because they have no voiced counterpart, the voiced [dʒ] and [ʁ] are phonemes that do not occur in coda position, and therefore do not meet the requirements of the process, either.

In featural accounts of phonological processes, the voiced obstruents of German are referred to with the phonological features [+voiced, −sonorant], and the voiceless obstruents with [−voiced, −sonorant]. A rule with features would thus look as follows:

(3)^f [+voiced, −sonorant] → [−voiced, −sonorant]/_ word boundary

This rule is hyper-inclusive in the sense that it theoretically affects all obstruents. Empirically, however, it is harmless as only the plosives and part of the fricatives occur in the relevant context and thus undergo the rule.

3.2.2 Regressive Nasal Assimilation

The nasal alveolar /n/ is often assimilated to its following context in German. The word /ankʊnft/ 'arrival', for instance, can be realized as [aŋkʊnft], and /a̯ɪnfaːɐt/ 'gateway' as [a̯ɪmfaːɐt]. This so-called regressive assimilation of nasals has to be formalized as two processes, velar and labial assimilation, distinguishing the two types of outcome:

(4) /n/ → [ŋ] / _ [k, g] *velar nasal assimilation*

(5) /n/ → [m] / _ [p, b, pf, f, v] *labial nasal assimilation*

The two contexts of nasal assimilation can be referred to with the features [+velar, −sonorant, −continuant] for [k, g], and the features [+labial, −sonorant] for [p, b, pf, f, v].

3.2.3 Glottal Stop Epenthesis

Vowel-initial words in German are always realized with a preceding glottal stop [ʔ], e.g. /aɪ̯/ 'egg' is realized as [ʔaɪ̯]. The distribution of the glottal stop is formalized in the following *epenthesis* rule, a rule that inserts segments:

(6) ∅ → [ʔ] / word boundary _ vowel

The regular and predictable occurrence of the glottal stop is the reason why it is not considered a phoneme of German by many phonologists. The class of vowels forming the context of glottal stop epenthesis can be referred to with the phonological feature [−consonantal].

3.2.4 Vowel Shortening

The German vowels [iː, yː, eː, øː, uː, oː], which are referred to as long and tense vowels, only occur in stressed position. Their short counterparts [i, y, e, ø, u, o], on the other hand, only appear in unstressed position, and are restricted to loanwords, e.g. the first vowel in [monaɐçiː] 'monarchy'. This complementary distribution led most phonologists to tread the short tense vowels as allophones of the long ones and to describe their distribution with a process as it is formalized in (7).

(7) /iː, yː, eː, øː, uː, oː/ → [i, y, e, ø, u, o] / $\_\_{\text{unstressed}}$

The long, tense vowels are usually referred to with the features [+long, +tense, −consonantal], and their short counterparts with [−long, +tense, −consonantal].

3.2.5 Palatal Fricative Assimilation

The palatal fricative /ç/ is articulated with a more backed tongue position, i.e. as the velar fricative [x], after the vowels [aː, a, uː, ʊ, oː, ɔ, aʊ̯]. This process is described in (8).

(8) /ç/ → [x] / [aː, a, uː, ʊ, oː, ɔ, aʊ̯] __

The context vowels for palatal fricative assimilation can be described with the phonological features [−consonantal, +low] for [aː, a], and [−consonantal, +back] for [uː, ʊ, oː, ɔ, aʊ̯]. This class is an example for a disjunct phonological context, where the sounds forming the context cannot be united under one feature description.

3.2.6 Umlaut

In German, we can observe a process of vowel change that has lost its phonological context. This process, called *umlaut*, changes the vowels /uː, ʊ, oː, ɔ, aː, a, aʊ̯/ into [yː, ʏ, øː, œ, ɛː, ɛ, ɔɪ̯], respectively, when a noun is set in the plural. The example [hant] – [hɛndə] 'hand (sg. – pl.)' in (1) illustrated this. Umlaut also occurs for diminutive forms of nouns and the comparative forms of many adjectives, see e.g. [21] (1996: 182f.). A formalization is given in (9), though the context is not specified because the process is morphologically conditioned.

(9) /uː, ʊ, oː, ɔ, aː, a, aʊ̯/ → [yː, ʏ, øː, œ, ɛː, ɛ, ɔɪ̯]

Such a process is familiar to Sanskrit scholars from the Sanskrit ablaut grades termed *guṇa* and *vṛddhi*.

For a description with phonological features, this process has to be divided into three subprocesses. The sounds [uː, ʊ, oː, ɔ] are [−consonant, −low, +back] and change to [yː, ʏ, øː, œ] with the feature specification [−consonant, −low, +front]. The sound class [aː, a] with the specification [−consonant, +low] changes to [ɛː, ɛ] with the specification [−consonant, −low, −high, +front, −tense]. And lastly, the single segment [aʊ̯] changes to [ɔɪ̯].

The six processes described in this section involve ten classes of sounds, namely the input and output of final devoicing, the context of labial and velar nasal assimilation, the context of glottal stop epenthesis, the input and output of vowel shortening, the context of palatal fricative assimilation, and the input and output of the umlaut process. For a featural description of these processes, a total of 16 features has to be used.[2] In the following section we will see how an account with pratyāhāras for the same processes looks.

---

[2] In order to make our analysis independent of the theoretical assumption that features are binary, we treat features like [+high] and [−high] as two distinct privative features.

4 *Pratyāhāra sūtras* of German

In section 3, we defined the collection of phonological processes and thereby implicitly the set of 10 phonological classes which we intend to represent as pratyāhāras. Hence, our task is to develop a list of the German sound segments interrupted by markers in the style of the Śivasūtras which allows the formation of a pratyāhāra for each phonological class of the collection. A simple but undesirable solution to the problem would be to line up the phonological classes in one single list and to put a marker behind each class. In such a list, the number of occurrences of a phonological segment would be equal to the number of phonological classes to which the phonological segment belongs. Since Pāṇini duplicates in his Śivasūtras only one segment, namely [h], it is obvious that the Śivasūtras are constructed more economically: Pāṇini aims at the minimization of the number of duplicated segments and the reduction of markers [9].

In [14,15] the general problem of generating economical *pratyāhāra sūtras* for given sets of sets has been tackled by applying methods from Formal Concept Analysis (FCA) [4]. In what follows we will apply those former results in order to construct adequate, economical *pratyāhāra sūtras* for German. A small example helps to clarify the required terminology of FCA:

	a	b	c	d
class1	×	×		
class2		×	×	
class3			×	×
class4	×	×	×	

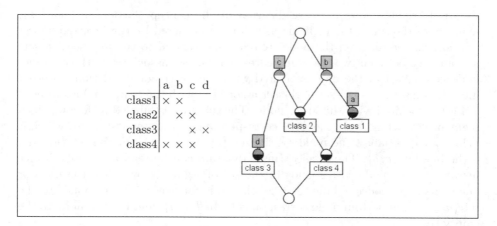

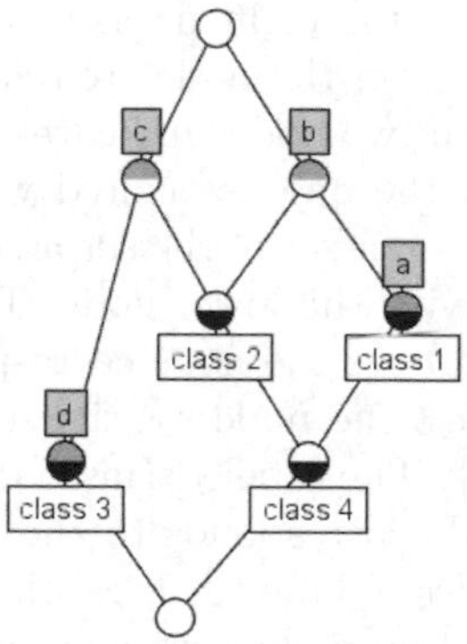

Fig. 4. Example formal context (left) with corresponding concept lattice (right)

In Figure 4 (left) an example set of four classes, namely $\{a, b\}$, $\{b, c\}$, $\{c, d\}$, and $\{a, b, c\}$, is given in form of a *formal context*. A cross in the table indicates that an element (top row) belongs to a class (first column). All four classes can be denoted as pratyāhāras of the list of *pratyāhāra sūtras*

$$d\, c\, M_1\, b\, M_2\, a\, M_3 \ \ \text{(pratyāhāras: } bM_3, cM_2, dM_1, cM_3\text{)}$$

Note that no element in the *pratyāhāra sūtras* occurs twice. The *concept lattice* of the formal context is the set of all intersections (ordered by the inverted subset

	p	t	k	f	s	ʃ	b	d	g	v	z	ʒ	ʏ	ɐ	ə	ɪ	ɛ	œ	ɔ	ʊ	a	i	e	o	u	y	ø	aː	iː	oː	uː	yː	øː	eː	aɪ	ɔɪ̯	aʊ̯	ɛː	pf
devoicing input							×	×	×	×	×	×																											
devoicing output	×	×	×	×	×	×																																	
vowel shortening in.																												×	×	×	×	×	×	×					
vowel shortening out.																					×	×	×	×	×	×	×												
umlaut input																			×	×	×							×		×	×						×		
umlaut output													×				×	×														×	×			×		×	
palatal assimilation																			×	×	×			×	×			×		×	×						×		
nasal assim. velar			×						×																														
nasal assim. labial	×			×			×			×																													×
glottal epenthesis													×	×	×	×	×	×	×	×	×	×	×	×	×	×	×	×	×	×	×	×	×	×	×	×	×	×	

Fig. 5. Formal context for the phonological classes of German (due to space limits, the segments [dʒ, tʃ, ts, ʁ, l, ŋ, n, m, j, ʔ, h, ç, x] which belong to no class are left out)

relation) which can be generated from the classes (plus the set of all elements). For our example, the set of all intersections is

$$\{\{\,\}, \{c\}, \{b\}, \{a, b\}, \{b, c\}, \{c, d\}, \{a, b, c\}, \{a, b, c, d\}\}.$$

Figure 4 (right) shows a Hasse diagram of the concept lattice, i.e., a Hasse diagram of the set of the eight intersection sets ordered by the inverted subset relation. In the diagram the nodes (circles) correspond to the intersection sets and an edge between two nodes indicates that the set associated with the upper node is a subset of the one associated with the lower node and that no other intersection set is a superset of the set associated with the upper and a subset of the set associated with the lower node. The diagram is labeled as follows: each segment is written above the node corresponding to the smallest set to which it belongs, e.g., c labels the node for the set $\{c\}$, b the node for the set $\{b\}$, and a the one for $\{a, b\}$. The labeled Hasse diagram can be read as an inheritance hierarchy: each node corresponds to the set of segments by which the node or one of its supernodes is labeled. E.g., the node for 'class 4' corresponds to the set $\{a, b, c\}$, the bottom node corresponds to $\{a, b, c, d\}$, and the top node to the empty set.

The formal context for our collection of phonological classes for German is given in Figure 5. [13] proves that it is impossible to order the phonological segments in *pratyāhāra sūtras* without a single duplication if the corresponding concept lattice is not planar, i.e., if it is impossible to draw a Hasse diagram without intersecting edges. A Hasse diagram of the concept lattice of the formal context for our phonological classes of German is given in Figure 6. In order to improve readability, most of the labels are left out. One part of the diagram, namely the one corresponding to the subsets of the phonological class 'glottal epenthesis' (i.e. the vowels), stands out, as it is plane. The question is whether it is possible to give a plane drawing, i.e. a drawing without intersecting edges, of the remaining part of the diagram. By Figure 7 we will argue that such a drawing is impossible.

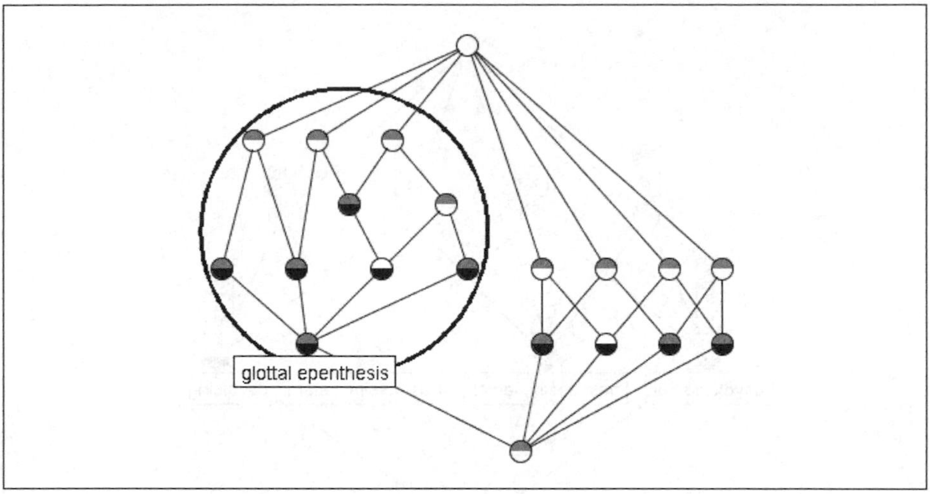

Fig. 6. Concept lattice for the phonological classes of German (the big circle indicates the planar part constituted by the vowel segments)

Figure 7 shows a Hasse diagram with intersecting edges of the concept lattice of the formal context constituted by the four classes 'devoicing output', 'devoicing input', 'nasal assimilation labial', and 'nasal assimilation velar'. According to [11] and [16], a lattice has no plane diagram if it is possible to gain the graph ⬠ (i.e. the complete graph K^5) from the graph of the diagram enlarged by an additional edge connecting the lowest and the top most node by removing some of the edges and contracting others. It is obvious that the graph in Figure 7 (top) can be constructed from the one in Figure 6 by leaving out some of the edges. The sequence of drawings at the bottom of Figure 7 starts with the graph from the top of the figure enlarged by an additional edge connecting the top and the bottom node. Each remaining graph is gained from its left neighbor by contracting the emphasized edge (thick grey edge). As the final graph is isomorphic to the graph ⬠, the sequence proves that it is impossible to draw a diagram of the concept lattice in Figure 6 without intersecting edges. Hence, the concept lattice of the formal context of our collection of phonological classes is not planar. It follows that it is impossible to construct *pratyāhāra sūtras* for the formal context in which each sound segment occurs only once. Hence, analogously to the Śivasūtras for Sanskrit we are forced to repeat at least one sound segment for our *pratyāhāra sūtras* for German.

Thus, the next step towards *pratyāhāra sūtras* for German is the identification of sound segments which are good candidates for duplication. Hence, we are interested in identifying those segments for which we can add a copy to our formal context in Figure 5 and distribute the crosses in the table between the two copies in such a way that the corresponding concept lattice gets planar. The aim thereby is to copy as few sound segments as possible. Note that the nodes

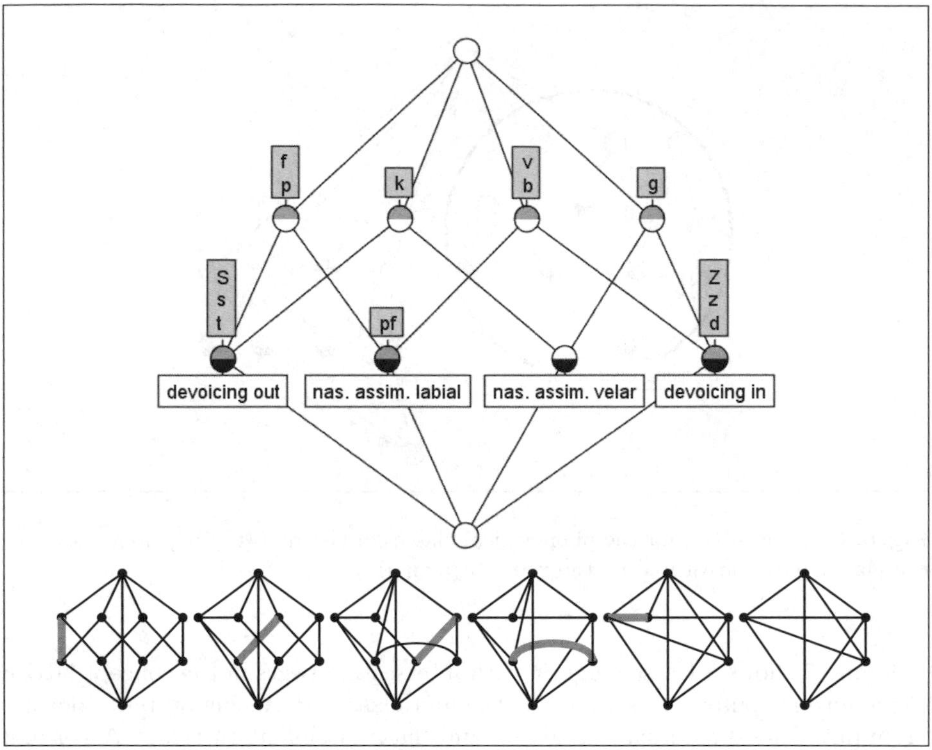

Fig. 7. Non-planar concept lattice for a selection of the phonological classes of German (key: S ↦ [ʃ], Z ↦ [ʒ])

of the four minimal nonempty sets ({f, p}, {k}, {v, b}, {g}) in Figure 7 do not differ structurally with respect to their position in the concept lattice. Thus, it is better to duplicate one of the segments [g] or [k] than one of the remaining four segments, as duplicating for example the segment [f] would force one to duplicate the segment [p] too, since these two segments are not distinguishable with respect to the chosen phonological classes. In what follows we will concentrate on the duplication of the segment [g]; the duplication of [k] would give analogous results.

If in the formal context in Figure 5 the segment [g] is replaced by two copies – one classified as 'devoicing input' and the other one as 'nasal assimilation velar' – a plane Hasse diagram of the resulting concept lattice can be drawn (cf. Figure 8). In the diagram in Figure 8, the top node of the concept lattice, which corresponds to the empty set, is left out. The boundary graph of the diagram is called the *S-graph* of the formal context. In [14] it has been proven that the S-graph of a formal context is unique up to isomorphism if the formal context can be encoded as *pratyāhāra sūtras* without duplicated elements. Furthermore, the main theorem on S-sortability [15,14] states that a formal context can be encoded as *pratyāhāra sūtras* without duplicated elements if and only if its

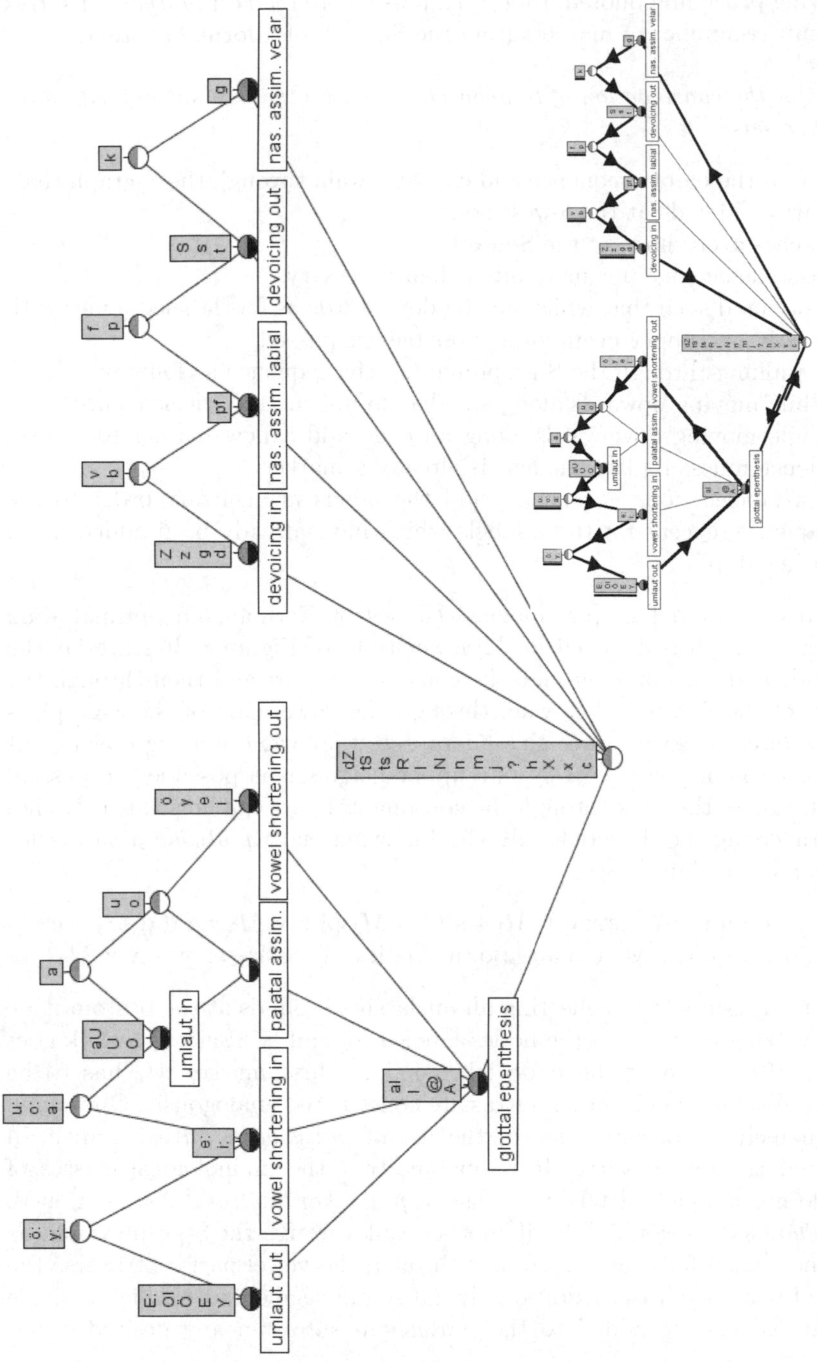

Fig. 8. Concept lattice without top node for the phonological classes of German (key: c ↦ [ç], S ↦ [ʃ], ? ↦ [ʔ], Z ↦ [ʒ], N ↦ [ŋ], R ↦ [ɾ], Y ↦ [ʏ], A ↦ [ɐ], @ ↦ [ə], I ↦ [ɪ], E ↦ [ɛ], Ö ↦ [œ], O ↦ [ɔ], U ↦ [ʊ], ö ↦ [ø], ö: ↦ [øː], aI ↦ [aɪ], OI ↦ [ɔɪ], aU ↦ [aʊ̯], E: ↦ [ɛː], tS ↦ [tʃ])

concept lattice is planar and its S-graph contains all nodes labeled by elements. The following procedure quoted from [15] allows one to read off *pratyāhāra sūtras* with a minimal number of markers from the S-graph of a formal context.

Procedure for the construction of S-alphabets (here: pratyāhāra sūtras) with minimal marker sets:

1. Start with the empty sequence and choose a walk through the S-graph that:
   - starts and ends at the lowest node,
   - reaches every node of the S-graph,
   - passes each edge not more often than necessary,
   - is oriented such that while moving downwards as few labeled nodes with exactly one upper neighbor as possible are passed.
2. While walking through the S-graph modify the sequence as follows:
   - While moving upwards along an edge do not modify the sequence.
   - While moving downwards along an edge add a new marker to the sequence unless its last element is already a marker.
   - If a labeled node is reached, add the labels in arbitrary order to the sequence, except for those labels which have already been added in an earlier step.

Applied to our context of phonological classes of German, an optimal walk through the S-graph is depicted in the lower right of Figure 8. It starts at the bottom node and runs first through the consonantal part and then through the vowel part of the S-graph. The walk through the vowel part of the S-graph is oriented counter-clockwise since this guarantees that while moving downwards as few labeled nodes with exactly one upper neighbor as possible are passed. The orientation of the walk through the consonantal part can be arbitrarily chosen. By traversing the depicted walk the following eight *pratyāhāra sūtras* for German can be read off:

dʒ tʃ ts ɾ l ŋ n m j ʔ h ç x g k M_1 t s ʃ f p M_2 pf b v M_3 ʒ z d g M_4 ə ɐ
ɪ aɪ ø y e i u o a M_5 ʊ ɔ aʊ̯ aː oː uː M_6 iː eː yː øː M_7 ɛː ɔɪ̯ œ ʏ ɛ M_8

The first sūtra results from collecting all unclassified sounds at the bottom node and then walking upwards to the nodes labeled ‘g’ and ‘k’. Since the walk goes downwards after reaching the node labeled ‘k’, a first marker M_1 has to be added to the sequence. The other sūtras are constructed analogously. One sound segment, namely [g] occurs twice in the list of *pratyāhāra sūtras*, namely in the first and the fourth sūtra. It is obvious that the phonological classes of German do not uniquely determine a list of *pratyāhāra sūtras*.[3] As mentioned, the *pratyāhāra sūtras* would differ if another walk through the S-graph would be chosen which could for example go first through the vowel part or traverse the consonantal part clockwise. Additionally, all sound segments by which a single node is labeled can be added to the *pratyāhāra sūtras* in any desired order.

[3] As mentioned in section 2.2, there are nearly 12 000 000 *pratyāhāra sūtras* of the same length for Sanskrit from which Pāṇini has chosen one sample (i.e., the Śivasūtras).

Finally, instead of duplicating the sound segment [g], the sound segment [k] could have been duplicated, resulting in different *pratyāhāra sūtras*.

The *pratyāhāra sūtras* given above yield the following pratyāhāras for the 10 phonological classes from our formal context in Figure 5:

gM_1 : Input to velar nasal assimilation
fM_3 : Input to labial nasal assimilation
bM_4 : Input to final devoicing
kM_2 : Output to final devoicing
uM_6 : Context of palatal fricative assimilation
aM_6 : Input to umlaut
yːM_8: Output to umlaut
aːM_7: Input to vowel shortening
øM_5 : Output to vowel shortening
əM_8 : Context of glottal epenthesis

In the remainder of this section we will demonstrate how the pratyāhāras for German can be employed for the description of the phonological rules discussed in section 3.2. The problem we are faced with is that phonological rules treat phonological classes not always as plain sets. In a phonological rule only the left and right contexts are unordered sets; the input and the output class has to be linearly ordered. The reason for this is that if a phonological rule is viewed as a rewriting rule then it has to be ensured that each segment of the input class has to be rewritten by its *corresponding* segment of the output class; e.g., the rule for final devoicing has to ensure that /d/ is rewritten as [t] and not as [k]. Pāṇini's Śivasūtras fulfill this constraint, his pratyāhāras are considered to be linearly ordered sets. But our formal model of the pratyāhāra technique so far does not take internally ordered sound classes into account. By our definitions pratyāhāras denote unordered sets. Our approach only guarantees that the resulting *pratyāhāra sūtras* offer the possibility to form for each phonological class a pratyāhāra which denotes the unordered set of the elements of the class. However, for the concrete example of phonological classes for German we were able to arrange the sounds in our *pratyāhāra sūtras* such that the order of the sounds in the input classes corresponds to the reversed order of the sounds in the output classes. Take for example the rule of final devoicing (rule (3) in section 3.2):

(3)' bM_4 → kM_2/_ word boundary *final devoicing*

Here, the pratyāhāra bM_4 denotes the class [b v ʒ z d g] and kM_2 the class [k t s ʃ f p]. Hence, kM_2 denotes the devoiced counterparts of the elements of bM_4 in reversed order. The remaining rules of section 3.2 can be stated in terms of pratyāhāras as follows:

(4)' /n/ → [ŋ] /_ gM_1 *velar nasal assimilation*

(5)' /n/ → [m] /_ fM_3 *labial nasal assimilation*

(6)' { } → [ʔ] / word boundary _ əM_8 *glottal stop epenthesis*

(7)' a:$M_7 \rightarrow øM_5$ / _unstressed *vowel shortening*

(8)' /ç/ → [x] / uM_6 _ *palatal fricative assimilation*

(9)' a$M_6 \rightarrow$ y:M_8 *umlaut*

Hence, all phonological processes of German described in section 3.2 can be rewritten with pratyāhāras of our proposed *pratyāhāra sūtras*.

5 Discussion

By describing the sound classes and phonological alternations of German with Pāṇini's pratyāhāra technique we have demonstrated that the pratyāhāra method is generalizable to the description of the phonological systems of other languages. Pāṇini's aim while constructing his Śivasūtras was to allow the formation of a pratyāhāra for every phonologically motivated class, i.e. for every class that is needed in the description of the phonological processes of Sanskrit. He did not construct pratyāhāras for all phonetically-based classes of sounds in a fashion that modern phonological features do. In the present article, we applied both methods, namely the description with *pratyāhāra sūtras* based solely on phonological processes and the description with features based on phonetic criteria, to a sample of phonological processes of German.[4] Our phonological system of German consisted of 52 sound segments. This yields a total of $2^{52} = 4\,503\,599\,627\,370\,496$ potential sound classes; just 10 of those classes are actually required for the description of our sample of phonological processes of German.

Any method for describing phonological classes which is not simply listing their elements overgenerates in the sense that it allows the formulation of classes which are needed in no phonological rule. The ratio of describable to actual classes constitutes a quality criterion for class-description devices [10]. Considering our example of German phonology, we get the following results: The pratyāhāras for the 10 required classes are given in section 4, but how many pratyāhāras can be formed with our *pratyāhāra sūtras*? Figure 9 lists the *pratyāhāra sūtras* for German and calculates for each single sūtra the number of pratyāhāras which can be build with its sound elements. For instance, each of the 5 sound elements of the second sūtra can be combined with any of the succeeding 7 markers $M_2 \ldots M_8$ in order to form a pratyāhāra. Altogether 268 pratyāhāras can be formed.[5] Although at first glance the ratio of describable to actual classes seems low for the pratyāhāra method, this method still performs better than the description with phonological features: In section 3 we used a total of 16 features to describe the phonological processes under consideration. But 16

[4] [13] combined the pratyāhāra account and the featural account in an analysis of the vowel system of German and a constructed language, by transferring the featural specifications for the vowels into pratyāhāras. It turned out that this demanded the duplication of disproportionately many sounds.

[5] Usually one would exclude pratyāhāras which are formed by the final sound segment and the marker of a *pratyāhāra sūtra* and which thus only denote single sound segments. Therefore, only 260 pratyāhāras of our *pratyāhāra sūtras* are well-formed.

dʒ tʃ ts ɾ l ŋ n m j ʔ h ç x g k M_1	15 × 8
t s ʃ f p M_2	5 × 7
pf b v M_3	3 × 6
ʒ z d g M_4	4 × 5
ə ɐ ɪ aɪ ø y e i u o a M_5	11 × 4
ʊ ɔ aʊ̯ aː oː uː M_6	6 × 3
iː eː yː øː M_7	4 × 2
ɛː ɔɪ̯ œ ʏ ɛ M_8	5 × 1

Fig. 9. The *pratyāhāra sūtras* for German in tabular form

features yield a total of $2^{16} = 65\,536$ classes which can be described by feature sets, thus far more than by pratyāhāras. Even if one drops the requirement that features should have acoustic, articulatory or auditory definitions and allows for unnatural features, a featural description of the 10 phonological classes cannot perform better than our pratyāhāra description. As none of our 10 classes can be described in terms of an intersection of some of the other classes, every featural description which is able to distinguish those 10 classes has to make use of at least 10 different features. A minimal featural description of the 10 classes would be to use the class identifiers (e.g., 'devoicing input', 'devoicing output') as features. Such a description would be minimal since none of the features would be reducible to other features. 10 features still yield a total of $2^{10} = 1\,024$ classes which can be described by feature sets; this is nearly four times more than by the pratyāhāras we employed.

One objection against our approach could be that we are only considering seven phonological processes while Pāṇini is describing many more processes in the *Aṣṭādhyāyī*. The main reason for this is that German exhibits far less Sandhi phenomena than Sanskrit. Furthermore, there is no standard description of the complete phonological system of German comparable to the *Aṣṭādhyāyī*. In this paper we refrained from testing how hyper-inclusive processes could yield a more economic description of German. The main reason for this is that our mathematical approach to the induction of the *pratyāhāra sūtras* is not yet able to automatically identify cases of harmless hyper-inclusivity. This is left for future research.

It is important to note that Pāṇini pursues a mixed strategy for the description of phonological classes: they are denoted by pratyāhāras (e.g., sūtra 6.1.77), they are referred to by the older phonetical *varga*-classification (e.g., sūtra 3.1.8) or their elements are simply listed (e.g., sūtra 1.1.24). In contrast to Pāṇini we restricted ourselves to the pratyāhāra-method.

References

1. Cardona, G.: Studies in Indian grammarians I: The method of description reflected in the Śiva-Sūtras. Transactions of the American Philosophical Society 59(1), 3–48 (1969)

2. Chomsky, N., Halle, M.: The Sound Pattern of English. Harper and Row, New York (1968)
3. Deshpande, M.M.: Ancient Indian phonetics. In: Koerner, E.F.K., Asher, R.E. (eds.) Concise History of the Language Sciences: From the Sumerians to the Cognitivists, pp. 72–77. Elsevier, Oxford (1995)
4. Ganter, B., Wille, R.: Formal Concept Analysis. Mathematical Foundations, Berlin (1999)
5. Hall, T.A.: Phonologie: Eine Einführung. Walter de Gruyter, Berlin (2000)
6. Hamann, S.: The Phonetics and Phonology of Retroflexes. Ph.D. thesis, Utrecht Institute of Linguistics (2003)
7. Jakobson, R., Fant, G., Halle, M.: Preliminaries to Speech Analysis: The Distinctive Features and their Correlates. MIT Press, Cambridge (1952)
8. Kenstowicz, M.: Phonology in Generative Grammar. Blackwell, Cambridge (1994)
9. Kiparsky, P.: Economy and the construction of the Śivasūtras. In: Deshpande, M.M., Bhate, S. (eds.) Pāṇinian Studies, Ann Arbor, Michigan (1991)
10. Kornai, A.: The generative power of feature geometry. Annals of Mathematics and Artificial Intelligence (8), 37–46 (1993)
11. Kuratowski, K.: Sur le problème des courbes gauches en topologie. Fundamenta Mathematicae 15, 271–283 (1930)
12. Misra, V.N.: The Descriptive Technique of Pāṇini. An Introduction. Mouton & Co., The Hague (1966)
13. Petersen, W.: A mathematical analysis of Pāṇini's Śivasūtras. Journal of Logic, Language, and Information 13(4), 471–489 (2004)
14. Petersen, W.: Zur Minimalität von Pāṇinis Śivasūtras – Eine Untersuchung mit Methoden der Formalen Begriffsanalyse. Ph.D. thesis, University of Düsseldorf (2008)
15. Petersen, W.: On the construction of Śivasūtras-alphabets. In: Kulkarni, A., Huet, G. (eds.) Sanskrit Computational Linguistics. LNCS (LNAI), vol. 5406, pp. 78–97. Springer, Heidelberg (2009)
16. Platt, C.R.: Planar lattices and planar graphs. Journal of Combinatorial Theory (B) 21, 30–39 (1976)
17. Staal, F.J.: A method of linguistic description. Language 38, 1–10 (1962)
18. Staal, F.J.: The Sanskrit of science. Journal of Indian Philosophy 23(1), 73–127 (1995)
19. Vennemann, T.: The German velar nasal: a case for abstract phonology. Phonetica 22(65-81) (1970)
20. Whitney, W.D.: Sanskrit Grammar. Harvard University Press, Cambridge (1889)
21. Wiese, R.: Phonology of German. Oxford University Press, Oxford (1996)

Building a Prototype Text to Speech for Sanskrit

Baiju Mahananda[1], C.M.S. Raju[2], Ramalinga Reddy Patil[2], Narayana Jha[1], Shrinivasa Varakhedi[3], and Prahallad Kishore[2]

[1] SCSVMV University, Kanchipuram
[2] International Institute of Information Technology, Hyderabad
[3] Sanskrit Acadamy, Osmania University, Hyderabad
{baijunanda,mouli.raju,patilrlreddy53,shrivara}@gmail.com,
kishore@iiit.ac.in
http://www.iiit.ac.in

Abstract. This paper describes about the work done in building a prototype text to speech system for Sanskrit. A basic prototype text-to-speech is built using a simplified Sanskrit phone set, and employing a unit selection technique, where prerecorded sub-word units are concatenated to synthesize a sentence. We also discuss the issues involved in building a full-fledged text-to-speech for Sanskrit.

Keywords: Text to Speech for Sanskrit, Festvox.

1 Introduction

Developing a text to speech (TTS) system for Sanskrit primarily involves studying its phonology and phonetics [2]. Sanskrit has a highly developed system of phonemic description, developed in ancient times mainly for preserving Vedic texts[3].

1.1 Goal

We try to develop a text to speech system for simple classical Sanskrit, using simplified phone set of Sanskrit. This uses Direct synthesis, in which the speech signal is generated by direct manipulation of its wave form representation. Wave form concatenation, is representative of this synthesis category. In this approach, several fundamental periods of pre-recorded phonemes are simply concatenated. The phonemes are then connected to form words and sentences [1].

1.2 Previous Work

Making a computer to talk in different languages has been attempted by many scholars from many years and they are successful. We are making an attempt to take forward the work done on "Making a Computer to talk Sanskit". The following are few studies in this path:

1. Vani - An Indian Language Text to speech Synthesizer for Sanskrit [15], is a speech synthesizer which employs formant synthesis, in which the basic

G.N. Jha (Ed.): Sanskrit Computational Linguistics, LNCS 6465, pp. 39–47, 2010.

assumption is that the vocal tract transfer function can be satisfactorily modeled by simulating formant frequencies and formant amplitudes. The synthesis thus consists of the artificial reconstruction of the formant characteristics to be produced. This is done by exciting a set of resonators by a voicing source or noise generator to achieve the desired speech spectrum, and by controlling the excitation source to simulate either voicing and voicelessness. The addition of a set of anti-resonators furthermore allows the simulation of nasal tract effects, fricatives and plosives [1].

2. Text to speech synthesis for Indian languages (Acharya), is a syllable level representation of the text and each syllable directly translates into a sound that can be synthesized or simply played from a prerecorded piece of audio [2].
3. Text to Speech conversion systems developed by C-DAC (Centre for Development of Advanced Computing) for various Indian languages supplementing the GIST Card [3].

Speech synthesis by concatenation of sub-word units (e.g. diphones) has become basic technology. It produces reliable clear speech and is the basis for a number of commercial systems. However with simple diphones, although the speech is clear, it does not have the naturalness of real speech[16].

In this work we are going to present you the general issues in building a TTS for Sanskrit, which could help in building an efficient Speaking system for Sanskrit.

2 Sanskrit Orthography

2.1 Nature of Sanskrit Scripts

The basic units of the writing system in Sanskrit are characters which are an orthographic representation of speech sounds [3]. A character in Indian language scripts is close to a syllable and can be typically of the form: C*VN, where C is a consonant, V is a vowel and N is anusvAra, visarha, jivhAmUllya e.tc. . There is fairly good correspondence between what is written and what is spoken [5]. Sanskrit has richer set of characters than many of the other Indian languages.

2.2 Phone Set

Most institutions and scholars propose "Pāṇinīya - Śikṣa" as a reference for written form of Sanskrit[8], in this book it is said that "The speech-sounds in Prakrit and Sanskrit are 63 or 64, according to their origin, has been said by Brahman (Svyambhū) himself." among which vowels are 21, and consonants are (included stop, approximent, sibilant, nasal, palatels) 43 [4]. We are using UTF-8 format to represent the text. In our system we use 8 vowels and 40 consonants including other symbols, which are given in the below picture (Fig. 2) with respective IT3 representations [1].

[1] Since the present computers can only process ASCII characters at machine level, we use a codding mechanism to represent each character (may belong to any language) in an intermediate format. This format was formulated by CMU.

Vowels

अ	आ	इ	ई	उ	ऊ	ए	ओ
a	aa	I	ii	u	uu	e	o

Consonants

क	ख	ग	घ	ङ	त	थ	द	ध	न
ka	kha	ga	gha	ng~a	t:	tha	da	dha	na
च	छ	ज	झ	ञ	प	फ	ब	भ	म
cha	chha	ja	jha	nj~a	pa	pha	ba	bha	ma
ट	ठ	ड	ढ	ण	श	ष	स	ह	
t:a	t:ha	d:a	d:h	nd~a	sha	shha	sa	ha	

Other symbols

ऋ	ॠ	ऌ	ऐ	औ	ं	ः	य	र	ल	व
rx	rx~	lx	ai	au	n:	h:	ya	ra	la	va

Fig. 1. Set of phones and their respective IT3 Format of Devanagari Script used in TTS

There has been many discussions on the character set of Sanskrit, but some say devanagari alphabets are developed for writing Sanskrit (which are descended from brahmi scripts) [9], however as we all can see Sanskrit can be written in multiple languages.

Letters not used in TTS - ऽ, जिह्वामूलीयः, उपध्मानीय, प्लुत, कुं, खुं, गुं, घुं

2.3 Suitability of Sikshaa Granthas for Building a TTS

Shiksha (Phonetics) explains the proper articulation and pronunciation of vedic texts. There are six parts of Shiksha letters (varnas), accent (swara), time consumed in articulating vowel (matra), effort (bala) Melodius chanting of mantras (sama) and conjugation of letters (sandhi). If some mistake is committed in any of the above six, instead of giving the desired result it can prove to be disastrous as well [8].

In order to make an efficient talking system all the parts of shiksha are to be met, however the present day systems can only add some parts of siksha, research has to be done to add the above six to the synthesized speech.

ystem Architecture

iskrit TTS system is implemented within the Festival/Festvox framework. egmentation, unit selection and synthesis are done using Festvox and Festival framework.

3.1 Why Festvox?

Frameworks are built to ease the software development. But the frameworks which can be customized for user need are mostly preferred. Festvox offers such a facility and it is an open software.

4 Letter to Sound Rule

Letter-to-sound rules are almost straight forward in Indian languages, as they are phonetic in nature. We write what we speak and we speak what we write, and hence generally the necessity of a pronunciation dictionary does not arise in our case. The pronunciation for a Sanskrit word such as chatura (clever) (the word chatura is the IT3 representation of the word "चतुर" using the chart specified above)in terms of phones marked with syllable boundaries can be written as ((ch a) 1) ((t u) 0) ((r a) 0). As the characters in Indian language are close to a syllable, clustering C*VN can be done easily taking into account a few exceptions [6].

In this work, simple syllabification rules are followed. Syllable boundaries are marked at the vowel position. If the number of consonants between two vowels is more than one, then first consonant is treated as coda of the previous syllable and the rest of the consonant cluster as the onset of the next syllable. For stress assignment, the primary stress is associated with the first syllable and secondary stress with the remaining syllables in the word. The integer 1 assigned to first syllable in the word chatura indicates the primary stress associated with it. Letter to sound rules, syllabification rules and assignment of stress patterns are different for different languages, which has to be specified accordingly. The architecture of Festival synthesis engine allows these rules to be written in Scheme, so that they get loaded at the run time, essentially avoiding recompilation of the core code for every new language.

5 Creation of Speech Database

A common trend in concatenative approach for TTS system is to use large database of phonetically and prosodically varied speech.

5.1 Text Selection for Sanskrit

The quality of data-driven synthesis approaches is inherently bound to speech database from which the units are selected. It is important to have an optimal text corpus balanced in terms of phonetic coverage and the diversity in the realizations of the units. Sanskrit has a vast literature starting from Vedic text to latest classical literature. In this work, the test corpus was generated from a set of simple Sanskrit sentences selected from daily life conversations . Some of the example sentences are given below. A common trend in concatenative approach for TTS system is to use large database of phonetically and prosodically varied speech.

हरिः ओम्! अस्ति किं गृहे ?

किं भोः, मन्त्री इव विलम्बेन आगच्छति?

मम नाम श्रीधरः।

कार्यक्रमः कथम् आसीत् ?

आगच्छतु! किम् आवश्यकम्?

5.2 Speech Data Collection

The number of Sentences recorded to generate Sanskrit database is 852. In these sentences we tried to cover the possible conversations related to our daily life. As it is not possible to complete the recording in a single session, to ensure consistency, the recordings were made at the same time every day.

6 Speech Segmentation

One of the most important tasks in building speech databases is the annotation of speech data with its contents (labeling) and the time alignment between labeling and speech (segmentation). Phoneme segmentation and labeling are highly desirable and useful for TTS as this information is used for classifying the speech units that helps to select and concatenate the right units in terms of linguistic and acoustic features.

The most precise way to annotate speech data is manually by linguistic experts. However, manual phoneme labeling and segmentation are very costly and require much time and effort. Even well trained, experienced phoneme labelers using efficient speech display and editing tools require about 200 times real time to segment and align speech utterances. To reduce this effort considerably and aid the phoneme labelers, an automated segmentation tool is required. For automatic phoneme segmentation, we have been using the most frequently used Tool EHMM based phoneme recognizer.

Spoken Sentence (UTF-8): आगच्छतु! किम् आवश्यकम्?
Spoken Sentence (IT3): aagachhatu! kim aavashyakam?

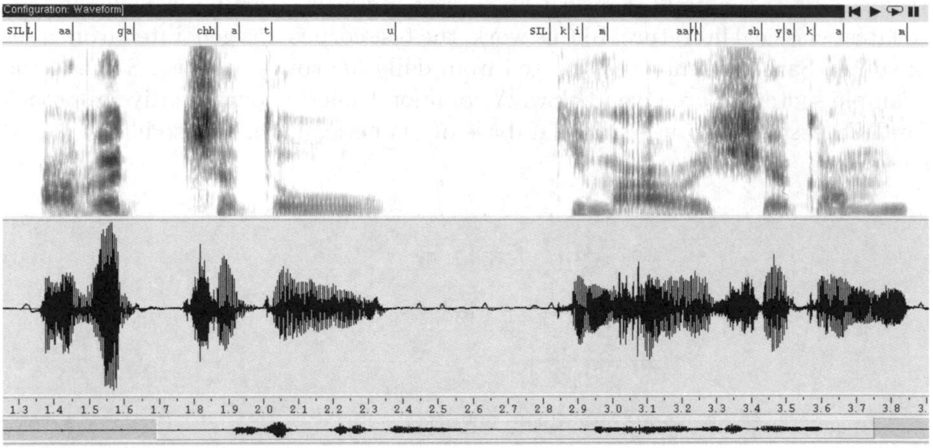

Fig. 2. Sample speech used for training with its labels, spectrogram and wave form

7 Unit Clustering and Synthesis

Given these segments, the unit selection algorithm (a Statistical Parametric Synthesizer) in the framework clustered the phones based on their acoustic differences. These clusters are then indexed based on higher level features such as phonetic and prosodic features. During synthesis, the appropriate clusters are sought using phonetic and prosodic features of the sentence. A search is then made to find a best path through the candidates of these clusters. Though the units used here are phones, the acoustic frame of previous unit is used during clustering as well as for concatenation.

Clustergen is a statistical parametric synthesizer released as part of the Festival distribution [11]. It predicts frame-based MFCCs clustered using phonetic, metrical, and prosodic contexts. Unlike CLUNITS, the unit size is one frame (5ms by default), and the signal is partitioned at the HMM-state size level (3 states per phone). The clustering, done via CART, optimizes the standard deviation of the frames in the cluster. The frames are 24 coefficient MFCCs plus F0. Clustergen offers a number of options for clusters which can be single frames, trajectories, or trajectories with overlap and add. We used the simplest model for our TTS. Synthesis is done by predicting the HMM-state duration, then predicting each frame with the appropriate CART tree. The track of MFCC plus F0 vectors is re-synthesized with the MLSA algorithm [13], as implemented in the HTS system and already implemented within Festival.

8 Evaluation

In order to evaluate we conducted subjective and objective evaluations. As part of subjective evaluation 10 samples were extracted from the database. These

Table 1. MOS Test Results

Sentence Name	Least Score	Best Score	Average Score
1	3	4	3.4
2	3	5	4.2
3	2	4	3.1
4	4	5	4.5
5	1	3	2.2
6	2	4	3.3
7	4	4	4
8	5	5	5
9	3	4	3.5
10	2	5	3.4
Total Average			3.67

samples were played to 10 Sanskrit speakers for obtaining mean opinion scores (MOS), i.e., score between 1 (worst) to 5 (best) [7]. The results were given in the Table 1.

For objective evaluation we have choose MCD. Mel cepstral distortion (MCD) is an objective error measure used to compute cepstral distortion between original and the synthesized MCEPs. Lesser the MCD value the better is synthesized speech. MCD is essentially a Euclidean Distance defined as

$$MCD = \frac{10}{\ln 10} * \sqrt{2 * \sum_{i=1}^{25} (mc_i^t * mc_i^e)^2} \quad (1)$$

where mcti and mcei denote the target and the estimated MCEPs, respectively. MCD is used as an objective evaluation of synthesized speech. Informally it is observed in [11] that an absolute difference of 0.2 in MCD values makes a difference in the perceptual quality of the synthesized signal and typical values for synthesized speech are in the range of 5 to 8.

Table 2. MCD Test Results

	Synthesis with Diphones
MCD	6.474

To compute MCD, we have taken 20 sentences from Sanskrit database and synthesized using diphone as unit.

9 Discussion and Future work

So far, we have discussed the build process of a prototype text-to-speech for Sanskrit. While we have demonstrated the usefulness of this prototype text-to-speech for Sanskrit by conducting listening tests, it should be noted that there exists several research issues that need to be addressed in building a full-fledged text-to-speech for Sanskrit. The following are a few research issues that need to be addressed in building a complete voice for Sanskrit.

1. Choice of Phoneset: In this work, we have used a simplified phoneset (mostly borrowed from Hindi) to develop this prototype text-to-speech system. A careful acoustic-phonetic study needs to be done to build a phone set for Sanskrit. Also, the relationship between Akshara and the sound is assumed to one-to-one in Sanskrit. However, it is often may not be the case. Typically it is known that the Sanskrit scholars in the Northern part of India drop Schwas at the end. For example, /raama/ is pronounced as /raam/. Hence a carefully analysis need to be done to derive a Akshara (written symbol) to phone (spoken sound) correspondence.
2. Choice of Unit: In this work, we have used diphone as a unit for concatenation. This type of unit is found to be useful for English. It is also well known that syllable is a better unit for Indian languages which are mostly derived from Sanskrit. Hence, it is important to study various levels of units (diphone, syllable, polysyllable) for the case of Sanskrit TTS.
3. Nature of Sanskrit Language: In this work, the prototype TTS is built for spoken on conversational form of Sanskrit. It is known that the Sanskrit comes in various flavors such as Vedic. Poetry prose, classical etc. and hence the nature of the Sanskrit language has to be studied and considered in developing a complete text-to-speech for Sanskrit.
4. Prosody: Prosody is often found to be manifested in intonation (rhythm), stress (energy), and duration patterns. Sanskrit is known for its richness in prosody. Hence, a study on prosodic aspects of Sanskrit TTS has to be conducted to identify suitable acoustic properties that can incorporated in TTS.
5. Incorporation of Sandhi Rules: Sanskrit is also known for Sandhi rules. The effect of these Sandhi rules has to be studied and understood in order to implement these rules in Sanskrit TTS.

Acknowledgments

We would like to thank all the persons who participated in conducting the subjective evaluation.

References

1. Formant Synthesis by Thomas Styger and Eric Keller, Laboratoire danalyse informatique de la parole (LAIP), Université de Lausanne, Switzerland
2. Acharya Project by IIT, Madras. Multilingual computing fo r Literacy and Education, `http://acharya.iitm.ac.in/disabilities/tts.php`
3. Ramani, S., Chandrasekar, R., Anjaneyulu, K.S.R. (eds.): KBCS 1989. LNCS, vol. 444. Springer, Heidelberg (1990)

4. पाणिनीय शिक्षा verse 3 and 4.
5. Sarkar, T., Keri, V., Yuvaraj, S., Prahalad, K.: Building Bengali Voice Using Festival. In: Proceedings of ICLSI 2005, Hyderabad, India (2005)
6. Kishore, S.P., Sangal, R., Srinvas, M.: Building Hindi and Telugu Voices using Festvox. In: Proceedings of International Conference on Natural Language Processing, ICON (2002)
7. Raghavendra, E.V., Desai, S., Yegnanarayana, B., Black, A.W., Prahallad, K.: Global Syllable Set for Building Speech Synthesis in Indian Languages. In: Proceedings of IEEE workshop on Spoken Language Technologies, Goa, India (December 2008)
8. About Sanskrit, `http://www.sanskrit.nic.in/ABOUTSANSKRIT1.htm`
9. Ager, S.: Omniglot writing systems and languages of the world, `http://www.omniglot.com/writing/devanagari.htm`
10. Veera Raghavendra, E., Prahallad, K.S.: Database Pruning for Indian Language Unit Selection Synthesizers. In: ICON-2009, Hyderabad, India (December 2009)
11. Black, A.W.: CLUSTERGEN: a statistical parametric synthesizer using trajectory modeling. In: Proceedings of Interspeech, pp. 1762–1765 (2006)
12. Black, A.W., Lenzo, K.: Building voices in the festival speech synthesis system (2000), `www.festvox.org/festvox/index.html`
13. Tokuda, K., Yoshimura, T., Masuko, T., Kobayashi, T., Kitamura, T.: Speech parameter generation algorithms for HMM-based speech. In: Proceedings of ICASSP, Istanbul, Turkey, pp. 1315–1318 (June 2000)
14. Black, A.W., Bennett, C.L., Blanchard, B.C., Kominek, J., Langner, B., Prahallad, K., Toth, A.R.: CMU Blizzard 2007: A Hybrid Acoustic Unit Selection System from Statistically Predicted Parameters. In: Blizzard Challenge 2007, Bonn, Germany (2007)
15. Jain, H., Kande, V., Desikan, K.: Vani - An India Language Text to speech Synthesizer. IIT, Mumbai
16. Black, A.W., Taylor, P.: Automatically Clustering Similar Units for Unit Selection in Speech Synthesis (1997)

Rule-Blocking and Forward-Looking Conditions in the Computational Modelling of Pāṇinian Derivation

Peter M. Scharf

Department of Classics,
Brown University
scharf@brown.edu

Abstract. Attempting to model Pāṇinian procedure computationally forces one to clarify concepts explicitly and allows one to test various versions and interpretations of his grammar against each other and against bodies of extant Sanskrit texts. To model Pāṇinian procedure requires creating data structures and a framework that allow one to approximate the statement of Pāṇinian rules in an executable language. Scharf (2009: 117-125) provided a few examples of how rules would be formulated in a computational model of Pāṇinian grammar as opposed to in software that generated speech forms without regard to Pāṇinian procedure. Mishra (2009) described the extensive use of attributes to track classification, marking and other features of phonetic strings. Goyal, Kulkarni, and Behera (2009, especially sec. 3.5) implemented a model of the asiddhavat section of rules (6.4.22-129) in which the state of the data passed to rules of the section is maintained unchanged and is utilized by those rules as conditions, yet the rules of the section are applied in parallel, and the result of all applicable rules applying exits the section. The current paper describes Scharf and Hyman's implementation of rule blocking and forward-looking conditions. The former deals with complex groups of rules concerned with domains included within the scope of a general rule. The latter concerns a case where a decision at an early stage in the derivation requires evaluation of conditions that do not obtain until a subsequent stage in the derivation.

1 Implementations of Sandhi and Inflection

Scharf and Hyman implemented Pāṇinian sandhi rules in a portable framework using modified regular expressions in an XML file to model Pāṇinian rules, as described in Scharf 2009: 118-120. Each rule is written as one or more XML rule tags each of which contains several parameters: source, target, lcontext, rcontext, optional, and c. The optional parameters lcontext and rcontext specify the left and right contexts for the replacement of the source by the target. The optional parameter `optional` specifies that the current state is to be duplicated and subsequent parallel paths created, one in which the rule is implemented and the other in which it is not. The parameter c (for comment) contains the

G.N. Jha (Ed.): Sanskrit Computational Linguistics, LNCS 6465, pp. 48–56, 2010.

number of the Pāṇinian rule implemented by the rule tag. The implementation utilizes the Sanskrit Library Phonetic encoding scheme SLP1, described in Scharf and Hyman 2010, Appendix B, in which Sanskrit sounds and common phonetic features such as tones and nasalization are each represented by a single character.

The rule syntax utilizes a number of macros that model Pāṇinian structures. Macros are used to model Pāṇinian sound classes: varṇa, varga, guṇa, vṛddhi, saṁprasāraṇa, etc.; to create pratyāhāras: अक्, अण्, इक्, यण्, etc.; and to group sounds with common phonetic features: aspirated sounds, unaspirated sounds, voiced sounds, unvoiced sounds, etc. For example, the macros @(f) and @(x) in 1.1.9 vt. represent the varnas ऋ and ऌ respectively. The macros @(eN) in 6.1.109 and @(ac) in 6.1.78 represent the pratyāhāras एङ् (monothongs) and अच् (vowels), respectively. Mappings are used to map sets of sounds onto corresponding sounds, such as short vowels onto long, and unvoiced stops onto voiced stops. Functions, such as `lengthen`, `guRate`, and `vfdDiize`, utilize the mappings to facilitate implementation of common operations, namely, the replacement of a vowel by its corrresponding long vowel, guṇa vowel, or vṛddhi vowel, respectively.

Rules are not pre-selected by hand; rather they are triggered by the data that meets the conditions for the application of the rule. Yet rules are arranged in sequence and placed in blocks to ensure appropriate application of general rules and exceptions. In particular, negations and identical replacements that are exceptions select paths that avoid the application of rules of which they are negations and exceptions. Hyman wrote a Perl program that converts the XML file of regular expressions to Perl executable code. The model succeeds in encoding Pāṇinian rules in a manner that allows the rules that come into play to be tracked. Rule tracking has valuable research and pedagogical applications. Hyman (2009) describes the procedure by which the XML vocabulary to express Pāṇini's sandhi rules was developed and how a series of stages converts the rules not only into executable Perl code, but also into a network, and a finite state transducer. The latter, being extremely fast, will permit realtime web use of the models.

Scharf and Hyman augmented the XML data structures created to model Pāṇinian sandhi to allow derivation of nominal stems, as described in Scharf 2009: 120-23. They introduced an additional parameter `morphid` in the XML rule tag that utilizes Scharf's (2002: 29-30) set of nominal inflection tags. They further enriched the XML structure utilized for nominal declension to model verbal conjugation. Scharf (2009: 123-125) described Scharf and Hyman's implementation of a computational model of Pāṇinian verbal inflection in a single cascade of rules that apply to whatever strings meet their conditions. The procedure begins with a single set of verbal terminations for all verbs, just as Pāṇini does, and introduces replacements on the basis of phonetic context. They widened the set of tags employed by the parameter `morphid` to utilize, in addition to nominal inflection tags, Scharf's (2002: 30-31) verbal inflection tags. They added two parameters to the rule tag: `lexid`, and `root`. The former allows reference to the class of the root in the Pāṇinian *Dhātupāṭha*. The latter allows reference to the original form of the root even when the previous rules have modified the input string. The implementations of Pāṇinian inflection include rule tracking so that a derivational history of the form can be provided.

2 Participles

In 2008, Scharf and Hyman enriched the XML tagset further to allow derivation of participle stems. A `vlexid` parameter was added to allow reference to the class of the verbal root in the *Dhātupāṭha* and an `affix` parameter to allow reference to the affix in the form in which it was originally introduced. The implementation of 1.2.21 demonstrates both parameters. 1.2.21 is a negation rule that optionally denies क्-marking to the affixes क्त and क्तवत् that have the initial augment इ, when they occur after a verbal root with a penultimate short उ.

```
<rule source="k" target=""
lcontext="@(begin)[@(al)]*u[@(hal)]#i:(?:ta|tavat);[@(it)]*"
rcontext="[@(it)]*$" vlexid="^vt?1" lexid="^(ppp|pap)$"
affix="@(nizWA)"
root="^[@(char)]*`?[@(char)]*u[@(hal)];?[@(char)]*$"
optional="yes"
c="1.2.21 udupaDAdBAvAdikarmaRoranyatarasyAm"/>
<!--Only roots with the stem-forming affix शप्, for now roots
of class 1 or 10, are subject to the negation of k-marking. Kā-
śikā: व्यवस्थितविभाषा चेयम्।तेन शब्विकरणानामेव भवति। Rule includes
vlexid.-->
```

The parameter `source` has the value of the क् to be deleted; the parameter `target` has the null string value. The parameter `lcontext` has the value of a regular expression matching strings preceding the `source` parameter. These include conditions specified in the ablative in the rule, namely, a string in which there is a penultimate short उ, as well as any markers that might occur prior to the marker क् to be deleted. The pound sign in the string marks the morpheme boundary between the stem and the affix. This is followed by the initial augment इ separated from the rest of the affix by a colon. The subsequent parenthesis matches either `ta` or `tavat` representing the phonetic strings त or तवत् respectively. A semicolon separates the phonetic form of the affix from its markers, regardless of whether they are initial or final. The value of the lcontext parameter ends with an expression including a macro reference that matches any possible marker.

```
<macro name="it" value="ufkNcYRtnprlS" c="1.3.2-8"/>
```

The `rcontext` parameter in the implementation of 1.2.21 again has the value of any marker that might follow क् represented in the `source` parameter. Now it is possible that the string त belongs to some affix other than क्त. It is also possible that an affix has a final marker क् rather than an initial one. In order to ensure reference to the desired affix, the original form of the affix is included as the value of the `affix` parameter. Here the location of markers as initial or final is preserved. The affixes in question are introduced with an initial marker क् and are termed निष्ठा. Initial and final markers are separated from the phonetic form of the affix by a grave accent and a semicolon respectively. A macro implements the classification rule as follows:

```
<macro name="nizWA" value="k`ta|k`tavat;u"
c="1.1.26 ktaktavatU nizWA"/>
```

The macro name `nizWA` is then employed in the implementation of 1.2.21, just as the technical term निष्ठा is in 1.2.19 from where it recurs. Recurring terms are explicitly stated in XML implementations of rules. The option parameter has the value `yes` which initiates two streams of derivation from this point forward, one in which the rule is applied and one in which it is not.

Now the *Kāśikā* on 1.2.21 states that the option is specifically distributed (व्यवस्थित) in that the rule applies only when the affixes occur after roots to which the stem-forming affix (*vikaraṇa*) शप् will be introduced (व्यवस्थितविभाषा चेयम्। तेन शब्विकरणानामेव भवति।). In order to limit the rule in accordance with the *Kāśikā*'s statement, the rule includes the parameter `vlexid` which has the value of the lexical tag assocated with the verbal root in our digital *Dhātupāṭha*. Pertinently, `vlexid` contains the roots' class. The example shown of the implementation of 1.2.21 demonstrates the utility of the two new parameters `vlexid` and `affix`.

3 Blocking

Scharf and Hyman enriched the structure of the xml file containing rules for participle derivation to allow complex blocking relations. Pāṇini formulates general rules (उत्सर्ग) and exceptions (अपवाद). Exceptions take precedence over their related general rule. Where the exception alters the string so that it no longer meets the conditions for the application of the general rule, it is easy to implement exceptions by simply ordering the rules in such a way that the exceptions have the opportunity to apply first. This is achieved by placing them earlier in the cascade of rules. However, rule ordering alone is insufficient to capture the structure of Pāṇini's blocking relations where the exceptions do not alter the string. This is particularly obvious where the exception takes the form of a negation. In previous versions of their framework, Scharf and Hyman rewrote general rules that had negative exceptions to reflect the narrower scope of application in the conditions of the general rule itself. Such a procedure makes it more difficult to implement accurate rule tracking. Hence we have enriched the framework of the XML rule file to reflect blocking.

Let's take, as an example, rules concerning the provision of the initial augment इ. 7.2.35 provides that the augment इ occurs as the initial part of an ārdhadhātuka affix that begins with a consonant other than य्.

```
<rule source="" target="i:" lcontext="#"
rcontext="[@(val)][@(al)]*@(anta)" affix="^(@(ArDaDAtuka))$"
c="7.2.35 ArDaDAtukasyeq valAdeH"/>
```

For the purposes of the participal derivation in this ruleset, the macro name ArDaDAtuka refers to any one of several affixes that form what are typically called participals as follows:

```
<macro name="ArDaDAtuka"
value="k`vas;u|k`Ana;c|k`ta|k`tavat;u|k`tvA|tum;un|@(kftya)"
c="3.4.114 ArDaDAtukaM SezaH, 3.4.115 liw ca"/>
```

The macro kftya again refers to what are called gerundives:

```
<macro name="kftya"
value="tavya|tavya;t|anIya;r|ya;t|k`ya;p|R`ya;t"
c="3.1.95 kftyAH prANRvulaH"/>
```

The general rule above accounts for the इ-augment in the infinitive लवितुम् (< लू 'cut' + इ-तुम्) and the gerundive stem लवितव्य (< लू 'cut' + इ-तव्य), for example.

Several negations that disallow the addition of the initial augment इ are exceptions to 7.2.35. Their domains are entirely included within the domain of the general rule; hence these negations would have no scope of application if they were not given precedence over 7.2.35. For example, 7.2.8 provides that the initial augment इ does not occur in affixes termed *kṛt* that begin with a voiced consonant other than य् or ह्. The domain of 7.2.8 is included within the domain of 7.2.35 because all kṛt-affixes that begin with a voiced consonant other than य् or ह् are also ārdhadhātuka affixes that begin with a consonant other than य्. Obviously the set of voiced consonants excluding य् and ह् is a subset of the set of consonants excluding य्. Moreover, most kṛt-affixes are ārdhadhātuka; only eight kṛt-affixes are termed *sārvadhātuka* rather than *ārdhadhātuka* due to being marked with श्, but they all begin with a vowel, in particular with अ or आ.[1] Since the domain of 7.2.8 is entirely included within the domain of 7.2.35, 7.2.8 is an exception to 7.2.35 and takes precedence over it. The XML rule 7.2.8 is formulated as follows:

```
<rule source="" target="" lcontext="#"
rcontext="[@(vaS)][@(al)]*;[@(it)]*k[@(it)]*$"
affix="^(@(kft))$" c="7.2.8 neqvaSi kfti"/>
```

The rule includes reference in the affix parameter to the macro kft, which lists several kṛt-affixes relevant to participle formation , including absolutive and non-Vedic infinitives.

[1] शतृ (3.2.124, 3.2.130), शानच् (3.2.124), शानन् (3.2.128), चानश् (3.2.129), खश् (3.2.28, 3.2.83), श (3.1.137, 3.3.100), and शध्यै and शध्यैन् (3.4.9). Note that the vikaraṇas श्नु (3.1.73, 3.1.82), श्नम् (3.1.78), and श्ना (3.1.81, 3.1.83) are taught before the heading धातोः in 3.1.91; hence they are not subject to being termed *kṛt* according to the *Kāśikā* on 3.1.93 which limits the rule to the range of the heading धातोः in 3.1.91 (अस्मिन् धात्वधिकारे, etc.). Although these vikaraṇas are not subject to the इ-augment's negation by 7.2.8, they are not subject to its provision by 7.2.35 or any other rule either.

```
<macro name="kft"
value="S`at;f|S`Ana;c|S`Ana;n|c`Ana;S|k`vas;u|k`Ana;c|k`ta|
k`tavat;u|k`tvA|tum;un|@(kftya)"
c="3.1.93 kfdatiN"/>
```

The exception 7.2.8 does nothing to change the conditions that would allow the general rule 7.2.35 from applying subsequently; it replaces nothing by nothing at the beginning of the affix. Now, if 7.2.8 were placed prior to 7.2.35 in the cascade of rules without any other restriction, 7.2.35 would proceed to add the इ-augment there. This is not desired. 7.2.8 should prevent 7.2.35 from applying. To achieve this, Hyman created an XML structure similar to an `otherwise` statement found in some programming languages. The structure groups rules within a block that contains two subsections: a `try` section and an `otherwise` section, as shown:

```
<block>
  <try>
    <rule/>
    <rule/>
    ...
  </try>
  <otherwise>
    <rule/>
    <rule/>
    ...
  </otherwise>
</block>
```

To effect the blocking of 7.2.35 by 7.2.8, then we write 7.2.8 in the `try` block and 7.2.35 in the `otherwise` block as follows:

```
<block>
  <try>
    ...
    <rule source="" target="" lcontext="#"
      rcontext="[@(vaS)][@(al)]*;[@(it)]*k[@(it)]*$"
      lexid="^(@(kft))$" c="7.2.8 neqvaSi kfti"/>
    ...
  </try>
  <otherwise>
    <rule source="" target="i:" lcontext="#"
      rcontext="[@(val)][@(al)]*@(anta)"
      affix="^(@(ArDaDAtuka))$"
      c="7.2.35 ArDaDAtukasyeq valAdeH"/>
  </otherwise>
</block>
```

All instances of exceptions whose domains are totally included within the domain of their related general rule can be handled similarly. In instances of partial blocking, the rule that contains a domain partially included within the domain of a related general rule must be split into a rule with a totally included domain and a rule with an excluded domain. Only the rule with the totally included domain will be placed in the `try` block.[2]

4 Look Ahead

Some rules in the *Aṣṭādhyāyī* apply only under conditions that are not created until subsequent rules apply. Let's consider the case of 7.2.67, necessary for the derivation of perfect active participles. Rules in the अङ्ग-section (6.4.1–7.4.97) apply before rules for doubling and related stem-internal changes (6.1.1 etc.). 1.1.59 द्विर्वचनेऽचि provides that the replacement of a vowel whose replacement is conditioned by a vowel-initial affix has the status of its substituend for the purpose of doubling in the section of rules headed by 6.1.1 एकाचो द्वे प्रथमस्य (See Cardona 1997: 60). This implies that the augment इ precedes vowel deletion which in turn precedes doubling. Likewise, rules providing the augment इ to the beginning of an affix apply before rules that change the stem prior to the augmented affix. Hence 7.2.67 applies to add the augment इ to the affix -क्वसु before 6.4.98 applies to delete the penultimate vowel of the root before vowel-initial affixes. The augment must be provided first because without it, the affix would not be vowel-initial. 6.4.98 in turn applies prior to 6.1.8 लिटि धातोरनभ्यासस्य. However, 7.2.67, the rule that provides the augment, includes among its conditions that the root be single-syllabled, and the *Kāśikā* explains that this refers to the root after doubling has applied. But doubling can't apply until after the augment इ is added, which conditions deletion of the penultimate vowel. It is not possible to evaluate the condition of being a single-syllabled doubled root at the time the augment is added. It is not possible to evaluate a state of affairs brought about subsequent to a rule as a condition at the time of application of that rule. Hence it is necessary to apply the rule tentatively, then trace the derivation subsequent to the application of the rule to the point where its conditions can be evaluated before deciding whether to apply the rule or not. Our XML framework implements look ahead to achieve this by deriving the form both with and without the rule application, then abandoning the rejected line of derivation at the point the decision is made.

Let's trace the derivation of the perfect active participle of the root गम् 'go' to illustrate our implementation. The Pāṇinian derivation is shown in Table 1; the Scharf-Hyman derivation in Table 2.

The Scharf-Hyman framework does not at present implement sthānivadbhāva; it does not reintroduce replaced sounds at certain points where they are required in the conditions of subsequent rules. Instead, for the time being, rules are modified to include the replacements in the conditions of the subsequent rule. Hence our derivation of the perfect active participle of गम् differs from the Pāṇinian

[2] See Scharf (forthcoming) for a thorough examination of principles for determining rule precedence in cases of conflict between rules that have independent as well as overlapping domains (विप्रतिषेध).

Table 1. Pāṇinian derivation

Perfect active participle of the root गम्

1	गम्(ल्ट)	MDhP 1.702	गमॢ सृपॢ गतौ
2	गम्-ल्(**इट्**)[लिट्]	3.2.115	परोक्षे लिट्
3	गम्-वस्(कु)	3.2.107	क्वसुश्च
4	गम्-इ:वस्(क्)	7.2.67	वस्वेकाजाद्घसाम्
5	ग्म्-इ:वस्(क्)	6.4.98	गमहनजनखनघसां लोपः क्ङित्यनङि
6a	गम्-इ:वस्	1.1.59	द्विर्वचनेऽचि
6	गम्ग्म्-इ:वस्	6.1.8	लिटि धातोरनभ्यासस्य
7	गग्म्-इ:वस्	7.4.60	हलादिः शेषः
8	जग्म्-इ:वस्	7.4.62	कुहोश्चुः
9	जग्मिवस्		Delete morpheme boundaries

Table 2. Scharf-Hyman derivation

Perfect active participle of the root गम्

1	गम्(ल्ट)	MDhP 1.702	गमॢ सृपॢ गतौ
2	गम्-ल्(**इट्**)[लिट्]	3.2.115	परोक्षे लिट्
3	गम्-वस्(कु)	3.2.107	क्वसुश्च
4	गम्-इ:वस्(क्)	7.2.67	वस्वेकाजाद्घसाम्
5	गम्गाम्-इ:वस्(क्)	6.1.8	लिटि धातोरनभ्यासस्य
6	गम्ग्म्-इ:वस्(क्)	6.4.98	गमहनजनखनघसां लोपः क्ङित्यनङि
7	गग्म्-इ:वस्	7.4.60	हलादिः शेषः
8	जग्म्-इ:वस्	7.4.62	कुहोश्चुः
9	जग्म्-इ:वस्	7.2.67	वस्वेकाजाद्घसाम्
10	जग्मिवस्		Delete morpheme boundaries

derivation. While Pāṇini implements penultimate vowel deletion by 6.4.98 (Table 1, step 5) prior to doubling by 6.1.8 (Table 1, step 6), we implement doubling by 6.1.8 first (Table 2, step 5) and then delete the penultimate vowel by 6.4.98 (Table 2 step 6), modifying the condition for deletion to accomodate the doubled root. To avoid dealing with the issue of sthānivadbhāva, deletion of the penultimate vowel is implemented after doubling by including the parameters to match a doubled root in the conditions for the application of 6.4.98. The `root` parameter compensates for the overbroad `lcontext` parameter to ensure the application of the rule only to the proper roots.

```
<rule source="a" target=""
lcontext="@(begin)(?:[ghjKG]a[mns][ghjKG])"
rcontext="[mns]#(?!a;N@(anta))[@(ac)][@(al)]*@(kNit)"
root="^(g\am;/x|h\an;/a|j/an;/I|j\an;/a|j/an;\I|K/an;^u|G/as;/x)$"
c="6.4.98 gamahanajanaKanaGasAM lopaH kNityanaNi"/>
```

In either derivation, 7.2.67 at step 4 must anticipate the state of the derivation at a subsequent step (at least step 6 in our derivation or step 7 in the Pāṇinian derivation). Our framework implements 7.2.67 in two XML rules to separate the single-syllable condition, which requires look-ahead, from conditions that don't require look-ahead. The single-syllable portion is written as follows:

```
<rule source="" target="i:"
lcontext="^(?:@(upasarga)-|)[@(hal)]*[@(ac)][@(hal)]*#"
rcontext="vas@(anta)" affix="^(k`vas;u|S`at;f)$"
root="@(DpAdi)[@(hal)]*[^/\][@(ac)][@(hal)]*@(Dpanta)"
c="7.2.67 vasvekAjAdGasAm"/>
```

At step 4 we begin two lines of derivation, one with the augment and one without. Only the one with is shown in Table 2. At step 9, we evaluate the condition in 7.2.67 and, finding that the string qualifies for augmentation, throw out the derivation without the augment.

We look forward to utilizing the enriched framework created for participle derivation in a revised, more faithful model of Pāṇinian inflection and hope to go on to implement derivational morphology generally.

References

1. Cardona, G.: Pāṇini: His Work and its Traditions. Vol. I, Background and Introduction, 2nd edn. Motilal Banarsidass, Delhi (1997)
2. Goyal, P., Kulkarni, A., Behera, L.: Computer Simulation of *Aṣṭādhyāyi*: Some Insights. In: Huet, G., Kulkarni, A., Scharf, P. (eds.) Sanskrit Computational Linguistics 2007/2008. LNCS (LNAI), vol. 5402, pp. 139–161. Springer, Heidelberg (2009)
3. Huet, G., Kulkarni, A., Scharf, P. (eds.): Sanskrit Computational Linguistics 2007/2008. LNCS (LNAI), vol. 5402. Springer, Heidelberg (2009)
4. Hyman, M.: From Pāṇinian Sandhi to Finite State Calculus. In: Huet, G., Kulkarni, A., Scharf, P. (eds.) Sanskrit Computational Linguistics 2007/2008. LNCS (LNAI), vol. 5402, pp. 253–265. Springer, Heidelberg (2009)
5. Mishra, A.: Simulating the Pāṇinian System of Sanskrit Grammar. In: Huet, G., Kulkarni, A., Scharf, P. (eds.) Sanskrit Computational Linguistics 2007/2008. LNCS (LNAI), vol. 5402, pp. 127–138. Springer, Heidelberg (2009)
6. Scharf, P. M.: *Rāmopākhyāna*–the Story of Rāma in the Mahābhārata: An Independent-study Reader in Sanskrit. Routledge Curzon, London (2002)
7. Scharf, P.: Modeling Pāṇinian Grammar. In: Huet, G., Kulkarni, A., Scharf, P. (eds.) Sanskrit Computational Linguistics 2007/2008. LNCS (LNAI), vol. 5402, pp. 95–126. Springer, Heidelberg (2009)
8. Scharf, P.: Rule selection in the *Aṣṭādhyāyi* or Is Pāṇini's grammar mechanistic? In: Proceedings of the 14th World Sanskrit Conference, Kyoto University, Kyoto, September 1-5 (2009) (forthcoming)
9. Scharf, P., Malcolm, H.: Linguistic issues in encoding Sanskrit. The Sanskrit Library; Providence. Delhi; Motilal Banarsidass (2010)

Sanskrit Compound Processor

Anil Kumar[1], Vipul Mittal[2], and Amba Kulkarni[1]

[1] Department of Sanskrit Studies, University of Hyderabad, India
anil.lalit22@gmail.com, apksh@uohyd.ernet.in
[2] Language Technologies Research Centre, IIIT, Hyderabad, India
vipulmittal@research.iiit.ac.in

Abstract. Sanskrit is very rich in compound formation. Typically a compound does not code the relation between its components explicitly. To understand the meaning of a compound, it is necessary to identify its components, discover the relations between them and finally generate a paraphrase of the compound. In this paper, we discuss the automatic segmentation and type identification of a compound using simple statistics that results from the manually annotated data.

Keywords: Sanskrit Compound Splitter, Sanskrit Compound Type Identifier, Sanskrit Compounds, Optimality Theory.

1 Introduction

In recent years Sanskrit Computational Linguistics has gained momentum. There have been several efforts towards developing computational tools for accessing Sanskrit texts ([12], [14], [23], [16], [7], [1]). Most of these tools handle morphological analysis and sandhi splitting. Some of them ([10], [13], [8]) also do the sentential parsing. However, there have been almost negligible efforts in handling Sanskrit compounds, beyond segmentation.

Sanskrit is very rich in compound formation. The compound formation being productive, forms an open-set and as such it is also not possible to list all the compounds in a dictionary. The compound formation involves a mandatory *sandhi*[1]. But mere *sandhi* splitting does not help a reader in identifying the meaning of a compound. Typically a compound does not code the relation between its components explicitly. To understand the meaning of a compound, thus, it is necessary to identify its components, discover the relations between them, and finally produce a *vigrahavākya*[2] of the compound. Gillon [6] suggests

[1] *Sandhi* means euphony transformation of words when they are consecutively pronounced. Typically when a word w_1 is followed by a word w_2, some terminal segment of w_1 merges with some initial segment of w_2 to be replaced by a "smoothed" phonetic interpolation, corresponding to minimising the energy necessary to reconfigure the vocal organs at the juncture between the words. [11]

[2] An expression providing the meaning of a compound is called a *vigrahavākya*.

G.N. Jha (Ed.): Sanskrit Computational Linguistics, LNCS 6465, pp. 57–69, 2010.

tagging of compounds by enriching the context free rules. He does so by specifying the vibhakti, marking the head and also specifying the enriched category of the components. He also points out how certain components such as *'na'* provide a clue for deciding the type of a compound. *Pāṇini* captures special cases and formulates rules to handle them. Implementing these rules automatically is still far from reality on account of the semantic information needed for the implementation. In the absence of such semantic information, statistical methods have proved to be a boon for the language engineers. These statistical tools use manually annotated data for automatic learning, and in turn develop a language model, which is then used for analysis. In what follows we discuss the automatic segmentation and type identification of compounds using simple statistics that result from the manually annotated data.

2 Sanskrit Compounds

The Sanskrit word *samāsaḥ* for a compound means *samasanam* which means "combination of more than one word into one word which conveys the same meaning as that of the collection of the component words together". While combining the components together, a compound undergoes certain operations such as loss of case suffixes, loss of accent, etc.. A Sanskrit compound thus has one or more of the following features ([17], [6]):

1. It is a single word (*ekapadam*).[3]
2. It has a single case suffix (*ekavibhaktikam*) with an exception of *aluk* compounds such as *yudhiṣṭiraḥ*, where there is no deletion of case suffix of the first component.
3. It has a single accent(*ekasvaraḥ*).[4]
4. The order of components in a compound is fixed.
5. No words can be inserted in between the compounds.
6. The compound formation is binary with an exception of *dvandva* and *bahupada bahuvrīhi.*
7. Euphonic change (*sandhi*) is a must in a compound formation.
8. Constituents of a compound may require special gender or number different from their default gender and number. e.g. *pāṇipādam, pācikābhāryaḥ*, etc.

Though compounds of 2 or 3 words are more frequent, compounds involving more than 3 constituent words with some compounds even running through pages is not rare in Sanskrit literature. Here are some examples of Sanskrit compounds involving more than 3 words.

1. वेदवेदाङ्गतत्त्वज्ञः[5]=वेद-वेदाङ्ग-तत्त्व-ज्ञः
2. प्रवरमुकुटमणिमरीचिमञ्जरीचयचर्चितचरणयुगल[6]=प्रवर-मुकुट-मणि-मरीचि-मञ्जरी-चय- चर्चित-चरण-युगल

[3] aikapadyam aikasvaryam ca samāsatvāt bhavati [Kāśikā 2.1.46].
[4] aikapadyam aikasvaryam ca samāsatvāt bhavati [Kāśikā 2.1.46].
[5] *Rāmāyaṇam 1-1-14.*
[6] *Pañcatantram in kathamukham.*

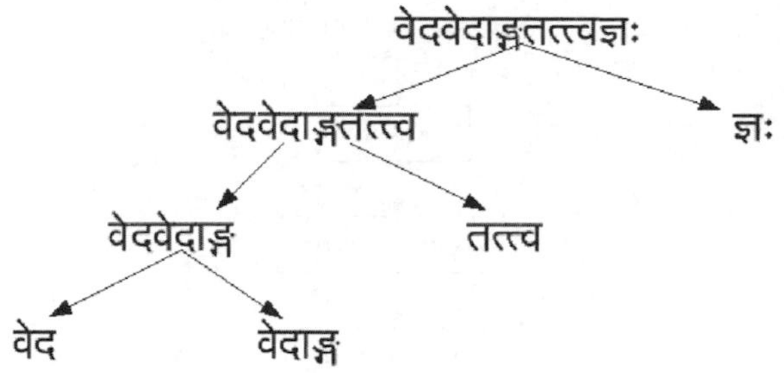

Fig. 1. Constituency representation of वेदवेदाङ्गतत्त्वज्ञ

3. जलादिव्यापकपृथिवीत्वाभावप्रतियोगिपृथिवीत्ववती[7]=जल - आदि - व्यापक - पृथिवीत्व - अभाव - प्रतियोगि - पृथिवीत्ववती

The compounds are formed with two words at a time and hence they can be represented faithfully as a binary tree, as shown in figure 1.

Semantically *Pāṇini* classifies the Sanskrit compounds into four major types:

- Tatpuruṣaḥ: (Endocentric with head typically to the right),
- Bahuvrīhiḥ: (Exocentric),
- Dvandvaḥ: (Copulative), and
- Avyayībhāvaḥ: (Endocentric with head typically to the left and behaves as an indeclinable).

This classification is not sufficient for generating the paraphrase. For example, the paraphrase of a compound *vṛkṣamūlam* is *vṛkṣasya mūlam* and *gṛhagataḥ* is *gṛham gataḥ*, though both of them belong to the same class of *tatpuruṣaḥ*. Based on their paraphrase, these compounds are further sub-classified into 55 sub-types. Appendix-A provides the classification and Appendix-B describes the paraphrase associated with each tag.

3 Compound Processor

Understanding a compound involves (ref Fig 2)

1. Segmentation (समासपदच्छेदः),
2. Constituency Parsing (समासपदान्वयः),
3. Compound Type Identification (समस्तपदपरिचायकः), and
4. Paraphrasing (विग्रह-वाक्यम्).

These four tasks form the natural modules of a compound processor. The output of one task serves as an input for the next task until the final paraphrase is generated.

[7] *Kevalavyatireki-prakaraṇam : Maṇikaṇa* [22].

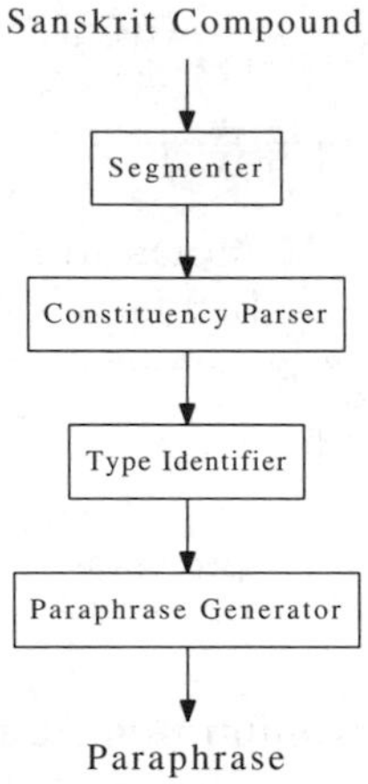

Fig. 2. Compound Analyser

3.1 Segmenter

The task of this module is to split a compound into its constituents. For instance, the compound

sumitrānandavardhanaḥ
is segmented as
sumitrā-ānanda-vardhanaḥ

Each of the constituent component except the last one is typically a compounding form (a bound morpheme)[8].

3.2 Constituency Parser

This module parses the segmented compound syntactically by pairing up the constituents in a certain order two at a time. For instance,

sumitrā-ānanda-vardhanaḥ
is parsed as
<***sumitrā***-<***ānanda-vardhanaḥ***>>

3.3 Type Identifier

This module determines the type on the basis of the components involved. For instance,

<*sumitrā*-<*ānanda-vardhanaḥ*>>
is tagged as
<***sumitrā***-<***ānanda-vardhanaḥ***>***T6***>***T6*** .

where T6 stands for compound of type *ṣaṣṭī-tatpuruṣa*. This module needs an access to the semantic content of its constituents, and possibly even to the wider context.

[8] With an exception of components of an *'aluk'* compound.

3.4 Paraphrase Generator

Finally after the tag has been assigned, the paraphrase generator [16] generates a paraphrase for the compound. For the above example, the paraphrase is generated as:

ānandasya vardhanaḥ = *ānandavardhanaḥ*,
sumitrāyāḥ ānandavardhanaḥ = *sumitrānandavardhanaḥ*.

4 Compound Segmenter

The task of a segmenter is to split a given sequence of phonemes into a sequence of morphologically valid segments [18]. The compound formation involves a mandatory sandhi. Each sandhi rule is a triple (x, y, z) where y is the last letter of the first primitive, z is the first letter of the second primitive, and x is the letter sequence resulting from the euphonic combination. For analysis, we reverse these rules of sandhi and produce $y + z$ corresponding to a x. Only the sequences that are morphologically valid are selected. The segmentation being non-deterministic, segmenter produces multiple splits. To ensure that the correct output is not deeply buried down the pile of incorrect answers, it is natural to prioritize solutions based on some scores. We follow GENerate-CONstrain-EVALuate paradigm attributed to the Optimality Theory [21] for segmentation. As is true of any linguistic theory, the Optimality Theory basically addresses the issue of generation. Nevertheless, there have been successful attempts ([4], [3]) to reverse the process of generation. It will be really challenging to implement the sandhi rules from *Aṣṭādhyāyī* as CONstraints and then reverse them for splitting. In this attempt however, we simply use the 'cooked' sandhi rules in the form of triplets. We describe below the GENerate-CONstrain-EVALuate cycle of the segmenter and the scoring matrix used for prioritizing the solutions.

4.1 Scoring Matrix

A parallel corpus of Sanskrit text in sandhied and unsandhied form is being developed as a part of the Sanskrit Consortium project in India. The corpus contains texts from various fields ranging from children stories, dramas, purāṇas to Ayurveda texts. From around 100K words of such a parallel corpus, 25K parallel instances of sandhied and unsandhied text were extracted. These examples were used to get the frequency of occurence of various sandhi rules. If no instance of a sandhi rule is found in the corpus, for smoothing, we assign the frequency of 1 to this sandhi rule.

We define the estimated probability of the occurrence of a sandhi rule as follows:

Let R_i denote the ith rule with f_{R_i} as the frequency of occurrence in the manually split parallel text. The probability of rule R_i is:

$$P_{R_i} = \frac{f_{R_i}}{\sum_{j=1}^{n} f_{R_j}}$$

where n denotes the total number of sandhi rules found in the corpus.

Let a word be split into a candidate S_j with k constituents as $< c_1, c_2, ..., c_k >$ by applying $k-1$ sandhi rules $< R_1, R_2, ..., R_{k-1} >$ in between the constituents. It should be noted here that the rules $R_1, ..., R_{k-1}$ and the constituents $c_1, ..., c_k$ are interdependent since a different rule sequence will result in a different constituents sequence. The sequence of constituents are constrained by a language model whereas the rules provide a model for splitting. We define two measures each corresponding to the constituents and the rules to assign weights to the possible splits.

Language Model. Let the unigram probability of the sequence $< c_1, c_2, ..., c_k >$ be PL_{S_j} defined as:

$$PL_{S_j} = \prod_{x=1}^{k} (P_{c_x})$$

where P_{c_x} is the probability of occurrence of a word c_x in the corpus.

Split Model. Let the splitting model PS_{S_j} for the sandhi rules sequence $< R_1, R_2, ..., R_{k-1} >$ be defined as:

$$PS_{S_j} = \prod_{x=1}^{k-1} (P_{R_x})$$

where P_{R_x} is the probability of occurrence of a rule R_x in the corpus.

Therefore, the weight of the split S_j is defined as the product of the language and the split model as:

$$W_{S_j} = \frac{PL_{S_j} * PS_{S_j}}{k}$$

where the factor of k is introduced to give more preference to the split with less number of segments than the one with more segments.

4.2 Segmentation Algorithm

The approach followed is GENerate-CONstrain-EVALuate. In this approach, all the possible splits of a given string are first generated and the splits that are not validated by the morphological analyser are subsequently pruned out.

Currently we apply only two constraints viz.

- C1 : All the constituents of a split must be valid morphs.
- C2 : All the segments except the last one should be valid compounding forms.

The system flow is presented in Figure 3.

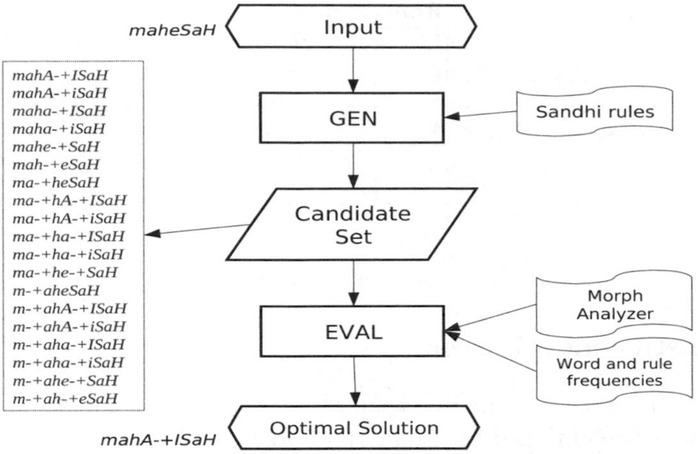

Fig. 3. Compound Splitter: System Data Flow

The basic outline of the algorithm is:

1. Recursively break a word at every possible position applying a sandhi rule and generate all possible candidates for the input.
2. Pass the constituents of all the candidates through the morph analyser.
3. Declare the candidate as a valid candidate, if all its constituents are recognised by the morphological analyser, and all except the last segment are compounding forms.
4. Assign weights to the accepted candidates and sort them based on the weights as defined in the previous subsection.
5. The optimal solution will be the one with the highest weight.

Results. The current morphological analyser[9] can recognise around 140 million words. Using 2,650 rules and a test data of around 8,260[10] words parallel corpus for testing, we obtained the following results:

- Almost 92.5% of the times, the first segmentation is correct. And in almost 99.1% of the cases, the correct split was among the top 3 possible splits.
- The precision was about 92.46% (measured in terms of the number of words for which first answer is correct w.r.t. the total words for which correct segmentation was obtained).
- The system consumes around 0.04 seconds per string of 15 letters on an average.[11]

[9] Available at http://sanskrit.uohyd.ernet.in/ anusaaraka/sanskrit/samsaadhanii/morph/index.html

[10] The test data is extracted from manually split data of *Mahābhāratam*.

[11] Tested on a system with 2.93GHz Core 2 Duo processor and 2GB RAM.

Table 1. Rank-wise Distribution

Rank	% of words
1	92.4635
2	5.0492
3	1.6235
4	0.2979
5	0.1936
>5	0.3723

The complete rank wise distribution is given in Table 1.

5 Constituency Parser

Segmenter takes a compound as an input and produces one or more possible segmentations conditioned by the morphological analyser and the sandhi rules. Constituency parser takes an output of the segmenter and produces a binary tree showing the syntactic composition of the compound corresponding to each of the possible segmentations. Each of these compositions show the possible ways various segments can be grouped. To illustrate various possible parses that result from a single segmentation, consider the segmentation a-b-c of a compound. Since a compound is binary, the three components a-b-c may be grouped in two ways as $< a < bc >>$ or $<< ab > c >$. Only one of the ways of grouping may be correct in a given context as illustrated by the following two examples.

1. < *baddha-* < *jāmbūnada- srajaḥ* >>
2. << *tapas-svādhyāya* >- *niratam* >

With 3 components, only these two parses are possible. But as the number of constituents increase, the number of possible ways the constituents can be grouped grows very fast. The constituency parsing is similar to the problem of completely parenthesizing *n+1* factors in all possible ways. Thus the total possible ways of parsing a compound with $n + 1$ constituents is equal to a *Catalan number,* C_n [13] where for $n \geq 0$,

$$C_n = \frac{(2n)!}{(n+1)!n!}$$

The task of the constituency parser is then to choose the correct way of grouping the components together, in a given context. To the best of our knowledge no work has been initiated yet that produces the best constituency parse for a given segmentation, in a given context.

6 Type Identifier

After getting a constituency parse of a compound, the next task in the compound analysis is to assign an appropriate operator (tag) to each non-leaf node. This operator will then operate on the leaf nodes to produce the associated meaning.

We use manually tagged corpus as a model for predicting the tags, given a pair of constituents of a compound. Manually tagged corpus consists of approximately 150K words which contain 12,796 compounds[12]. These texts were tagged using the tagset given in appendix-I. All these compounds are thus tagged 'in context' and contain only one tag. This corpus formed the training data. Another small corpus with 400 tagged compounds from totally different texts, was kept aside for testing.

Some features of the manually tagged data

1. Around 10% of the compounds were repeated.
2. The 12,796 tokens of compounds contain 2,630 types of left word and 7,106 types of right word.
3. The frequency distribution of highly frequent tags is shown in Table 2. To study the effect of fine-grained-ness we also merged the sub-types. Table 3 gives the frequency distribution of major tags, after merging the sub-types.

Table 2. Distribution of Fine-grain-Tags

Tag	% of words
T6	28.35
Bs6	12.45
K1	9.63
Tn	8.56
Di	7.23
U	5.73

Table 3. Distribution of Coarse-Tags

Tag	% of words
T	52.43
B	18.96
K	12.04
D	8.84
U	5.73

We define,

$$P(T/W_1 - W_2) = \text{probability that a compound } W_1 - W_2 \text{ has tag T.}$$

Assuming that occurrence of W_1 and W_2 are independent,

$$P(T/W_1 - W_2) = P(T/W_1) * P(T/W_2)$$

where $P(T/W_i) = \frac{P(T*W_i)}{P(W_i)}$, i= 1,2.

Since the data is sparse, to account for smoothing, we define, for unseen instances,

$$P(T.W_i) = \tfrac{1}{F},$$
$$P(W_i) = \tfrac{1}{F},$$

where F is the total number of manually tagged compounds.

[12] Only compounds with 2 components were selected.

6.1 Performance Evaluation

The test data of 400 words are tagged 'in context'. While our compound tagger does not see the context, and thus suggests more than one possible tag, and ranks them. Normally a tool is evaluated for its coverage and precision. In our case, the tool always produces tags with weights associated with them. The coverage and precision therefore are evaluated based on the ranks of the correct tags. Table4 shows results of 400 words with a coarse as well as fine grained tagset.

Table 4. Precision of Type Identifier

	with 55 tags		with 8 tags	
Rank	no of words	% of words	no of words	% of words
1	252	63.0	291	72.7
2	44	10.9	53	13.2
3	29	7.2	38	9.5

Thus, if we consider only the 1st rank, the precision with 8 tags is 72.7% and with 55 tags, it is 63.0%. The precision increases substantially to 95.4% and 81.1% respectively if we take 1st three ranks into account.

The performance of the type identifier can be further improved by using semantically tagged lexicon. There are around 200 rules in the *Aṣṭādhyāyī* which provide semantic contexts for various compounds. One such example is the *Pāṇini*'s sūtra *annena vyañjanam* (2.1.34). This sūtra gives a condition for forming *tṛtīyā tatpuruṣaḥ* compound. Thus what is required is a list of all words that denote eatables. A lexicon rich with such semantic information would enhance the performance of the tagger further. The compound processor with all the functionality described above is available online at `http://sanskrit.uohyd.ernet.in/samAsa/frame.html`

References

1. Bharati, A., Kulkarni, A.P., Sheeba, V.: Building a wide coverage Sanskrit Morphological Analyser: A practical approach. In: The First National Symposium on Modelling and Shallow Parsing of Indian Languages, IIT-Bombay (2006)
2. Bhat, G.M.: Samāsaḥ. Samskrita Bharati, Bangalore, Karnataka (2006)
3. Fortes, F.C.L., Roxas, R.E.O.: Optimality Theory in Morphological Analysis. In: National Natural Language Processing Research Symposium (January 2004)
4. Fosler, J.E.: On Reversing the Generation Process in Optimality Theory. In: Proceedings of the Association for Computational Linguistics (1996)
5. Gillon, B.S.: Exocentric Compounds in Classical Sanskrit. In: Proceeding of the First International Symposium on Sanskrit Computational Linguistics (SCLS 2007), Paris, France (2007)
6. Gillon, B.S.: Tagging Classical Sanskrit Compounds. In: Kulkarni, A., Huet, G. (eds.) Sanskrit Computational Linguistics. LNCS (LNAI), vol. 5406, pp. 98–105. Springer, Heidelberg (2009)

7. Hellwig, O.: Sanskrit Tagger: A Stochastic Lexical and POS Tagger for Sanskrit. In: Huet, G., Kulkarni, A., Scharf, P. (eds.) Sanskrit Computational Linguistics 2007/2008. LNCS (LNAI), vol. 5402, pp. 266–277. Springer, Heidelberg (2009)
8. Hellwig, O.: Extracting Dependency Trees from Sanskrit Texts. In: Kulkarni, A., Huet, G. (eds.) Sanskrit Computational Linguistics. LNCS (LNAI), vol. 5406, pp. 106–115. Springer, Heidelberg (2009)
9. Hoeks, J.C.J., Hendriks, P.: Optimality Theory and Human Sentence Processing: The Case of Coordination. In: Proceedings of the 27th Annual Meeting of the Cognitive Science Society, Erlbaum, Mahwah, NJ, pp. 959–964 (2005)
10. Huet, G.: Shallow syntax analysis in Sanskrit guided by semantic nets constraints. In: Proceedings of International Workshop on Research Issues in Digital Libraries, Kolkata (2006)
11. Huet, G.: Lexicon-directed Segmentation and Tagging of Sanskrit. In: XIIth World Sanskrit Conference, Helsinki, Finland (August 2003); Tikkanen, B., Hettrich, H. (eds.) Themes and Tasks in Old and Middle Indo-Aryan Linguistics, Motilal Banarsidass, Delhi, pp. 307–325 (2006)
12. Huet, G.: Formal structure of Sanskrit text: Requirements analysis for a Mechanical Sanskrit Processor. In: Huet, G., Kulkarni, A., Scharf, P. (eds.) Sanskrit Computational Linguistics 2007/2008. LNCS (LNAI), vol. 5402, pp. 162–199. Springer, Heidelberg (2009)
13. Huet, G.: Sanskrit Segmentation. In: South Asian Languages Analysis Roundtable XXVIII, Denton, Ohio (October 2009)
14. Jha, G.N., Mishra, S.K.: Semantic Processing in Pāṇini's Kāraka System. In: Huet, G., Kulkarni, A., Scharf, P. (eds.) Sanskrit Computational Linguistics 2007/2008. LNCS (LNAI), vol. 5402, pp. 239–252. Springer, Heidelberg (2009)
15. Kulkarni, A.P., Shukla, D.: Sanskrit Morphological Analyser: Some Issues. To appear in Bh.K Festschrift volume by LSI (2009)
16. Kulkarni, A.P., Kumar, A., Sheeba, V.: Sanskrit compound paraphrase generator. In: Proceedings of ICON 2009: 7th International Conference on Natural Language Processing. Macmillan Publishers, India (2009)
17. Mahavira: Pādnini as Grammarian (With special reference to compound formation). Bharatiya Vidya Prakashan, Delhi (June 1978)
18. Mittal, V.: Automatic Sanskrit Segmentizer using Finite State Transducers. In: Proceeding of Association for Computational Linguistics - Student Research Workshop (2010)
19. Murty, M.S.: Sanskrit Compounds-A Philosophical Study. Chowkhamba Sanskrit Series Office, Varanasi, India (1974)
20. Ishvarachandra, P.: Aṣṭādhyāyī. Chaukhamba Sanskrit Pratisthan, Delhi (2004)
21. Prince, A., Smolensky, P.: Optimality Theory: Constraint Interaction in Generative Grammar. In: RuCCS Technical Report 2 at Center for Cognitive Science, Rutgers University, Piscataway (1993)
22. Sarma, E.R.S.: Maṇikaṇa: A Navya-Nyāya Manual. The Adyar Library and research Centre, Madras (1960)
23. Scharf, P.M.: Levels in Pāṇini's Aṣṭādhyāyī. In: Kulkarni, A., Huet, G. (eds.) Sanskrit Computational Linguistics. LNCS (LNAI), vol. 5406, pp. 66–77. Springer, Heidelberg (2009)
24. Yuret, D., Biçici, E.: Modeling Morphologically Rich Languages Using Split Words and Unstructured Dependencies. In: ACL-IJCNLP, Singapore (2009)

A Appendix-A : Sanskrit Compound Type and Sub-type Classification

अव्ययीभावः		कर्मधारयः	
Compound sub-types	Tags	Compound sub-types	Tags
अव्यय-पूर्वपदः	A1	विशेषण-पूर्वपद-कर्मधारयः	K1
अव्यय-उत्तरपदः	A2	विशेषण-उत्तरपद-कर्मधारयः	K2
तिष्ठद्गुप्रभृति	A3	विशेषण-उभयपद-कर्मधारयः	K3
संख्यापूर्वपद-नद्युत्तरपद	A4	उपमान-पूर्व-पद-कर्मधारयः	K4
नद्युत्तरपद-अन्यपदार्थे संज्ञायाम्	A5	उपमान-उत्तर-पद-कर्मधारयः	K5
संख्यापूर्वपद-वंश्योत्तरपदः	A6	अवधारणापूर्वपदः-कर्मधारयः	K6
पारे-मध्ये-पूर्वपद षष्ठ्युत्तरपद	A7	सम्भावनापूर्वपद-कर्मधारयः	K7
		मध्यमपदलोपी-कर्मधारयः	Km
तत्पुरुषः		बहुव्रीहिः	
Compound sub-types	Tags	Compound sub-types	Tags
प्रथमातत्पुरुषः	T1		
द्वितीयातत्पुरुषः	T2	द्वितीयार्थ-बहुव्रीहिः(समानाधिकरणः)	Bs2
तृतीयातत्पुरुषः	T3	तृतीयार्थ-बहुव्रीहिः(समानाधिकरणः)	Bs3
चतुर्थीतत्पुरुषः	T4	चतुर्थ्यर्थ-बहुव्रीहिः(समानाधिकरणः)	Bs4
पञ्चमीतत्पुरुषः	T5	पञ्चम्यर्थ-बहुव्रीहिः(समानाधिकरणः)	Bs5
षष्ठीतत्पुरुषः	T6	षष्ठ्यर्थ-बहुव्रीहिः(समानाधिकरणः)	Bs6
सप्तमीतत्पुरुषः	T7	सप्तम्यर्थ-बहुव्रीहिः(समानाधिकरणः)	Bs7
नञ्तत्पुरुषः	Tn	दिग्वाचक-बहुव्रीहिः(समानाधिकरणः)	Bsd
समाहार-द्विगुः	Tds	प्रहरणविषयक-बहुव्रीहिः(समानाधिकरणः)	Bsp
तद्धितार्थद्विगुः	Tdt	ग्रहणविषयक-बहुव्रीहिः(समानाधिकरणः)	Bsg
उत्तरपदद्विगुः	Tdu	अस्त्यर्थ-मध्यमपदलोपी(नञ्)-बहुव्रीहिः	Bsmn
गतिसमासः	Tg	प्रादि-बहुव्रीहिः	Bvp
कुसमासः	Tk	संख्योभयपद-बहुव्रीहिः(समानाधिकरणः)	Bss
प्रादिसमासः	Tp	उपमानपूर्वपद-बहुव्रीहिः(समानाधिकरणः)	Bsu
मयूरव्यंसकादिः	Tm	व्यधिकरण-बहुव्रीहिः	Bv
तत्पुरुषः बहुपदः	Tb	सङ्ख्योत्तरपदः व्यधिकरण-बहुव्रीहिः	Bvs
तत्पुरुषः उपपदः	U	सहपूर्वपद-व्यधिकरण-बहुव्रीहिः	BvS
		उपमानपूर्वपद-व्यधिकरण-बहुव्रीहिः	BvU
		बहुपदः-बहुव्रीहिः	Bb
द्वन्द्वः		केवल	
Compound type:	Tags	Compound type:	Tags
इतरेतरयोग-द्वन्द्वः	Di	केवल समासः	S
समाहार-द्वन्द्वः	Ds	द्विरुक्तिः	
एकशेष	E	द्विरुक्तिः	d

A : Rules for Paraphrase Generation

	अव्ययीभावः
1	<x-y>A1 => y{6} f{x} where f maps x to the noun with same semantic content. A function f needs to be defined.
2	<x-y>A2 => x{3} विपरीतम् वृत्तम्
3	<x-y>A4 => x{6} y{6} समाहारः
4	<x-y>A5 => x'{1} y{1} यस्मिन् देशे x' has same gender as that of y'.
5	<x-y>A6 => x{6} y{6} समाहारः if x = द्वि, both x and y will be in द्विवचनम्
6	<x-y>A7 => x{6} y
	तत्पुरुषः
7	<x-y> T_n => x{n} y $2 \leq n \leq 7$
8	<x-y>Tn => x{n} y
9	<x-y>Tds => x{6;ba} y{6;ba} समाहारः
10	<x-y-z>Tb = x{1} y{1} z{1}
	कर्मधारयः
11	<x-y>K1 => x{1} तत् y{1} च
12	<x-y>K2 => x{1} च y{1} च
13	<x-y>K3 => x{1} च असौ y{1} च
14	<x-y>K4 => x{1} इव y{1}
15	<x-y>K5 => x{1} y{1} इव
16	<x-y>K6 => x{1} एव y{1}
17	<x-y>K7 => x{1} इति y{1}
	बहुव्रीहिः
19	<x-y>Bsd => x{6} च y{6} च यदन्तरालम्
20	<x-y>Bsp => x{3} च y{3} च प्रहृत्य इदम् युद्धम् प्रवृत्तम्
21	<x-y>Bsg => x{7}-y{7}+ गृहीत्वा इदम् युद्धम् प्रवृत्तम्
22	<x-y>Bsmn => x'-y{1} यस्य
23	<x-y>Bss = > x{1} वा y{1} यस्य
24	<x-y>Bsu => x{1} इव y{1} यस्याः
25	<x-y>Bv => x y{1} यस्य
26	<x-y>Bvs => y{6} x' ये सन्ति ते
27	<x-y>BvS => y{3} सह
28	<x-y>BvU => x{6} इव y यस्य
	द्वन्द्वः
29	<x-y>Di => x{1} च y{1} च
30	<x-y>Ds => x{1} च y{1} च
31	<x-y>S => y{1} x{1}
32	<x-y>d => x y

Designing a Constraint Based Parser for Sanskrit

Amba Kulkarni, Sheetal Pokar, and Devanand Shukl

Department of Sanskrit Studies,
University of Hyderabad,
Hyderabad
apksh@uohyd.ernet.in {sjpokar,dev.shukl}@gmail.com

Abstract. Verbal understanding (*śābdabodha*) of any utterance requires the knowledge of how words in that utterance are related to each other. Such knowledge is usually available in the form of cognition of grammatical relations. Generative grammars describe how a language codes these relations. Thus the knowledge of what information various grammatical relations convey is available from the generation point of view and not the analysis point of view. In order to develop a parser based on any grammar one should then know precisely the semantic content of the grammatical relations expressed in a language string, the clues for extracting these relations and finally whether these relations are expressed explicitly or implicitly. Based on the design principles that emerge from this knowledge, we model the parser as finding a directed Tree, given a graph with nodes representing the words and edges representing the possible relations between them. Further, we also use the *Mīmāṃsā* constraint of *ākāṅkṣā* (expectancy) to rule out non-solutions and *sannidhi* (proximity) to prioritize the solutions. We have implemented a parser based on these principles and its performance was found to be satisfactory giving us a confidence to extend its functionality to handle the complex sentences.[1]

Keywords: Sanskrit, Constraint Based Parser, Information coding, *ākāṅkṣā*, *sannidhi*.

1 Introduction

Śābdabodha is the understanding that arises from a linguistic utterance. The three schools of Śābdabodha viz. *Vyākaraṇa*, *Nyāya* and *Mīmāmsā* mainly differ in the chief qualificand of the Śābdabodha. Nevertheless to begin with, all these three schools need an analysis of an utterance. This analysis expresses the relations between different meaningful units involved in an utterance. The utterance may be as small as a single word or as big as a complete novel. In what follows, however, we take a sentence[2] as a unit, and as such we discuss only the relations of words within a sentence and do not deal with the discourse analysis.

[1] Thanks to Gérard Huet and Peter Scharf for their valuable remarks.
[2] Roughly *ekatiṅ vākyam (vārttika* on *tvāmau dvitīyāyāḥ 8.1.23, halantapuṃlliṅga-prakaraṇam).*

G.N. Jha (Ed.): Sanskrit Computational Linguistics, LNCS 6465, pp. 70–90, 2010.

A generative grammar of any language provides rules for generation. For analysis, we require a mechanism by which we can use these rules in a reverse way. The reversal in some cases is easy and also deterministic. For example, subtraction is an inverse operation of addition and is deterministic. The reversal may not always be deterministic. Let us see a simple example of non-deterministic reversal with which all of us are familiar. The multiplication tables or simple method of repetitive addition provides a mechanical way for multiplication. Given a product, to find its factors is a reverse process. Multiplication of two numbers, say, 4 and 3 produces a unique number 12. But its decomposition into two factors is not unique. 12 may be decomposed into two factors as either {6,2} or {4,3} in addition to a trivial decomposition {12,1}. Thus the inverse process may at times involve non-determinism. Depending upon the context, if one factor is known, the other factor gets fixed. For example, if you are interested in distributing 12 apples among 2 children, then one of the factors being 2, the other factor, viz. 6, is determined uniquely.

This is true of a generative grammar as well. To give an example, look at the following two *sūtras* of Pāṇini.

- *anabhihite* (2.3.1)
- *kartṛkaraṇayos tṛtīyā* (2.3.18)

These two *sūtras* together, in case of a passive voice (karmaṇi prayogaḥ), assign third case[3] to both the *kartā* as well as *karaṇam kāraka* as in

(1) *rāmeṇa bāṇena vāliḥ hanyate.*

Now, when a hearer (who knows Sanskrit grammar) listens to this utterance, he notices two words ending in the third case suffix and that the construction is in passive voice. But unless he knows that *rāma* is the name of a person and *bāṇa* is used as an instrument, he fails to get the correct reading. In the absence of such an 'extra-linguistic' knowledge, there are two possible interpretations viz. either *rāma* is *kartā* and *bāṇa* is *karaṇam*, or *bāṇa* is *kartā* and *rāma* is *karaṇam* leading to a non-determinism.[4]

The process of analysing a sequence of words to determine the underlying grammatical structure with respect to a grammar is parsing. There are two distinct ways of developing a parser for a language. The first method which has gained recent predominance is to use statistical machine learning techniques to learn from a manually annotated corpus. This requires a large human annotated corpus. Second method is to use the grammar rules of generation to 'guess' the possible solutions and apply constraints to rule out obvious non-solutions. There have been notable efforts in developing parsers by both the statistical methods as well as grammar based methods for various languages (Lin,1998; Marneffe, 2006; Sleator,1993). A parser based on Paninian Grammar Formalism for modern

[3] The word 'case' is used for *vibhakti*.

[4] There are two more possibilities, since both have the same gender, number, and *vibhakti*, one can be an adjective of the other.

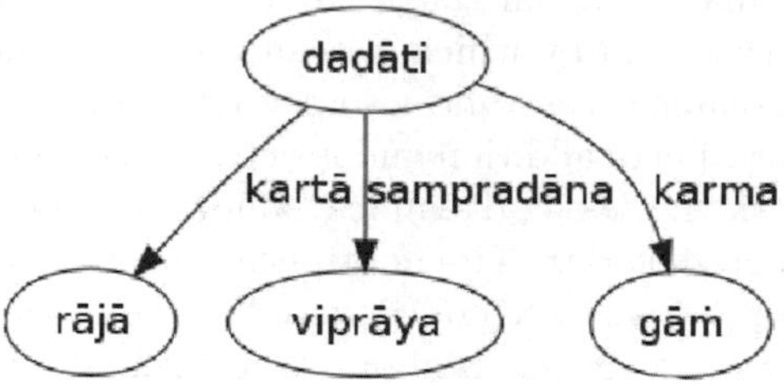

Fig. 1. Semantic relations

Indian languages is described in Bharati, et. al. (1995; 85-100). This parser is modeled as a bipartite graph matching problem. A statistical parser for analysing Sanskrit is described in Hellwig (2008). The shallow parser of Huet (2006, 2009) uses bare minimum information of transitivity of a verb as a sub-categorisation frame and models it as a graph-matching algorithm. The main purpose of this shallow parser is to filter out non-sensical interpretations. It is therefore natural for Huet to develop small tools such as 'ca' handler with more priority to rule out non-grammatical solutions (rather than to develop a full-fledged parser).

While designing a grammar based parser, two major design issues[5] one has to address are: a) what should be the level of semantic analysis, and b) which relations to represent in the parsed output. In order to decide on these issues, in what follows, we first look at the Sanskrit grammar to see what kind of semantic relations can be extracted from a language string, precisely where is the information about these relations coded, and whether the extracted relations are from primary sources or secondary. Later we discuss the issues the mechanical processing throws up, and the possible ways to handle them. Based on these observations, we decide various design parameters. The next section discusses mathematical formulation of the problem, its implementation and finally its performance analysis.

2 Encoding of Grammatical Relations in Sanskrit

Parsing unfolds a linear string of words into a structure which shows explicitly the relations between words. For example, the parse of

(2) *rājā viprāya gāṃ dadāti.*

may be described as in Figure 1.

The task of a parser involves identifying various relations between the words. So the parser developer should decide on the nature of relations and the means

[5] The issues in the development of a statistical parser are totally different. They are related to the size of the annotated corpus, the number of annotated tags used, their fine-grained-ness, etc.

to identify the relations. Sanskrit has the unique privilege of having an extant grammar in the form of Aṣṭādhyāyī. It has been demonstrated (Bharati, forthcoming) that Pāṇini had given utmost importance to the information coding and the dynamics of information flow in a language string. In what follows we look at the information coding in Sanskrit from the point of view of designing a parser.

2.1 Semantic Content of the Relations

Though the correspondence between the semantic relations and the *kāraka* relations is duly stated in the grammar, what is encoded in words is only the *kāraka* relations. There is no one-to-one relation between thematic and *kāraka* relations. One *kāraka* relation may correspond to more than one thematic relation and one thematic relation may be realised by more than one *kāraka* relations (Kiparsky, 2009: 49). What can be extracted from a language string alone without using any extra-linguistic information are the syntactico-semantic relations or the *kāraka* relations and not the pure semantic relations. We give below some examples in our support.

Svatantraḥ kartā. The *vārttikas* under Pāṇini's *sūtra kārake* (1.4.23) go like this[6]

> In the sentence *devadattaḥ pacati*, the activity of cooking refers to the activity of *devadattaḥ* viz. putting a vessel on the stove, pouring water in it, adding rice, supplying the fuel etc. and this activity refers to the activity of the *pradhāna kartā.* In the sentence *sthālī pacati*, the cooking activity refers to holding the rice and water till the rice cooks and this activity is that of a vessel. In the sentence *edhāḥ pakṣyanti*, the cooking activity refers to the supply of sufficient heat by a piece of firewood and thus refers to the activity of an instrument.

In real world, *devadattaḥ*, *sthālī* and *edhāḥ* are the agent, locus and the instrument respectively. But what is expressed by these language strings is just the *kartṛtva* of the *pradhāna kartā, adhikaraṇam* and *karaṇam* respectively and NOT the agent, locus and instrument.

śeṣe. Similarly the relation between *vṛkṣa* and *śākhā*, *pitṛ* and *putra*, and *rājan* and *puruṣa* in the phrases *vṛkṣasya śākhā*, *pituḥ putraḥ* and *rājñaḥ puruṣaḥ* is marked by the genitive case suffix, and Pāṇini groups all of them under the sūtra *śaṣṭhī śeṣe* (2.3.50). Semantically however the first is *avayava-avayavī-bhāva* (part-whole-relation), the second one is *janya-janaka-bhāva* (parent-child-relation), and the third one is *sva-svāmi-bhāva* (owner-possession-relation).

---

[6] *adhiśrayaṇodakāsecanataṇḍulāvapanaidho'pakarṣaṇakriyāḥ pradhānasya kartuḥ pākaḥ* || (ma. bhā. 1.4.23. vā 8) ||
droṇaṃ pacatyāḍhakaṃ pacatīti saṃbhavanakriyā dhāraṇakriyā cādhikaraṇasya pākaḥ || (ma. bhā. 1.4.23.vā 9)||
edhāḥ pakṣyantyā viklitter jvaliṣyantīti jvalanakriyā karaṇasya pākaḥ || (ma. bhā. 1.4.23.vā 10) ||.

***adhiśīṅsthāsāṃ karma* (1.4.46).** In the sentences

(3) *hariḥ vaikuṇṭham adhiśete.*
(4) *muniḥ śilāpaṭṭam adhitiṣṭhati.*
(5) *sādhuḥ parvatam adhyāste.*

vaikuṇṭha, śilāpaṭṭa and *parvata* are in the second case, and Pāṇini assigns them a karma role. However, semantically, all of them are the loci of the activities of the associated verbs viz. *adhi-śīṅ, adhi-sthā,* and *adhi-ās.* Hence the naiyāyikas, who want to map the 'world of words' to the real world, find it difficult to accept the *karmatva* of these words and they qualify this *karmatva* on the second case ending as *ādhārasya anuśāsanika-karmatva* (Dash, 1991;141). Thus, there is a deviation between the real world and what is expressed through the words.

***sahayukte'apradhāne* (2.3.19).** In the sentence,

(6) *mātrā saha bālakaḥ āgacchati.*

the agreement of the verb is with *bālakaḥ*, and not with *mātarā.* According to the *sūtra* (2.3.19), '*saha*' is used with the *apradhāna* (sub-ordinate) *kāraka.* Thus in this example, *mātā* is sub-ordinate and *bālaka* is the main *kartā.* However, at another level of semantic analysis, the situation is reversed. It is *mātā* who carries the child in her arms and thus *bālaka* is *apradhāna* and *mātā* is the *pradhāna kāraka.* Thus again there is a mismatch between the reality and what sentence actually codes in terms of grammatical relations.

From all the above examples, it is clear that the world of words (*śabda-jagat*) is different from the real world. To match the extracted relations with the experience of the real world, extra-linguistic information is needed. Since the extra-linguistic information is not easily accessible, and is open ended, we would extract only syntactico-semantic relations that depend solely upon the linguistic/ grammatical information in a sentence.

2.2 Clues for Extracting the Relations

Sanskrit being inflectionally rich, we know that suffixes mark the relation between words. Similarly certain indeclinables mark some grammatical relations. Agreement between the words also indicate certain grammatical relations. We discuss below these cases with examples.

1. Abhihitatva
   The Paninian *sūtra 'anabhihite'* (2.3.1) (if not already expressed) is an important *sūtra* that governs the *vibhakti* assignment to the nominals. The *vārttika*[7] on this *sūtra* explains *abhihita* as the one which is expressed either by *tiṅ* (a finite verbal suffix), *kṛt* (a non-finite verbal suffix), *taddhita* (derivational nominal suffix) or *samāsa* (compound). E.g. in the sentence

[7] *tiṅkṛttaddhitasamāsaiḥ parisaṃkhyānam* (ma. bhā. 2.3.1. vā.)

(7) *rāmaḥ vanaṃ gacchati.*

the verb being in the active voice (*kartari prayogaḥ*), the verbal suffix '*ti*' expresses the *kartā*, while in the following sentence in passive voice (*karmaṇi prayogaḥ*)

(8) *rāmeṇa vanaṃ gamyate.*

the *karma* is expressed by the verbal suffix. As such, in both cases, the one which is expressed (*kartā* and *karma* respectively) is in the nominative case and shows number and person agreement with the verb form.
Some of the *kṛt* suffixes also express the *kārakas*. For example, in

(9) *dhāvan aśvaḥ.*

the *kṛt* suffix in '*dhāvan*' expresses the relation of *kartā* (*kartari kṛt* (3.4.68)).

2. *Vibhakti*
   The verbal as well as nominal suffixes in Sanskrit are termed *vibhaktis*. We have already seen that verbal suffixes (*tiṅ*), through *abhihitatva*, mark the relations between words. Now we consider the nominal suffixes. They fall under two categories.
   (a) *vibhakti* indicating a *kāraka* relation
   This marks a relation between a noun and a verb known as a *kāraka* relation. Sanskrit uses seven case suffixes to mark six *kāraka* relations viz. *kartā*, *karma*, *karaṇam*, *sampradānam*, *apādānam* and *adhikaraṇam*. The genitive suffix, in addition to marking a *kāraka* relation[8], is predominantly used to mark the noun-noun relation. There is no one-to-one mapping between the case suffixes and the *kāraka* relations, which makes it difficult to determine the relation on the basis of *vibhakti* alone.

   (b) *upapada vibhakti*
   In addition to the noun-noun relations expressed by the sixth case, there are certain words, most of them indeclinables called *upapadas*, which also mark a special kind of noun-noun relation. These indeclinables, mark a relation of a noun with an another noun, and in turn demand a special case suffix for the preceding noun. For example, the *upapada* '*saha*' demands a third case suffix for the preceding noun as in

   (10) *rāmeṇa saha sītā vanaṃ gacchati.*

3. Indeclinables (*avyaya*)
   The indeclinables mark various kinds of relations such as negation, adverbial (manner adverbs only), co-ordination, etc. Sometimes they also provide information about interrogation, emphasis, etc. We distinguish the *upapadas* from the *avyayas*, mainly because, though most of the *upapadas* are also

[8] *kartṛkarmaṇoḥ kṛti* (2.3.65).

indeclinables, they demand a special case suffix on the preceding word, whereas it is not so with indeclinables.

For example, the relation of '*na*' with '*gacchati*' in the sentence

(11) *rāmaḥ gṛhaṃ na gacchati.*

is that of 'negation (*niṣedha*)'. Similarly, the relation of '*mandam*' with '*calati*' in the sentence

(12) *rāmaḥ mandaṃ calati.*

is that of 'adverbial (*kriyāviśeṣaṇa*)'. The relation of '*eva*' with '*rāma*' in the sentence

(13) *rāmaḥ eva tatra upaviṣṭati.*
is that of 'emphasis (*avadhāraṇa*)'.

4. *Samānādhikaraṇa*
Agreement in gender, number and case suffix marks *samānādhikaraṇa* (having the same locus), or the modifier-modified relation between two nouns as in

(14) *śvetaḥ aśvaḥ dhāvati.*
(15) *aśvaḥ śvetaḥ asti.*

In (14) as well as (15), the words *aśvaḥ* and *śvetaḥ* have the same gender, number and *vibhakti* indicating *samānādhikaraṇa*. However, there is a slight difference between the information being conveyed. In (15), the word *śvetaḥ* is a predicative adjective (*vidheya viśeṣaṇa*), while in (14) it is an attributive adjective.

2.3 Explicit versus Implicit Relations

Relations need not always be encoded directly through suffixes or morphemes. Sometimes the information is coded in the 'Language Convention'. The *sūtra*

samānakartṛkayoḥ pūrvakāle (3.4.21)

states that the suffix *ktvā* is used to denote the preceding of two actions that share the same *kartā*. Then the question is what relation does *ktvā* suffix mark? - the relation of *kartṛtva* or the relation of *pūrvakālīnatva*? or both?

Bhartṛhari in vākyapadīyam states (3.7.81-82),

pradhānetayoryatra dravyasya kriyayoḥ pṛthak
śaktirguṇāṣrayā tatra pradhānamanurudhyate 3.7.81

pradhānaviṣayā śaktiḥ pratyayenābhidhīyate
yadā guṇe tadā tadvad anuktāpi prakāśate. 3.7.82

i.e., in case X is an argument of both the main verb as well as the subordinate verb, it is the main verb which assigns the case and the relation of X to the sub-ordinate verb gets manifested even without any other marking.

From the sentences

(16) *rāmaḥ dugdhaṃ pītvā śālām gacchati.*
(17) *rāmeṇa dugdhaṃ pītvā śālā gamyate.*

it is clear that the *vibhakti* of *rāma* is governed by the main verb *gam*. And hence, the information that *rāma* is also the *kartā* of the verb *pā* is not expressed through any of the suffixes. The *ktvā* suffix expresses only the precedence relation (*pūrvakālīnatva*).

Similarly the *sūtra*

samānakartṛkeṣu (icchārtheṣu) tumun (3.3.158)

states that in case of verbs expressing desire, the infinitive verb in the subordinate clause will have the same *kartā* as that of the verb it modifies. Here also the primary information available from the non-finite verbal suffix *tumun* is the relation of purpose.[9]

The sharing in case of *ktvā* and *tumun* suffixes is the result of the preconditions *samānakartṛkayoḥ* or *samānakartṛkeṣu* in 3.4.21 and 3.3.158 respectively which act as Language Conventions.

3 Factors Useful for *Śābdabodha*

As mentioned above, the generation problem is a direct problem, and the analysis problem is a reverse problem, and is non-deterministic. This problem was well recognised by the *mīmāṃsakas* who proposed four conditions viz. *ākāṅkṣā* (expectancy), *yogyatā* (mutual compatibility), *sannidhi* (proximity) and *tātparya* (intention of the speaker) as necessary conditions for proper verbal cognition. With the help of examples, we explain below, how the first three factors play an important role in the rejection of non-solutions from among the several possibilities. We have not discussed the importance of the fourth factor, since the kind of analysis it involves is out of the scope of the present discussion.

[9] *tumunṇvulau kriyāyām kriyārthyāyām* (3.3.10).

3.1 *Ākāṅkṣā* (Expectancy)

In the sentence,
(18) *rāmaḥ vanaṃ gacchati.*

each of the 3 words in this sentence has multiple morphological analyses.
rāmaḥ = rāma {gender=m, case=1, number=sg},
= rā[10] {lakāra=laṭ, person=1, number=pl, voice=active, parasmaipadī}.

vanaṃ = vana {gender=n, case=1, number=sg},
= vana {gender=n ,case=2, number=sg}.

gacchati = gam {lakāra=laṭ, person=3, number=sg, voice=active, parasmaipadī},
= gacchat (gam śatṛ) {gender=m, case=1, number=sg},
= gacchat (gam śatṛ) {gender=n, case=1, number=sg}.

This may lead to the following two possible sentential analysis:

- *rāma* = kartā of the action indicated by *gam*,
  vana = karma of the action indicated by *gam*.

- *vana* = *karma* of the action indicated by *rā*,
  gacchati = simultaneity of the actions indicated by *rā* and *gam*,
  vayam = *kartā* of the action indicated by the verb *rā* (not expressed explicitly, but through the verbal suffix).[11]

Of these two analyses, the second analysis can be ruled out on the basis of non-fulfilment of *kartā* and karma expectancies of the verb *gam*, and the *sampradānam* expectancy of the verb *rā*. The first analysis being complete in itself, it is preferred over the second one.

3.2 *Yogyatā* (Compatibility)

Consider the sentence,
(19) *śakaṭaṃ vanaṃ gacchati.*

The possible morphological analyses of each of the three words are given below.
śakaṭam = śakaṭa {gender=n, case=1, number=sg},
= śakaṭa {gender=n, case=2, number=sg}.

vanaṃ = vana {gender=n, case=1, number=sg},
= vana {gender=n, case=2, number=sg}.

---

[10] *rā* in the sense of *dāne* from the second (*adādi*) gaṇaḥ.

[11] The sentence is interpreted as - (*tasmin*) *gacchati (sati), vayaṃ vanaṃ rāmaḥ* As (he) goes, let us give the forest (to somebody).

gacchati = gam {person=3, lakāra=laṭ, number=eka, voice=active, parasmaipadī},
= gacchat (gam+śatṛ) {gender=m, case=1, number=sg},
= gacchat (gam+śatṛ) {gender=n, case=1, number=sg}.

Now, more than one word can't have the same *kāraka* role unless it is already expressed (*abhihita*). This leads to the following possible sentential analyses[12]:

- *śakaṭa* = kartā of the action indicated by *gam*,
  vana = *karma* of the action indicated by *gam*.

- *vana* = *kartā* of the action indicated by *gam*,
  śakaṭa = karma of action indicated by *gam*.

- *vana* = *kartā* of the action indicated by *gam*,
  śakaṭa = modifier of *vana*.

- *vana* = *karma* of the action indicated by *gam*,
  śakaṭa = modifier of *vana*.

Out of these, the last two do not fulfil all the mandatory expectancies of a verb. Among the first two, the first one is preferable over the second one, since *śakaṭa* has an ability to move while *vana* can not move. Hence *śakaṭa* is preferable as a *kartā* of the verb *gam* than *vana*. Thus the *yogyatā* or the competency of the nouns to be eligible candidates for the *kāraka* roles plays an important role here. However, the context may overrule the condition of *yogyatā*. It is possible to have a reading where, all the residents of vana are going to see the new *śakaṭa*, and thus *vana* qualifies to be a *kartā. The yogyatā* and the context thus compete with each other and hence one needs discourse analysis to prune some of the possibilities.

3.3 *Sannidhi* (Proximity)

Consider,
(20) *rāmaḥ dugdham pītvā śālām gacchati.*

Here the possible analyses are:

- *rāma* = *kartā* of *gam*,
  dugdha = *karma* of *pā*,
  śālām = *karma* of *gam*,
  pā = preceding action with respect to *gam*.

- *rāma* = *kartā* of *gam*,
  dugdha = *karma* of *gam*,
  śālām = *karma* of *pā*,
  pā = preceding action with respect to *gam*.

[12] Assuming that the modifier is to the left, which need not be true in case of poetry.

A competent speaker rules out the second solution on account of non-compatibility of the arguments viz. *dugdha* and *śālā* do not have semantic competence to be the karma of *gam* and *pā* respectively.

The arguments in the correct solution are closer. We mark the words by their positions, and define the proximity measure of a relation as the distance between its two arguments, and the proximity measure of a solution as the sum of the proximity measures of the various relations in the parse. The proximity measure of the above two parses is

- *rāma* = *kartā* of *gam*
  (dist = position of *gam* - position of *rāma* = 5 -1 = 4)
  dugdha = *karma* of *pā* (dist = 3-2 = 1)
  śālām = *karma* of *gam* (dist = 5-4 = 1)
  pā = preceding action with respect to *gam* (dist = 5-3 = 2)
  Thus the total distance = 4 + 1 + 1 + 2 = 8

- *rāma* = *kartā* of *gam* (dist = 5-1 = 4)
  dugdha = *karma* of *gam* (dist = 5-2 = 3)
  śālām = *karma* of *pā* (dist = 4-3 = 1)
  pā = preceding action with respect to *gam* (dist = 5-3 = 2)
  Thus the total distance = 4 + 3 + 1 + 2 = 10

The one with greater proximity (or smaller distance) is preferred as a solution. Though Sanskrit is a free-word-order language, the following sentence with exchange of the karmas is not acceptable.

(21) **rāmaḥ śālām pītvā dugdhaṃ gacchati.*

Equally unacceptable prose orders are

(22) **rāmaḥ pītvā śālām dugdhaṃ gacchati.*
(23) **rāmaḥ dugdham śālām pītvā gacchati.*

which involve crossing of links expressing the relations. A small pilot study of *anvaya* of Saṃkṣepa Rāmāyaṇa (Kutumbashastri, 2002) sentences show no evidence of crossing of links.

It is worth exploring the Calder mobile model suggested by Staal (1967) and further worked out by Gillon (1993) in the light of the *mīmāṃsā* principle of *sannidhi*. It may result in a better computational criterion for *sannidhi*.

4 Design Principles

The foregoing discussions lead to the following design principles for the constraint-based parser.

1. The relations will be marked as *kāraka* relations.
   [Using these *kāraka* relations and extra-linguistic knowledge, the semantic analysis may be carried out in the next level of processing.]

2. Only those relations that are marked directly by the morphemes will be extracted.

   [No relations that require some post-processing, or are based on secondary information will be extracted in the first step. The next level of processing will use this information to mark the unspecified or shared relations, if any.]
3. To prioritize the solutions, only the conditions of *ākāṅkṣā* and *sannidhi* will be used.

   [The condition of *yogyatā* will be used as and when the information is available in machine usable form, with the understanding that this knowledge may not be relied on completely.]
4. While dealing with prose, it will be assumed that there is no cross-linking of the relations between the words.

5 Mathematical Model

Let each word in a sentence be represented as a node in a graph, and the nodes be connected by the directed labelled edges. Then the problem of parsing a sentence may be modelled as

> Given a Graph G with n nodes, the task is to find a sub-graph T which is a directed Tree.[13]

Assuming that the words can be partitioned into two classes viz. the words which have an expectancy called demand words and the words which satisfy the demand called source words, Bharati et. al. reduced the parsing problem to matching a bipartite graph (Bharati 1995: 96). But in reality, the words can not be partitioned into two classes. We come across words which can be demand words in some context and source words in some other context, or in the same context a *kṛdanta* (primary derivative), e.g. can be both a demand word as well as a source word. Bharati et. al. (1995: 91) also needed the requirement of *kārakas* and their optionality for each verb. But then, a parser based on such information will fail to parse sentences with ellipsis, or the real corpus where we come across sentences with incomplete information.

With a robust parser, that produces at least partial solution in case of ellipsis, as an aim, we relax the above conditions. So we give away the constraint that a word can be exclusively either a demand or a source word. Further we treat all *kārakas* at the same level, irrespective of whether they are mandatory or optional, and assign penalty to lower the priority of those solutions which do not satisfy the mandatory expectancies.
We divide the problem into three parts:

1. For a given sentence, draw all possible labeled directed edges among the nodes.

[13] A tree is a graph in which any two vertices are connected by exactly one simple path.

2. Identify a sub-graph T of G such that T is a directed Tree which satisfies the given constraints.
3. Prioritize the solutions, in case there is more than one possible directed Tree.

In what follows we describe our model.

A matrix is a convenient way of representing the graphs for computing purpose. In our case, each word represents a node of a graph, and with each pair of nodes is associated zero or more labels, indicating the possible relations between these nodes. The strong constraint on these relations is that there can be at the most one label associated with a pair of nodes. This then naturally suggests a 3D matrix representation, whose elements are either 0 or 1, where the 3 dimensions represent two nodes and a relation label. The task of the parser is to prune out all those relations which do not satisfy the given constraints, and finally rank them based on their weights. Further, each word has one or more morphological analyses. Hence, corresponding to each node there exists a record with one or more cells, each cell representing one morphological analysis of the word. Let the jth analysis of the ith node be represented by $[i, j]$. Thus the address of a typical element of the 3D matrix is $([i, j], R, [l, m])$. The first pair of letters i and j correspond to the source word analysis, while the second pair of letters l and m represent the demand word analysis. R is the name of the relation of the lth word to the ith word. j indicates the morphological analysis of the ith word responsible for this relation, and m indicates the morphological analysis of the lth word that triggers this relation. In short the tuple $([i, j], R, [l, m])$ represents a relation R due to the mth morphological analysis of the lth word to i^{th} word due to its jth morphological analysis. For ease of representation, we represent the tuple as (i, j, R, l, m). Thus, the initial graph with all possible relations between various nodes is represented as 5D matrix C such that

$$C[i, j, R, l, m] = 1, \text{ if such a relation exists,}$$
$$= 0, \text{ otherwise.}$$

Task 1. Based on the available information in a given sentence in the form of *abhihitatva, vibhakti, sāmānādhikaraṇya,* and the expectancies the matrix C is populated with 0s and 1s.

Here are sample rules (just enough to illustrate an example), expressed in English.

Rule 1:
If the sentence has
 a noun(say ‘s’) in prathamā vibhkati,
 a verb(say ‘t’) in kartari prayogaḥ, in 3rd person, and
 ‘s’ and ‘t’ agree in number,
then ‘s’ is possibly a kartā of ‘t’.

Rule 2:
If the sentence has
a noun(say 's') in dvitīyā vibhkati,
a verb(say 't') in kartari prayogaḥ, and is sakarmaka (roughly transitive)
then 's' is possibly a karma of 't'.

Rule 3:
If the sentence has
a noun(say 's') in saptamī vibhkati, and
a verb(say 't'),
then 's' is possibly an adhikaraṇa of 't'.

Now consider the sentence
(24) *rāmaḥ vanaṃ gacchati.*

The analyses of various words are numbered as follows:
[1, 1]: rāma {gender=m, case=1, number=sg},
[1, 2]: rā {gaṇaḥ=*adādi*, lakāra=laṭ, person=1, number=pl, prayogaḥ=kartari, parasmaipadī}.

[2, 1]: vana {gender=n, case=1, number=sg},
[2, 2]: vana {gender=n ,case=2, number=sg}.

[3, 1]: gam {lakāra=laṭ, person=3, number=sg, voice=active, parasmaipadī},
[3, 2]: gacchat (gam śatṛ) {gender=m, case=1, number=sg},
[3, 3]: gacchat (gam śatṛ) {gender=n, case=1, number=sg}.

The above 3 rules with this input then produce the following output showing all possible relations between various analses:

[2, 2] is a possible karma of [3, 2]
[2, 2] is a possible karma of [3, 3]
[2, 2] is a possible karma of [1, 2]
[2, 2] is a possible karma of [3, 1]
[2, 1] is a possible kartā of [3, 1]
[1, 1] is a possible kartā of [3, 1]

The resulting graph is shown in Figure 2.

Task 2. In order to get a Tree from this graph, we impose the following constraints.

1. A morpheme(*vibhakti*) marks only one relation.
   I.e., a node can have one and only one incoming arrow.
   $\sum_{j,R,k,l} C[i,j,R,k,l] = 1, \forall i.$

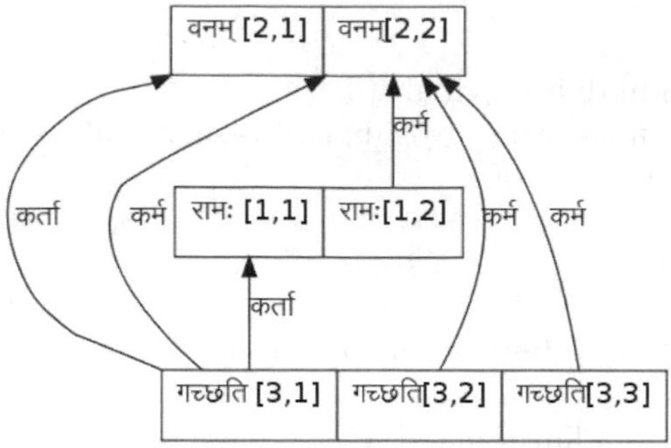

Fig. 2. Graph showing all possible relations

2. Each *kāraka* relation is marked by a single morpheme.
There can not be more than one outgoing arrow with the same label from the same cell, if the relation marks a *kāraka* relation,[14] i.e. there can not be two words satisfying the same *kāraka* role of the same verb.

   $\sum_{i,j} C[i,j,R,k,l] = 1$, for each tuple (R,k,l).

3. A morpheme does not mark a relation to itself.
A word can't satisfy its own expectancy. i.e. a word can't be linked to itself[15]. Or there can not be self loops in a graph.

   $\sum_{j,R,k} C[i,j,R,i,k] = 0, \forall i.$

4. Only one valid analysis of every word per solution
   (a) If a word has both an incoming arrow as well as an outgoing arrow, they should be through the same cell.

   $\forall i \forall j \sum_{R,l,n} C[i,j,R,l,n] + \sum_{a,b,R,k!=j} C[a,b,R,i,k] \leq 1.$

   (b) If there is more than one outgoing arrow through a node, then it should be through the same cell.
   if, for some i,j,R,l,m C[i,j,R,l,m] = 1,

   then $\forall a \forall b \forall R \sum_{a,b,R,k!=j} C[a,b,R,l,k] = 0.$
5. All the words in a sentence should be connected.[16]

---

14 *adhikaraṇam* is treated as an exception since one can have more than one *adhikaraṇam* as in
rāmaḥ adya pañca vādane gṛham agacchat.

15 In case of some of the *taddhita* suffixes which are in *svārtha*, there will be self loops. But we do not consider the meaning of *taddhita* suffixes in the first step, and thus can avoid the self loops.

16 This condition is not yet implemented.

6. There are no crossing of links
   If all the nodes are plotted in a straight line, then they should not intersect each other. i.e.,
   if $C[i, j, R, k, l] = 1$, then
   $\forall v \forall y C[u, v, w, x, y] = 0$, if i < x < k and u < i or u > k.

The resultant graph is a Tree provided:

1. It is connected[17].
2. It has n-1 edges.
   The fact that only sup / tiṅ suffix in every word marks a relation with some other word in a sentence, and abhihita kāraka is not expressed by any sup suffix, it is guaranteed that there are exactly n-1 edges.

Task 3. The solutions are prioritized using the conditions specified below. For each of the solutions, the cost is calculated as
Cost $= \sum_{i,R,j} c_{iRj}$, where
i) $c_{iRj} = |j - i| * wt_R$, if $C[i, a, R, j, b] = 1$ for some a and b.
$= 0$ otherwise.
ii) $wt_R = rank(R)$, if R is a *kāraka* relation (appendix I shows the ranking)
$= 100$, otherwise.
This cost ensures the following:

1. *ākāṅkṣā* (*kāraka* relation) is preferred over other relations (rank[18] of the relations takes care of this.).
2. The ranking of the solutions on the basis of distance-based weights takes care of *sannidhiḥ*.

6 Implementation

The first task demands the inputs from grammar, whereas the second and the third tasks are purely mathematical ones, which can be handled by a constraint solver. The separation of tasks into three sub-tasks makes it not only modular, but also easy for a grammarian to test his/her rules independently. For the first task, an expert shell CLIPS is being used, whereas for the second task, a constraint solver MINION is being used. The system is available at

`http://sanskrit.uohyd.ernet.in/~anusaaraka/sanskrit/MT/test_skt.html`

There is no specific reason behind using these special software tools except the familiarity and the availability under the General Public License. No special efforts were put in towards the efficiency of the system since the main purpose of this exercise is to have a proof of the concept.

17 Since, this condition is not yet implemented, the resulting graph need not be a Tree.
18 Better ranking scheme needs to be developed to take care of default word order.

7 Performance

The current system allows only *padaccheda-sahita-eka-tiṅ-gadya-vākyam.* To measure the performance of this parser, we used hand tagged data. Around 110 sentences with single finite verb were selected from a school book (see appendix A for a sample). These sentences were tagged manually showing the relation of each word in the context. The sentences being simple, each sentence had a single possible parse in the context. There were 525 token words. The average length of the sentences was approximately 5, with a maximum length of 14 words. Morphological analyser is a pre-requisite for a parser. In order to avoid the cascading effect of errors due to non-availability of the morphological analysis, before running the parser, we ensured that the correct morphological analysis of all the words is being produced. Thus, given all possible correct analyses of the words, the task of the parser was to come up with a correct parse. Though the parser produces multiple parses, for the evaluation purpose, we chose only the first parse. Among the 113 sentences, 97 (86%) sentences had the first parse correct and 16 (14%) sentences had one relation wrong. Out of these 16, 10 relations had wrong label, 3 had wrong attachments and 3 went wrong in both the label as well as attachments.

The analysis of wrong results showed that most of the wrong relations were due to non-availability of appropriate knowledge to make the fine-grained distinction. For example, manually tagged corpus makes a distinction between *kāla-adhikaraṇa* and *deśa-adhikaraṇa*, *gauṇa* and *mukhya karma* in case of *dvi-karmaka* (di-transitive) verbs, *hetu* and *karaṇam*, to name a few. Another cause of ambiguity was the verbs in the curādi (10th) *gaṇa*. For most of the verbs in this class, the causative and non-causative forms are the same. This then leads to a wrong parse, since we also allow elipsis. In case there are n (> 1) adjectives, there can be more than one possible way these adjectives can group with the following noun. But we produce a single parse where the adjectives are linked as a chain with the rightmost adjective qualifying the noun directly. This chain just indicates a chunk, and the internal grouping of these adjectives and also their relation with the head noun is left to the user for interpretation.

A sample output of a sentence

bālyakāle rāmaḥ daśarathasya ājñayā viśvāmitrasya yajñam rākṣasebhyaḥ rakṣitum vanam agacchat.

is produced in Figure 3.

8 Challenges

The result with limited test cases is encouraging. The real corpus, even with small children's stories involves much more complex constructions, not necessarily confining to '*eka tiṅ vākyam*'. The constructions involve co-ordination between two or more verbs, sentence connectives such as '*yadā-tadā, yathā-tathā,*

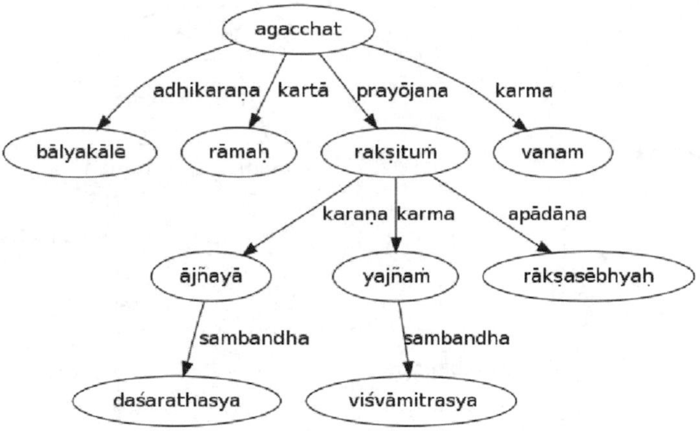

Fig. 3. Sample parse output

atha, tasmāt,' etc. Thus, even at the level of simple texts, one can not do away with discourse analysis.

Another important problem that needs to be addressed is to handle a little more semantics than can be handled with syntactico-semantic relations. For example, it would be desirable to distinguish between *hetu* and *karaṇa* at least, though not between *mukhya karma* and *gauṇa karma.*

Third problem is regarding the *upapadas. Upapada* acts more like a function word (*dyotaka*) than a content word (vācaka). So in case of *upapadas*, it would be desirable to group the *upapada* together with the content word in the *vibhakti* it demands and then mark its relation with other content word. Thus e.g. in the sentence *rāmaḥ muninā saha vanaṃ agacchat*, it is desirable to parse it as in figure 4 than as in figure 5. This means an *upapada* should be treated as a function word, and as such should not be represented by a node.

The vibhaktis, as we know, denote more than one meaning. For example, the second case suffix denotes the meaning of *kriyāviśeṣaṇa* (manner), *kāla* (time) or *adhvan* (path) in addition to the *karma.* To decide an appropriate role, now what one requires is the knowledge of yogyatā. In other words, our e-dictionaries should be rich with semantic properties of the words such as whether it denotes time, path or the manner, etc.

Since the parser does the analysis 'mechanically', it detects the problems of 'violation' of the rules more easily. We give just one example (more examples can be found in Gillon, 2002) from the *anvaya* of 'Saṃkṣepa Rāmāyaṇam'.

> *guhena lakṣmaṇena sītayā ca sahitaḥ rāmaḥ vanena vanaṃ gatvā bahūdakāḥ nadīḥ tīrtvā bharadvājasya śāsanāt citrakūṭam anuprāpya vane ramyam āvasathaṃ kṛtvā devagandharvasaṅkāśāḥ te trayaḥ ramamāṇāḥ sukhaṃ nyavasan. (Śloka* 30-32)

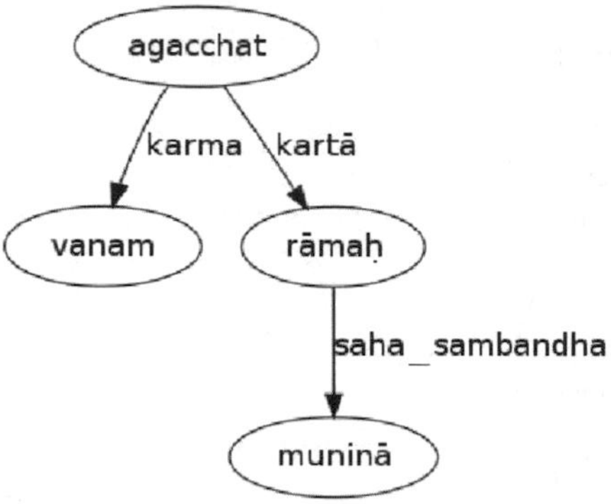

Fig. 4. Saha-function

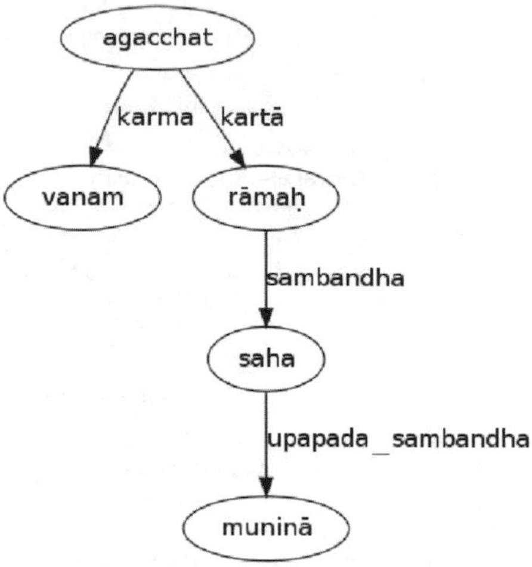

Fig. 5. Saha-content

This sentence poses the following problems:

a) Whom does the phrase '*te trayaḥ*' refer to?
b) *rāmaḥ* does not agree with the finite verb *nyavasan*. Is it not a violation of *samānakartṛkayoḥ pūrvakāle*?
c) Does *gatvā* precede *tīrtvā* or *nyavasan*?
d) In case of *vanena vanaṃ* what should be the meaning of the third case?

In spite of these problems, this parser can act as a tool to discover various kinds of semantic knowledge necessary to build a semantic parser.

Acknowledgement

This work is a part of the Sanskrit Consortium project entitled 'Development of Sanskrit computational tools and Sanskrit-Hindi Machine Translation system' sponsored by the Government of India.

References

1. Bharati, A., Sangal, R.: A Karaka Based Approach to Parsing of Indian Languages. In: COLING 1990: Proc. of Int. Conf. on Computational Linguistics, Helsinki, vol. 3. ACL, NY (August 1990)
2. Bharati, A., Chaitanya, V., Sangal, R.: NLP A Paninian Perspective. Prentice Hall of India, Delhi (1994)
3. Cardona, G.: Pāṇini and Pāṇinīyas on śeṣa Relations. Kunjunni Raja Academy of Indological Research, Kochi (2007)

4. Dash, A.: The syntactic role of adhi in the Pāninian *kāraka* system. In: Deshpande, M.M., Bhate, S. (eds.) Paninian Studies Prof. S. D. Joshi Felicitation Volume, Center for South and Southeast Asian Studies, University of Michigan, U.S.A. (1991)
5. Gent, I.P., Jefferson, C., Miguel, I.: MINION: A Fast, Scalable, Constraint Solver. In: The European Conference on Artificial Intelligence 2006, ECAI 2006 (2006)
6. Gillon, B.S.: Word Order in Classical Sanskrit. Indian Linguistics 57(1), 1–35 (1996)
7. Gillon, B.S.: Bhartṛhari's rule for unexpressed kārakas: The problem of control in Classical Sanskrit. In: Deshpande, H. (ed.) Indian Linguistic Studies, Festschrift in Honor of George Cardona, Motilal Banarasidass, Delhi (2002)
8. Hellwig, O.: Extracting Dependency Trees from the Sanskrit Texts. In: Huet, G., Kulkarni, A., Scharf, P. (eds.) Sanskrit Computational Linguistics Symposium. Springer, Heidelberg (2009)
9. Huet, G.: Formal Structure of Sanskrit Text: Requirements Analysis for a Mechanical Sanskrit Processor. In: Huet, G., Kulkarni, A., Scharf, P. (eds.) Sanskrit Computational Linguistics 2007/2008. LNCS (LNAI), vol. 5402, pp. 162–199. Springer, Heidelberg (2009)
10. Huet, G.: Shallow syntax analysis in Sanskrit guided by semantic nets constraints. In: Majumder, Mitra, Parui (eds.) Proceedings of International Workshop on Research Issues in Digital Libraries, ACM Digital Library (December 2006)
11. Jigyasu, B.: Ashtadhyayi (Bhashya) Prathamavrtti, Three volumes. Ramlal Kapoor Trust Bahalgadh, Sonepat, Haryana (1979) (in Hindi)
12. Joshi, S.D. (ed.): Patanjali's Vyakarana Mahabhashya(several volumes). Univ. of Poona, Pune (1968)
13. Joshi, S.D., Roodebergen, J.A.F.: The Aśṭādhyāyīof Pāṇini (several volumes). Sahitya Akademi, Delhi (1998)
14. Kiparsky, P.: On the Architecture of Panini's Grammar. In: Huet, G., Kulkarni, A., Scharf, P. (eds.) Sanskrit Computational Linguistics 2007/2008. LNCS (LNAI), vol. 5402, pp. 33–94. Springer, Heidelberg (2009)
15. Kutumbashastri, V.: Saṃkṣepa Ramāyaṇam. Teach Yourself Samskrit Series. Rashtriya Sanskrit Samsthanam, New Delhi (2002)
16. Lin, D.: Dependency-based evaluation of MINIPAR. In: Workshop on the evaluation of Parsing Systems, Granada, Spain (1998)
17. Marneffe, M., MacCartney, B., Manning, C.D.: Generating Typed Dependency Parses from Phrase Structure Parses. The Fifth International Conference on Language Resources and Evaluation, LREC 2006, Italy (2006)
18. Pande, G.D.: Vaiyākaraṇa Siddhāntakaumudī of Bhattojidikshita (Text only), Reprint Edition. Chowkhamba Vidyabhavan, Varanasi (2000)
19. Ramakrishnamacharyulu, K.V.: Annotating Sanskrit Texts based on Śābdabodha systems. In: Huet, G., Kulkarni, A., Scharf, P. (eds.) Sanskrit Computational Linguistics 2007/2008. LNCS (LNAI), vol. 5402, Springer, Heidelberg (2009)
20. Ramanujatatacharya, N.S.: Śābdabodha Mīmāṃsā. Institute Francis De Pondicherry (2005)
21. Sharma, R.: Vākyapadīyam, Part III With commentary Prakāśa by Helaraja and Ambakartri. Varanaseya Sanskrit Visvavidyalaya, Varanasi (1974)
22. SK: Siddhāntakaumudī See Pande
23. Staal, J.F.: Word Order in Sanskrit and Universal Grammar. In: Foundations of Language. Supplementary Series, vol. 5. Reidal, Dordercht (1967)
24. Sleator, D.D., Temperley, D.: Parsing English with a link grammar. In: Third international Workshop on Parsing Technologies (1993)

A Sample Story

nadyāḥ taṭe ekaḥ vṛkṣaḥ asti| vṛkṣasya samīpam ekā śilā asti| vṛkṣasya śākhāsu nīḍāḥ santi| nīḍeṣu vihagāḥ vasanti| nīḍāḥ vihagān rakṣanti| vṛkṣasya adhaḥ vānarāḥ santi| kapayaḥ gṛham na racayanti| te sarvadā itastataḥ bhramanti| ekasmin divase śītam tān pīḍayati| te śītāt trāṇāya agnim icchanti| kutrāpi te agnim na vindanti| ekaḥ guñjāyāḥ phalāni paśyati| guñjāyāḥ phalāni raktāni santi| saḥ agneḥ sadṛśāni guñjāyāḥ phalāni ānayati| tāni guñjāyāḥ phalāni śilāyām saṃharati| te sarve guñjā-phalam paritaḥ upaviśanti| agneḥ icchayā te mukhaiḥ tāni dhamanti| te agnim na vindanti| te vānarāḥ analāya vṛthā āyāsam kurvanti| teṣām śītam na naśyati| kapayaḥ mūrkāḥ santi|

B Relations

The relations used, along with their ranks are given in Table 1.

Table 1. Relations

(0)	upapada vibhakti	(12)	kāla-adhikaraṇaṃ
(1)	kartā	(13)	viṣaya-adhikaraṇaṃ
(2)	prayojaka kartā	(14)	kartā-samānādhikaraṇaṃ
(3)	prayojya kartā	(15)	viśeṣaṇaṃ
(4)	karma	(16)	kriyā-viśeṡaṇaṃ
(5)	reserverd for gauṇakarma	(17)	tādarthya
(6)	reserverd for mukhyakarma	(18)	pūrvakālīna
(7)	karaṇaṃ	(19)	sambandha
(8)	sampradānaṃ	(20)	kāraka_ṣaṣṭhī
(9)	apādānaṃ	(21)	niṣedha
(10)	adhikaraṇaṃ	(22)	sambodhana
(11)	deśa-adhikaraṇaṃ		

Generative Graph Grammar of Neo-Vaiśeṣika Formal Ontology (NVFO)

Rajesh Tavva and Navjyoti Singh

Center for Exact Humanities,
International Institute of Information Technology,
Gachi Bowli, Hyderabad, India
vrktavva@research.iiit.ac.in, navjyoti@iiit.ac.in

Abstract. NLP applications for Sanskrit so far work within computational paradigm of string grammars. However, to compute 'meanings', as in traditional *śābdabodha prakriyā*-s, there is a need to develop suitable graph grammars. Ontological structures are fundamentally graphs. We work within the formal framework of Neo-Vaiśeṣika Formal Ontology (NVFO) to propose a generative graph grammar. The proposed formal grammar only produces well-formed graphs that can be readily interpreted in accordance with *Vaiśeṣika* Ontology. We show that graphs not permitted by *Vaiśeṣika* ontology are not generated by the proposed grammar. Further, we write Interpreter of these graphical structures. This creates computational environment which can be deployed for writing computational applications of *Vaiśeṣika* ontology. We illustrate how this environment can be used to create applications like computing *śābdabodha* of sentences.

Keywords: Formal Ontology, *Vaiśeṣika*, Punctuator, NVFO, *śābdabodha*, Graph Grammar, Generative Grammar.

1 Introduction

Natural Language processing for Sanskrit like sandhi-splitting and parsing etc. are written on top of string grammars. Syntax of languages can be computationally handled with operations on generative string grammars. 'Meaning' of language, however, is a richer structure than the syntax of language. Words and inflexions therein can be linearly handled in terms of symbolic string manipulation and matching, whereas meanings of words are related to elements outside the linear sentence. Typically, specification of the meaning of a sentence will involve insertion of information about classes, properties and events etc. in the middle of the symbolic string. Such information is a graph. Thus, we need graph grammar of symbols to specify meaning of sentences. Such grammars should provide for embedding and paraphrasing of graphs in the middle of strings. This is exactly what is done in the traditional Sanskrit semantics. To analyze meaning of Sanskrit sentences *śābdabodha prakriyā*-s have been traditionally evolved based on the contending perspectives of (1) *Vyākaraṇa* , (2) *Nyāya*-Vaiśeṣika and (3)

G.N. Jha (Ed.): Sanskrit Computational Linguistics, LNCS 6465, pp. 91–105, 2010.

Mīmāṃsā. In these analytical models, a sentence is uniquely paraphrased into a graph which is the meaning of sentence.

In fact, our goal is much bigger than just solving the problem of *śābdabodha* where one only tries to analyze the meaning of a sentence. Our goal is to construct a computational ontology which is foundational as well as generative in nature. We are aiming at generative graph grammar on which comprehensive ontology interpreter can be constructed. Once that task is accomplished, one could apply it to solve domain-specific problems like *śābdabodha* . Hence the main focus of the paper is on the building of graph grammar suitable for ontology applications. Idea is to construct computational environment such that applications like *śābdabodha* can be built.

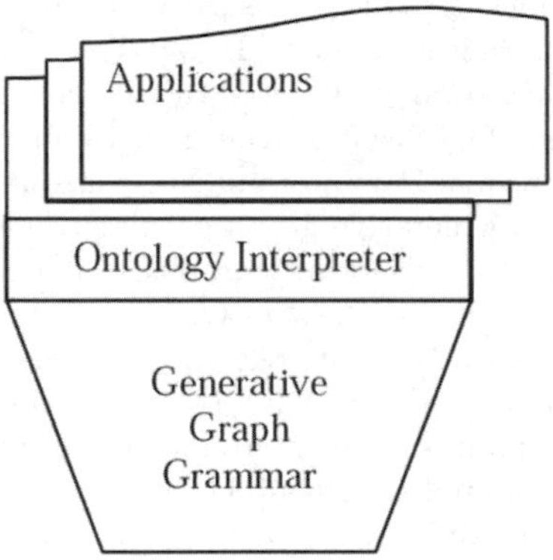

Fig. 1. Computational environment

Most of the existing ontologies follow top-down approach they start with the names of the entities (real as well as imaginary) existing in the universe, they try to classify these names and in turn try to classify the actual objects in the universe. But this approach has a problem since in the process of classifying names, one cannot help placing a name under two classes. In other words one name/class might have to have two parent classes. This is called *crossing*. For instance if one has the classes 'flying objects' and 'horses' in one's classification and now if one encounters a new object called 'flying horse', in which of the above classes should he place this object? Obviously, both classes should be its parents since it has the features of both the classes, and hence there is crossing. A good ontological classification is one in which there are no crossings. Having said this, we now elucidate an alternative bottom-up approach which we take to build not just any ontology but a formal, foundational and generative ontology. For this we utilized formalization of *Vaiśeṣika* ontology which is a comprehensive foundational ontology. While formalizing it, there was some difference noted with classical *Vaiśeṣika* traditions, hence our approach has been called Neo-Vaiśeṣika Formal Ontology (NVFO).

There are many proposed foundational ontologies based on Aristotle's ontology like BFO [1], GFO [2], OCHRE [3], DOLCE [4] etc. These ontologies are not generative, they are depictive; they declare categories and not formally derive categories. In contrast, NVFO is generative as its categories are formally derived from the first principle and thus generative graph grammar for NVFO can be done as has been shown in the paper.

2 Neo-Vaiśeṣika Formal Ontology (NVFO)

Vaiśeṣika ontology, due to Kaṇāda [5], Praśastapāda [6] and Udayana [7] has been formalized by Navjyoti [8, 9, 10, 11]. The formalization is based on the idea of an ontological form which is recursive. This form is called punctuator. Using punctuators categories of *Vaiśeṣika* ontology can be derived.

2.1 Punctuator

To understand differentiating features of categories, we introduce a form punctuator which takes contents strictly as its exteriors quite unlike forms such as 'proposition' or 'set'. Two examples of this form are 'locus-located' punctuator (*ādhāra-ādheya bhāva*) and 'predecessor-successor' punctuator (*kārya-kārañа bhāva*), which tell apart two entities or rather make two entities contiguants in the context of locational and temporal contiguum respectively. We shall develop this form to arrive at calculus that yields categories.

The idea of punctuation can be generalized as a form 'punctuator' *which vacuously tells apart two content-full entities that are otherwise-bound by relational context.* Generalized idea of punctuator holds for experience acquired through all sensory modalities and not just auditory modality. Rather it holds for all experience. Force behind this generalization is based on the insight that no real entity is given in experience without a punctuator. Any presentation in experience minimally has a punctuator and the entities it punctuates.

Punctuator is thus a general form that vacuously weaves two exterior entities as *contiguants* in the context of relational *contiguum.* Relational contiguum is made up of the rest of entities in the world. It is the rest of the entities in the world that go in making two punctuated entities contiguant in the first place. Relational contiguum is *determinable* in terms of *invariants* that arrest infinite regress in the system of entities and punctuators. Using these invariants, *categories* of entities can be characterized. Further, mutually distinguishing features of categories and shared features of *categories* can be exactly articulated. With these categories, well-formed punctuators with determinate relational context become *knowable* and *pronounceable.*

Punctuator between two entities e_x and e_y is represented as $p_r(e_x|e_y)$ where r is an underlying contiguum.

2.2 Fundamental Types of Punctuators

In *Vaiśeṣika*, there are three basic relations enumerated - *svarupa*, *samavāya*, and *saṃyoga-vibhāga*. These will form the relational contexts for our three basic punctuators (a) Self-linking (b) Inseparable punctuator and (c) Separable punctuator respectively.

Self-linking Punctuator. Punctuator $p_\phi(e_x|e_y)$ with structure $\langle e_x|e_y, \phi\rangle$ is called *self-linking punctuator* where ϕ is an empty relational context. It can be represented using directed graph notations by representing entities as circular nodes and punctuator as line.

Inseparable Punctuator. *Inseparable punctuator* $p_\sigma(e_x|e_y)$ with structure $\langle e_x|e_y, p_\phi(e_x|S) - p_\phi(S|e_y)\rangle$ where S is an inert entity (the top one in the following figure) called *inherence* that inseparably binds e_x and e_y. It can be represented using directed graph notations by representing entities as nodes and punctuator as line with arrow. The full structure of inseparable punctuator will be as depicted below.

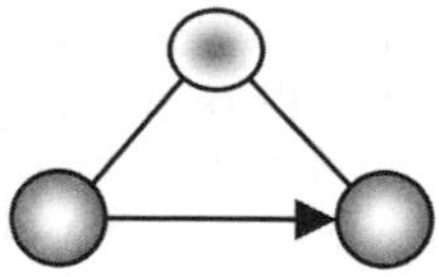

Separable punctuator. *Separable punctuator* $p_\tau(e_x|e_y)$ can be represented using directed graph notations by representing entities as nodes and punctuator as line with arrow. $p_\tau(e_x|e_y)$ is read as 'conjunction/disjunction with e_y inheres in e_x and conjunction/disjunction with e_x inheres in e_y' or simply as 'e_x is in conjunction/disjunction relation with e_y'. Unlike inherence relation, conjunction/disjunction relations are mutually contingent like a switch device. The full structure of separable punctuator is depicted below. Here conjunct 'J' is represented by the top black entity, and disjunct 'D' is represented by the top white entity. At any instant of time only one of them can inhere in both e_x and e_y since those entities can only be in conjunct or disjunct at a time but not both.

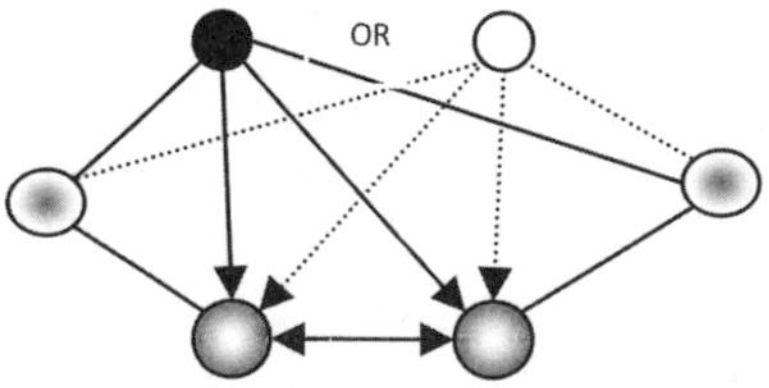

One can further generalize this and go on to construct any arbitrary punctuator of complex shape consisting of these three basic punctuators. This can be done if one has a grammar to combine these fundamental punctuators. Our claim is that any such structure represents some portion of reality. Hence one can identify each of the entities of that structure as one of the fundamental categories of *Vaiśeṣika* . This is possible if one has interpretation rules to interpret the structure generated by the above grammar. Thus the main agenda of this paper is to give the grammar rules (to generate arbitrary complex punctuator from the basic punctuators) as well as interpretation rules (to interpret this structure) after which one can deploy this knowledge for various applications like *śābdabodha* etc.

3 Graph Grammars

Graph Grammars are more generalized versions of string grammars where the alphabet are nodes and edges. As we have observed above that the 3 fundamental punctuators are graphs, and they are sufficient to construct any complex punctuator, one would naturally get a doubt 'How?' For that we need to define the rules for combining the basic punctuators in different ways.

Firstly, our grammar consists of a set of rules to generate a graph of arbitrary size composed of basic punctuators. Then we write an interpreter another layer of rules to identify the substructures of the graph with the above categories. Before we list the rules, let's have a look at some of the technical details necessary to understand them.

3.1 Definitions

Graph. Given two fixed alphabets Ω_V and Ω_E for node and edge labels, respectively, a (labelled) graph (over (Ω_V, Ω_E)) is a tuple G=$(G_V, G_E, s^G, t^G, lv^G, le^G)$, where G_V is a set of *vertices* (or *nodes*), G_E is a set of *edges* (or *arcs*), s^G, t^G : $G_E \longrightarrow G_V$ are the *source* and *target* functions, and $lv^G : G_V \longrightarrow \Omega_V$ and $le^G : G_E \longrightarrow \Omega_E$ are the *node* and *edge labeling* functions, respectively. [12]

Total graph morphism. A (total) graph morphism (mapping) $f : G \longrightarrow G'$ is a pair $f = (f_V : G_V \longrightarrow G'_V, f_E : G_E \longrightarrow G'_E)$ of functions which preserve sources, targets, and labels, i.e., which satisfies f_V o $t^G = t^{G'}$ o f_E, f_V o $s^G = s^{G'}$ o $f_E, lv^{G'}$ o $f_V = lv^G$, and $le^{G'}$ o $f_V = le^G$. [12]

Partial graph morphism. A subgraph S of G, written as $S \subseteq G$ or $S \hookrightarrow G$, is a graph with $S_V \subseteq G_V, S_E \subseteq G_E, s^S = s^G|S_E, t^S = t^G|S_E, lv^S = lv^G|S_V, le^S = le^G|S_E$. A (partial) graph morphism g from G to H is a total graph morphism from some subgraph $dom(g)$ of G to H, and $dom(g)$ is called the domain of g. A partial morphism can also be considered as a pair of partial functions. [12]

Production. A production $p : (L \xrightarrow{r} R)$ consists of a production name p and of an injective partial graph morphism r, called the production morphism. The graphs L and R are called the left-hand side (LHS) and the right-hand side (RHS) of p, respectively [12].

Graph Grammar. A graph grammar G is a pair $G = ((p : r)_{p \in P}, G_0)$ where $(p : r)_{p \in P}$ is a family of production morphisms indexed by production names, and G_0 is the start graph of the grammar. [12]

The above details can be concisely captured in the following diagram.

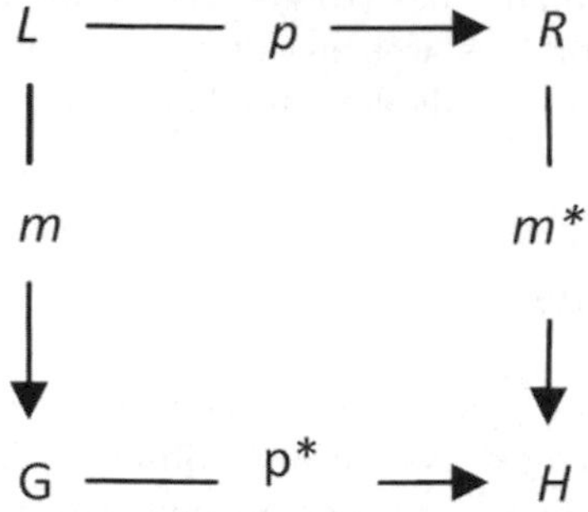

If the *match* m fixes an occurrence of L in a *given graph* G, then $G \overset{p,m}{\Rightarrow} H$ denotes the direct derivation where p is applied to G leading to a derived graph H. Intuitively, H is obtained by replacing the occurrence of L in G by R.

Negative Application Conditions (NACs). Rules can have exceptions - it may not be likely to apply a rule in some particular cases. Those cases/conditions can also be depicted in the form of graphs. Hence each rule may (or may not) have one or more NACs. So a rule will be applied only when its LHS matches with some subgraph of the host graph, and none of its NACs matches with any subgraph of the host graph.

4 Generation Rules of NVFO Graph Grammar

In NVFO, as mentioned earlier, there are rules for generating graphs as well as rules for interpreting these generated graphs. For the generation rules, $\Omega_V = \{G, X, E, S, J, D\}$ where

G – start node
X – variable node
E – anonymous entity (this will be identified as one of the categories during interpretation)
S – inherence entity (samavāya)
J – conjunct entity (saṃyoga)
D – disjunct entity (vibhāga)

And the set of edge labels, $\Omega_E = \{\text{in, con, dis}\}$ where

in – inherence
con – conjunct/contact
dis – disjunct

The edge representing self-linking relation has no label. It is recognised as such. Though it is directed it can be considered as undirected edge since for a self-linking edge direction doesn't matter.

The start graph, G_0, consists of a single node labelled 'G'. It is the same as the LHS of Rule 1 listed below. The following rules can be applied in any order, not necessarily in the order in which they are listed.

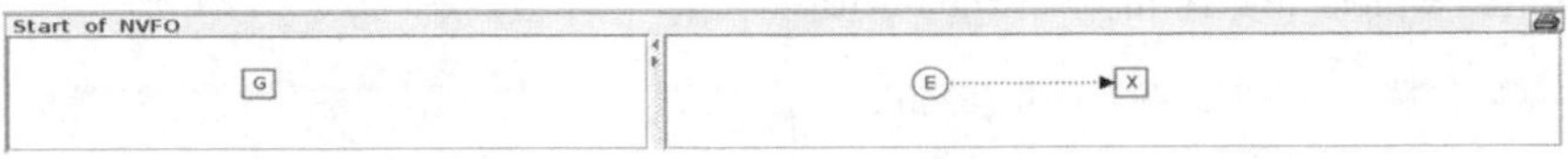

Rule 1 (Start). We start by replacing the start node, 'G', with anonymous entity, 'E', and variable node, 'X', which are self-linked.

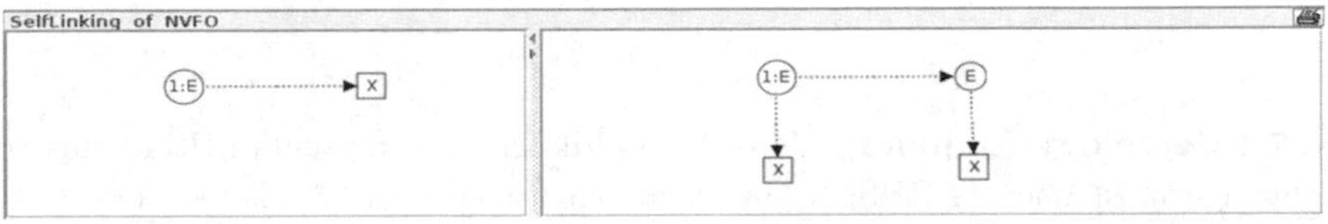

Rule 2 (SelfLinking). Here LHS is replaced by a punctuator which self-links already existing *'E'(1:E)* with another *'E'*. *'X'*s here are variable nodes.

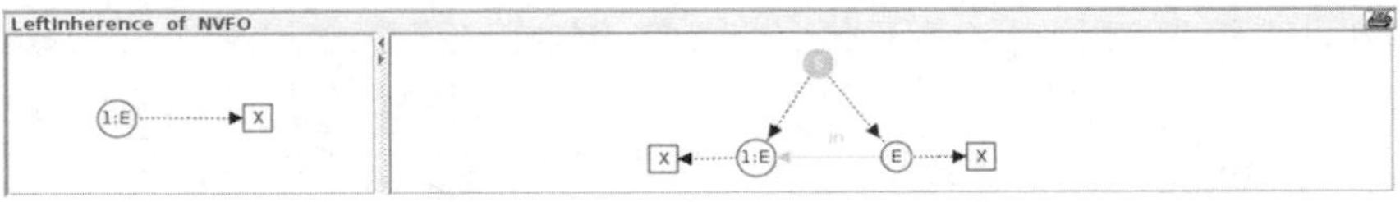

Rule 3 (LeftInherence). Here LHS is replaced by Inherence punctuator where the new entity (E) inheres in the old entity (*1:E*). In the process, 'S', is generated which is self-linked to both the entities.

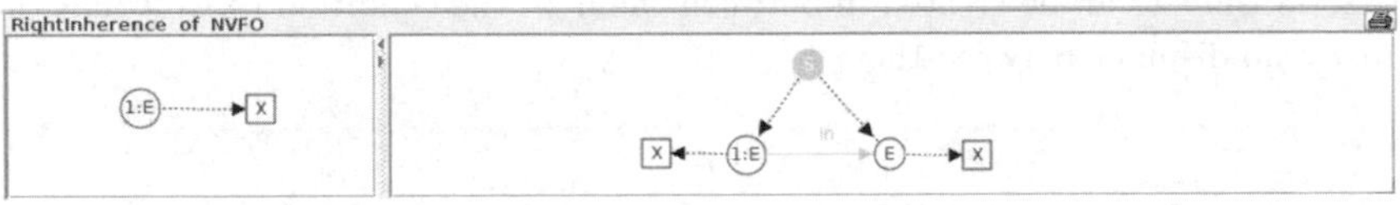

Rule 4 (RightInherence). Here LHS is replaced by Inherence punctuator where the old entity (*1:E*) inheres in the new entity (E).

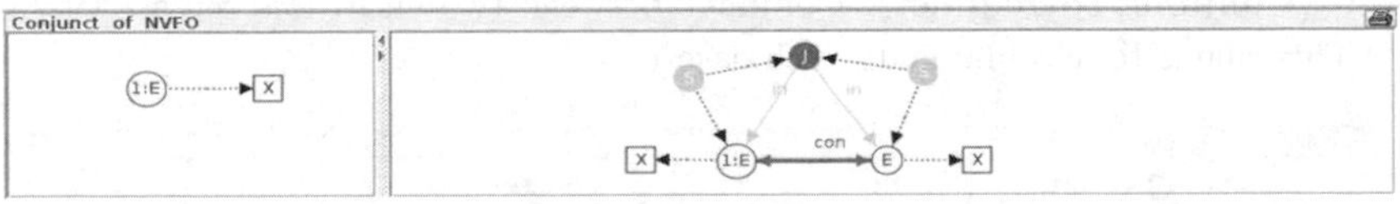

Rule 5 (Conjunct). LHS is replaced by separable punctuator where the new entity (E) and the old entity (*1:E*) are in contact with each other. In the process 'J' is generated which inheres in both the entities.

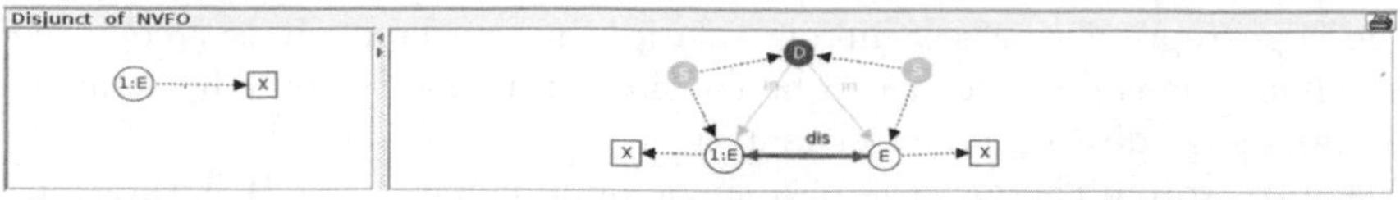

Rule 6 (Disjunct). LHS is replaced by separable punctuator where the new entity (*E*) and the old entity (*1:E*) are in disjunct with each other. In the process '*D*' is generated which inheres in both the entities.

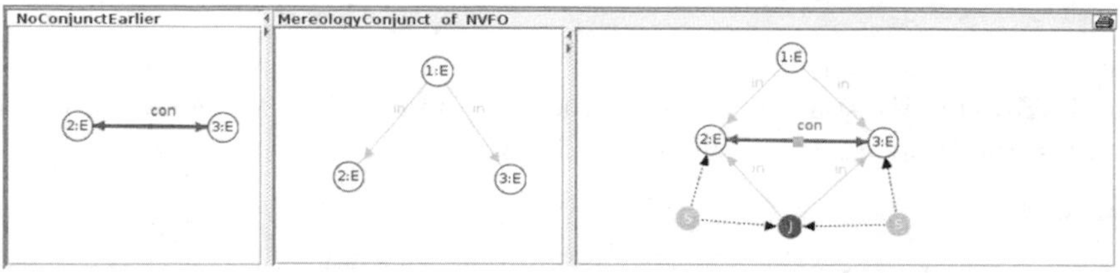

Rule 7 (MereologyConjunct). Here the middle graph represents LHS of this rule, and the rightmost one, its RHS. It says that when some entity inheres in two other entities, a conjunct can be created in between them on the condition (NAC) that there is already no conjunct between them, which is represented by the leftmost graph in the above image. This rule is to model the conjunct between parts of a whole.

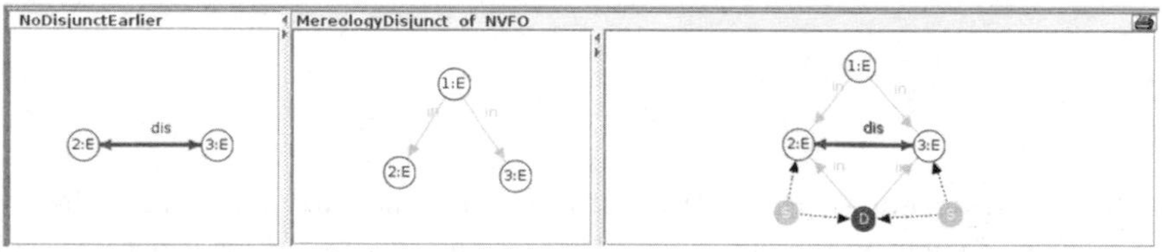

Rule 8 (MereologyDisjunct). It says that when some entity inheres in two other entities, a disjunct can be created in between them on the condition (NAC) that there is already no disjunct between them.

Rule 9 (ConjunctToDisjunct). Conjunct between two entities is replaced by disjunct. This rule is to accommodate such cases.

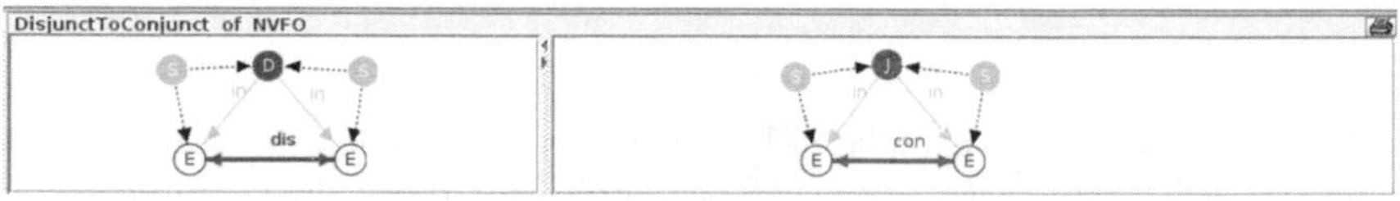

Rule 10 (DisjunctToConjunct). Disjunct between two entities is replaced by conjunct. This rule is to accommodate such cases.

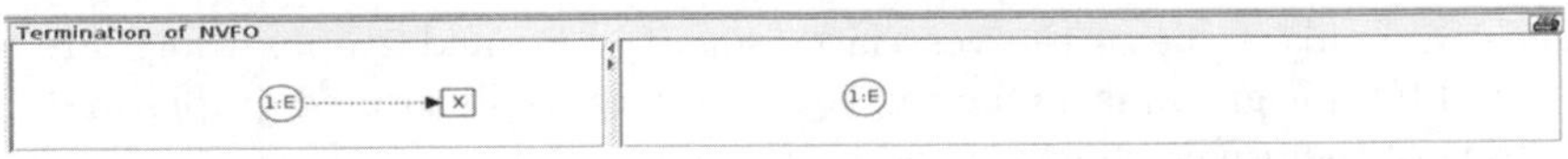

Rule 11 (Termination). This rule is applied when no more new entities need to be generated.

5 Testing If Ill-Formed Graphs Are Constructed

Our claim is that the above rules generate only well-formed graphs (those which comply with *Vaiśeṣika* Ontology) and not a single ill-formed graph. For instance one can show that the above rules do not generate the following ill-formed graphs. In NVFO grammar the rules in which only 'S' or 'E' appear (no 'J' or 'D') are rules 1,2,3,4 and 11. Let's show that, using these rules, the following graphs cannot be generated (we will ignore the node 'X' and its edges for the time being since it can always be done away with the 'Termination' rule (Rule 11)).

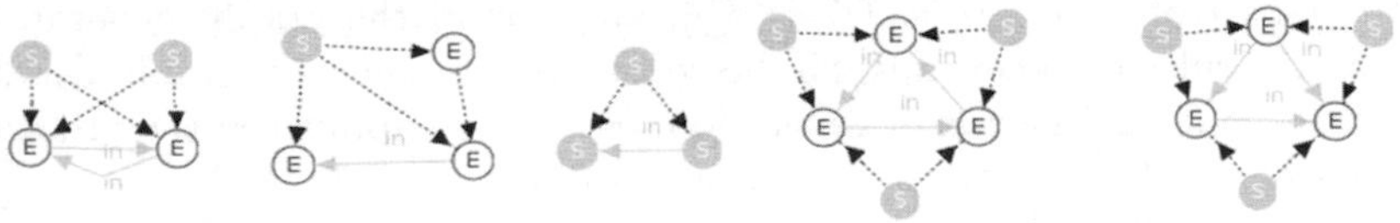

To generate the first graph, one can try applying the above rules in the following order: 1,3 (or 1,4). They would generate two 'E's and one 'S'. But if one further tries to apply one more rule which produces a new 'S' one cannot do it without producing a new 'E'. Hence there can be no graph with two 'E's and two 'S's. One needs minimum three 'E's to have two 'S's. This proves *Vaiśeṣika* theorem that same pair of entities cannot inhere in each other, and that inherence is unidirectional.

One can get closest to the second graph by applying the rules 1,2,4 in order. But here the self-linking edge between 'S' and the top-right 'E' would be missing. There is no other rule with which this edge can be created. Hence this graph too cannot be generated with this grammar. It's an ill-formed graph according to *Vaiśeṣika* because the same 'S' can never be self-linked to more than two entities at a time.

The third graph can be discarded right away since there is no single rule in our grammar which has only 'S's in it. In other words, any given graph, generated using these rules, has to contain minimum one or more 'E'. This is an ill-formed graph according to *Vaiśeṣika* since two inherence entities cannot have an inherence relation between them this would lead to infinite regress.

The 4th and 5th graphs can be rejected using similar arguments. One can get closest to them by applying rules 1,3,3 (or 1,3,4 or 1,4,3 or 1,4,4) in order. But at the end of none of these combinations does one get a cycle where the third 'E' inheres in the first 'E' or vice-versa. This chain of inherences always remains open, never closed. This proves another theorem of *Vaiśeṣika* that there can never be a closed chain consisting of inherence relations alone.

These are only some of the very large set of ill-formed graphs which are not generated by this grammar. One can test many other ill-formed graphs and find that none of them will be generated by this grammar.

6 Interpretation Rules of NVFO Graph Grammar

Here we interpret the above generated graphs to identify each of the anonymous entities as one of the *Vaiśeṣika* categories.

For the interpretation rules, $\Omega'_V = \Omega_V \bigcup \{U, US, V, SW, SA, Q, A\}$ where

U – Universal (*sāmānya*)
US – Ultimate Substance (*antya dravya*)
V – Ultimate Differentiator (*viśeṣa*)
SW – Substantial Whole (*avayavi*)
SA – Substantial Atom (*paramāṇu*)
Q – Quality (*guṇa*)
A – Action (*karma*)

And the set of edge labels, $\Omega'_E = \Omega_E$; and any of the graphs generated by applying the rules of generation can be considered as the start graph, G_0, here. These rules, unlike the rules of generation, need to be applied in the order they are listed below (the reasons are mentioned inline).

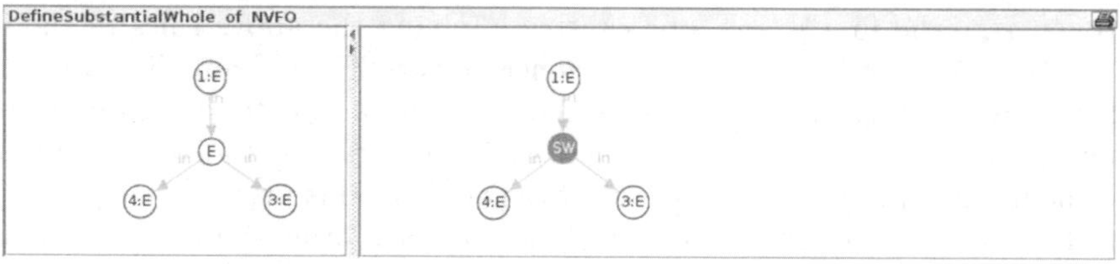

Rule 1 (DefineSubstantialWhole). A substantial whole is one in which something inheres, and also it itself inheres in some other entities. Hence 'E' is replaced by 'S_w' if at least one entity inheres in it, and it itself inheres in minimum two other entities (It has to inhere in minimum two entities for it to be differentiated from Qualities and Actions which can inhere only in one entity).

In *Vaiśeṣika* , Universal means a class which has minimum two instances of it. Now Universal is defined in three rules since an entity can be identified as a Universal in three different cases. Each of these rules has the same NACs (4 in number) which are mentioned at the end of these three rules.

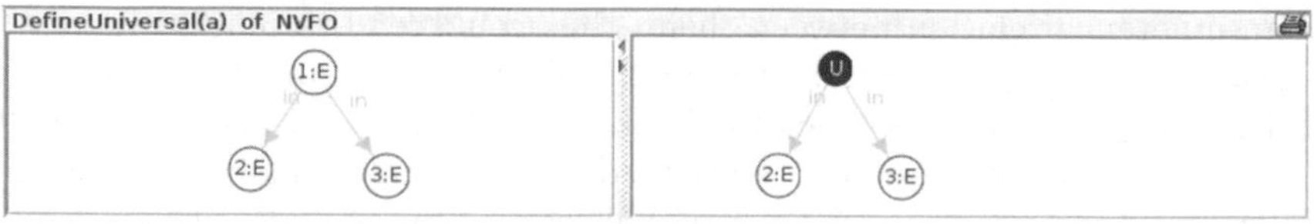

Rule 2 (DefineUniversal(a)). Universal inhering in two anonymous entities which haven't yet been identified.

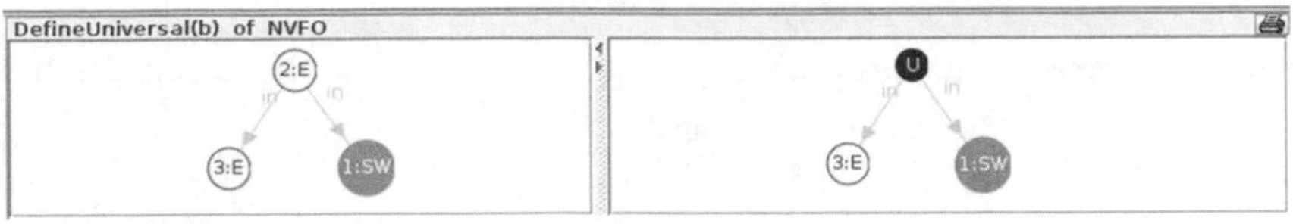

Rule 3 (DefineUniversal(b)). Universal inhering in one anonymous entity and one whole.

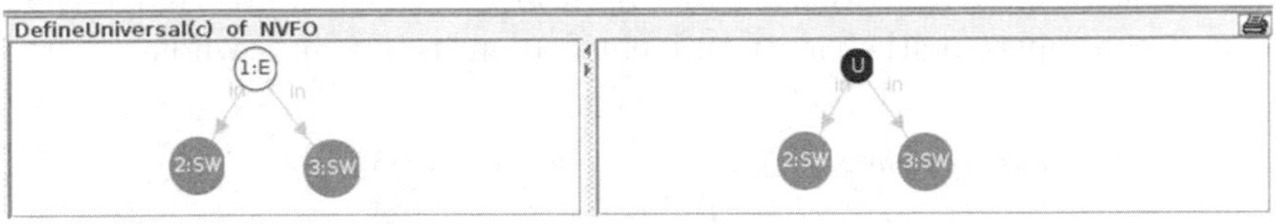

Rule 4 (DefineUniversal(c)). Universal inhering in two wholes.

None of the above 3 rules can be applied if at least one of the following 4 NACs is true which say that nothing can inhere nor be in contact with a Universal neither anonymous entity nor a whole.

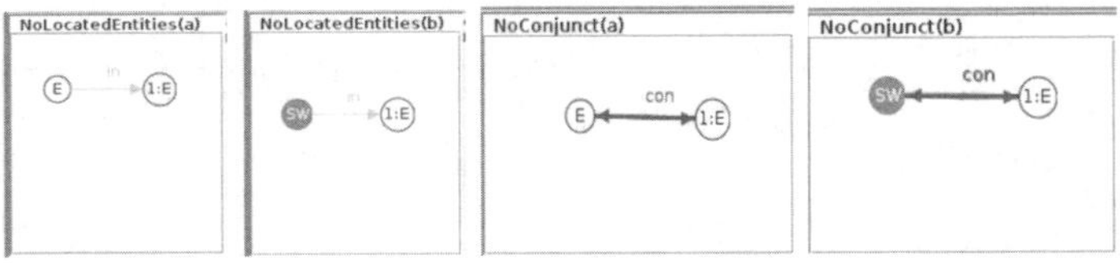

In Vaiśeṣika, Ultimate Substance is the one which has no locus, but it itself can serve as a locus for other entities. Here it is defined in three rules. Each of these rules has 2 NACs.

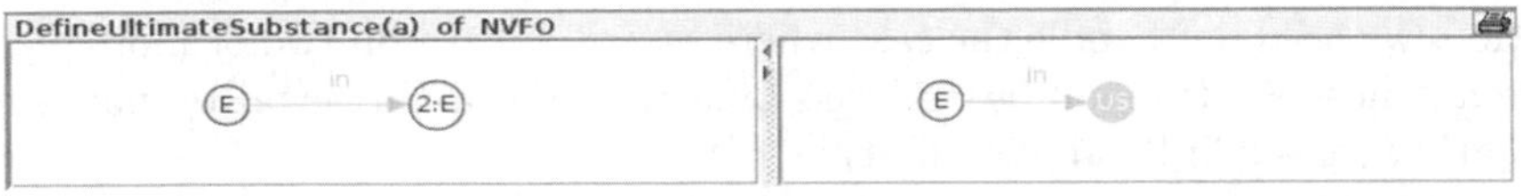

Rule 5 (DefineUltimateSubstance(a)). An anonymous entity can inhere in Ultimate Substance.

Rule 6 (DefineUltimateSubstance(b)). A Universal can inhere in Ultimate Substance.

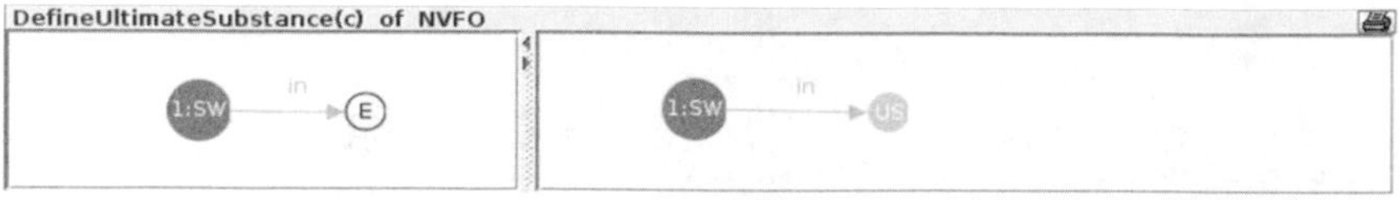

Rule 7 (DefineUltimateSubstance(c)). A whole can inhere in Ultimate Substance.

The following 2 NACs (for each of the above 3 rules) state that Ultimate Substance can inhere neither in an unidentified entity nor in a whole.

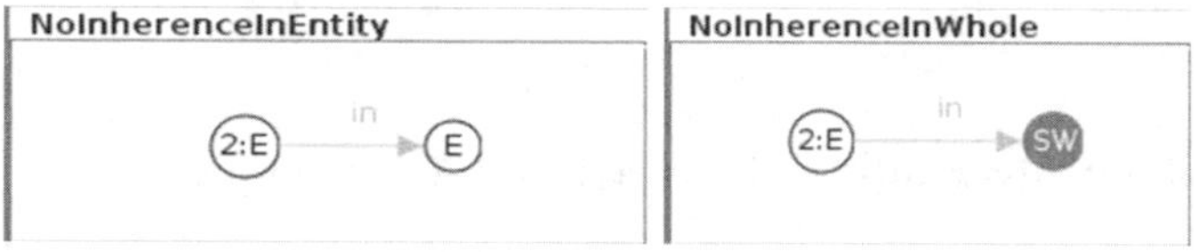

Viśeṣa, in *Vaiśeṣika* , can inhere in one and only one entity, that too Ultimate Substance, and nothing else inheres in *viśeṣa* . These Ultimate Differentiators give volume to the Universe. It is defined in one rule which has 6 NACs.

Rule 8 (DefineVisesa). If an entity inheres in Ultimate Substance then it can be identified as viśeṣa under the 6 NACs.

The 6 NACs state that neither an unidentified entity nor a Universal nor a whole can inhere or be in contact with *viśeṣa* .

(Here we need not handle the case where *viśeṣa* inheres in two or more places since all those entities will be replaced by a whole or a Universal by now since the earlier rules will be applied first).

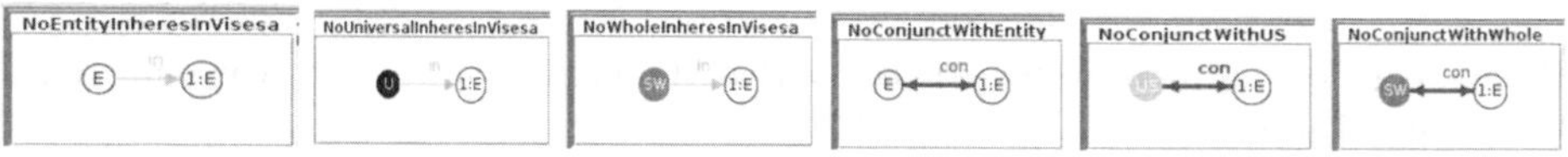

The only remaining *padārtha*-s are Quality and Action. We defined them as those entities in which at least one Universal inheres, but they themselves can inhere in one and only one whole or one Ultimate Substance, not more. Hence when an entity satisfies these conditions, it can be identified as either a Quality or an Action non-deterministically. Hence Quality and Action are defined in two rules each. Each of the following 4 rules has 4 NACs.

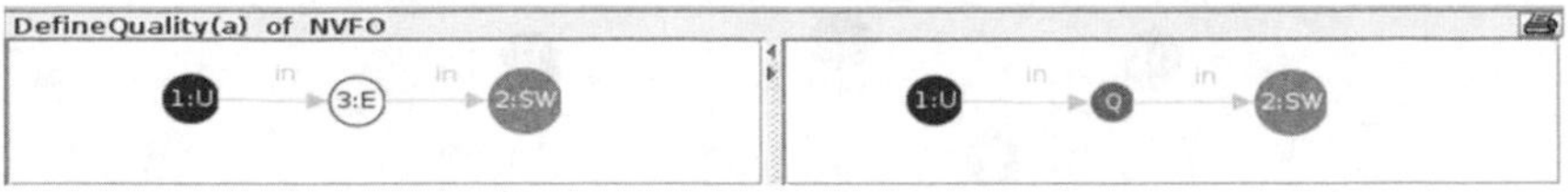

Rule 9 (DefineQuality(a)). In a Quality at least one Universal inheres, and it itself inheres in one and only one whole (its inherence in two or more wholes is handled in NACs).

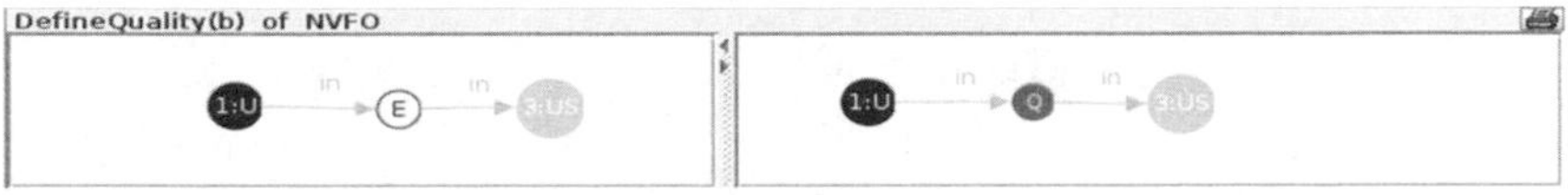

Rule 10 (DefineQuality(b)). In a Quality at least one Universal inheres, and it itself inheres in one and only one Ultimate Substance.

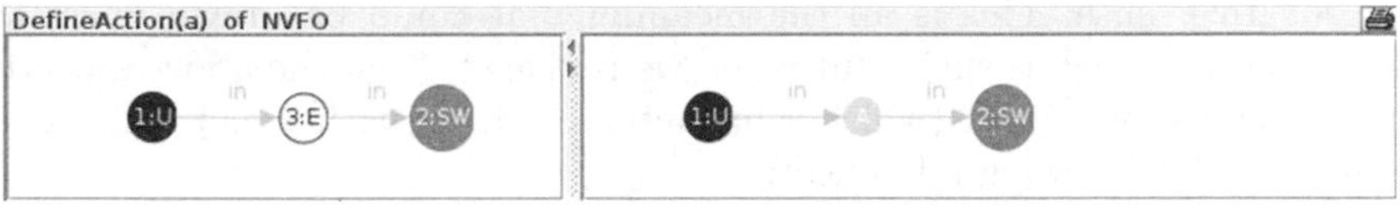

Rule 11 (DefineAction(a)). In an Action at least one Universal inheres, and it itself inheres in one and only one whole.

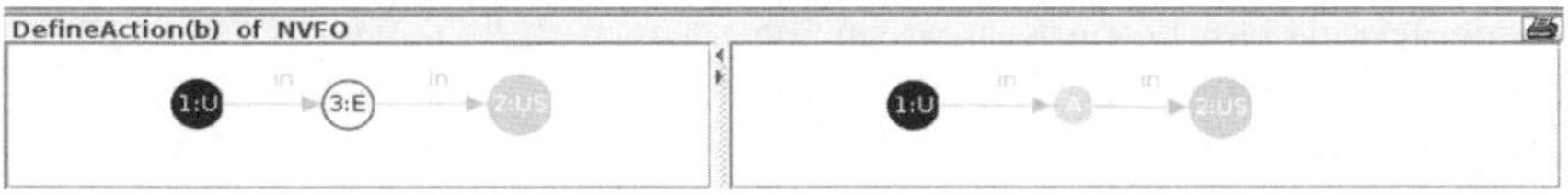

Rule 12 (DefineAction(b)). In an Action at least one Universal inheres, and it itself inheres in one and only one Ultimate Substance.

The 1st NAC says that quality/action cannot inhere in any entity other than whole or Ultimate Substance. The remaining 3 NACs state that they cannot inhere in two or more entities, whatever they may be.

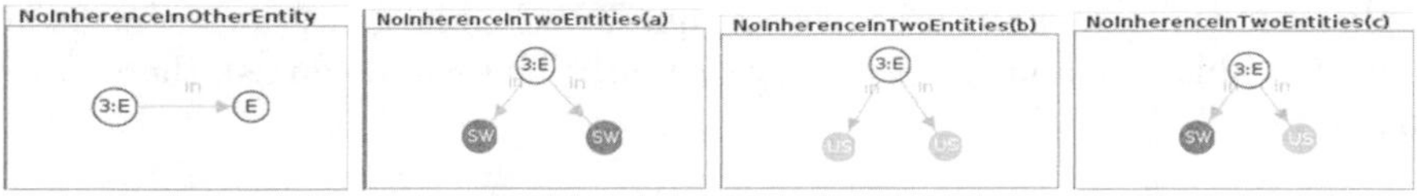

7 Application to *śābdabodha*

Now one can easily verify that the above rules generate the following well-formed graph which represents the structure of an arbitrary Substantial Whole (here inherence entity 'S' and its edges are ignored). This is just one of the many well-formed graphs generated by this grammar. In fact, our claim is that each and every substructure of a graph generated by these rules represents some portion of reality or the other. But for now, we will be using the structure of a substantial whole to do *śābdabodha* .

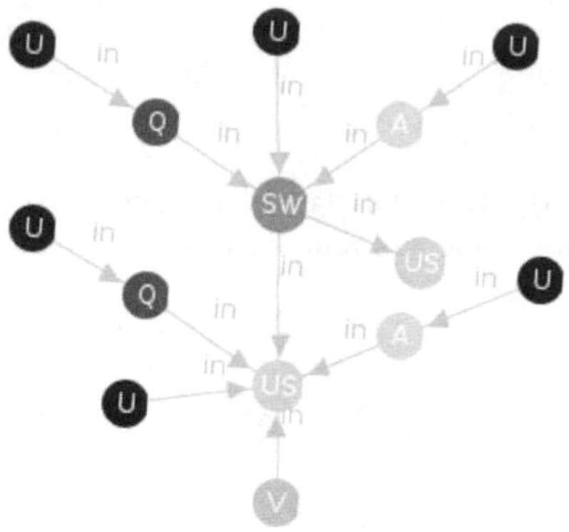

Let's consider the sentence, 'Jug is on the table'. Using standard NLP techniques, one can identify jug and table to be some objects and that table is used in the locative sense. This is all the meaning that could be drawn when this sentence is considered as such. But when we replace 'S_w' in the above structure with 'jug', and another 'S_w' (which is in contact with the earlier S_w) with 'table', we can draw the following inferences:

- Jug is in contact with table and vice-versa.
- Jug has some qualities.
- Table has some qualities.
- Some actions can be performed on jug.
- Some actions can be performed on table.
- Jug has some parts.
- Table has some parts.
- Jug belongs to some class 'jugness' which indicates that there are probably some other jugs in this Universe which may also have same qualities, and which may also be in contact with some table.
- Similarly, table belongs to the class 'tableness' which indicates that there are probably some other tables in this Universe.

This kind of reasoning has many applications. For instance one can write a query application which asks the user relevant queries when he gives the sentence, 'Jug is on the table', as input to the query system. The system can ask the following questions:

- What is the color of the jug?
- Are there some other jugs on the table?
- What is the length of the table?
- What is the texture of the surface of the table?
- ...

8 Conclusion

Śābdabodha is only one of the many applications of this grammar. Since *Vaiśeṣika* is not a domain ontology but a foundational ontology, our grammar is a foundational framework upon which any domain ontology can be modeled, not just *śābdabodha* . The power of this grammar lies in its generative nature which none of the other existing ontologies have.

References

1. BFO (Basic Formal Ontology), http://www.ifomis.org/bfo
2. GFO (General Formal Ontology), http://www.onto-med.de/ontologies/gfo/
3. OCHRE (Object-Centered High-level REference Ontology), http://ifomis.uni-leipzig.de/
4. DOLCE (Descriptive Ontology for Linguistic and Cognitive Engineering), http://www.loa-cnr.it/DOLCE.html
5. Kaṇāda: The Vaiśeṣika sutras of Kaṇāda, with the commentary of Śamkara Misra and extracts from the gloss of Jayanarayana. Translated in English by Nandalal Sinha. Allahabad (1911); 2nd edn. Revised and enlarged, Allahabad (1923); Reprinted New York (1974), Delhi (1986)
6. Praśastapāda: Padārthadharmasaṃgraha with Nyāyakandalī of Sridhara. English Translation by Ganganatha Jha. Chowkhamba, Varanasi (Reprint 1982)
7. Udayana: Lakshaṇāvali. In Musashi Tachikawa: The Structure of the World in Udayana's Realism: A study of the Lakshaṇāvali and Kiraṇāvali. Springer, Heidelberg (1982)
8. Singh, N.: Comprehensive Schema of Entities Vaiśesika Category System. Science Philosophy Interface 5(2), 1–54 (2001)
9. Singh, N.: Formal Theory of Categories through the Logic of Punctuator (2002) (unpublished)
10. Singh, N.: Theory of Experiential Contiguum. Philosophy and Science Exploratory Approach to Consciousness, pp. 111–159. Ramakrishna Mission Institute of Culture, Kolkata (2003)
11. Singh, N.: Foundations of Ontological Engineering. Lecture slides of course at IIIT Hyderabad (2008)
12. Rozenberg, G., et al. (eds.): Foundations. Handbook of Graph Grammars and Computing by Graph Transformation, vol. 1. World Scientific, Singapore (1997)
13. The Attributed Graph Grammar System, http://user.cs.tu-berlin.de/~gragra/agg/

Headedness and Modification in Nyāya Morpho-Syntactic Analysis: Towards a Bracket-Parsing Model

Malhar Kulkarni, Anuja Ajotikar, Tanuja Ajotikar, Dipesh Katira, Chinmay Dharurkar, and Chaitali Dangarikar

Indian Institute of Technology, Bombay
{malhar,anuja,tanuja,dipesh,chinmay,chaitali}@hss.iitb.ac.in

Abstract. The paper aims to develop a parsing model using the *nyāya*-morpho-syntactic analysis using the two terms namely, *prakāratā* and *viśeṣyatā*. The idea is that *prakāratā* and *viśeṣyatā* are to be seen as ***modification** (modifiedness)* and ***headedness*** respectively. Several representative sentences have been exemplified using the method developed. *prakāratā* and *viśeṣyatā* not only come through to give a thorough analysis at word level, but may be extended, as it has been shown in this paper, to get a thorough analysis at syntactic and discourse level, as well.

Keywords: headedness, modification, head-modifier relationship, *prakāratā* and *viśeṣyatā*.

1 Introduction

Many systems that provide the morphological information of Sanskrit are available online. They are built by following scholars at following institutions- Gérard Huet, INRIA, Paris,[1] Amba Kulkarni, University of Hyderabad,[2] Peter Scharf, Brown University[3] and Girish Nath Jha, JNU.[4] Ramakrishnamacharyulu (2009) presented a tagset to be used for the processing of Sanskrit sentences. We base ourselves on these existing systems and aim to provide with a model to parse Sanskrit sentences with head and modifier information. In order to do so we draw our material from the framework of the *Navya-Nyāya* school of Indian Philosophy. Several devices in Pāṇini's grammar like *abhidhāna, kāraka* have been employed to develop the analysis proposed. Several studies like Akshara Bharti (1995) and Pederson (2004), show use of Paninian framework in dependency parsing. They do not however, explicitly use the *Navya-Nyāya* technique of representing head and modifier relation within a word and concentrate mainly on intra-word relations. Further, they try to adapt the Paninian framework to the parsing of other Indian and non-Indian languages and not Sanskrit.

[1] `http://sanskrit.inria.fr/DICO/reader.html`
[2] `http://sanskrit.uohyd.ernet.in/lang-tech/lang-tech.html`
[3] `http://sanskritlibrary.org/`
[4] `http://sanskrit.jnu.ac.in/index.jsp`

G.N. Jha (Ed.): Sanskrit Computational Linguistics, LNCS 6465, pp. 106–123, 2010.

1.1 Heading towards *viśeṣyatā*

The idea of head across the syntactic theories has been residing in the basic idea of category, i.e. head as the unit that gives category to a phrase. It is as if the phrasehood of a phrase depends on the head.

The head of X is the part which determines the category of X. Example: The category of *ate the apple* (VP) is determined by the verb 'ate', not by any other element. This explains why it coordinates with other phrases which begin with a verb, but not with phrases that share the nominal part: *peeled the apple and ate the apple, *ate the apple and over the apple.*

Head in X-bar theory is an offshoot of the basic idea of phrase which is the centre of formalization. Here, though we would be using "head" as the English equivalent, it is strictly in the sense of *viśeṣya* and so we would distance this *viśeṣya* (head) from the head as in the linguistic theories.

The distinct nature of this headedness is that, save verb, for all the *padas*, it is the suffixes that hold the *viśeṣyatā*. Here, head-modifier relation is not of the kind that exists in X-bar theory. The *prakṛti-pratyaya* analysis is used to speak in terms of *prakāratā* and *viśeṣyatā*. So, in the analysis presented here (which takes inputs from *nyāya, vyākaraṇa* etc.), it is the *pratyaya* that is mostly the head and the *prakṛti* that is the modifier, and it is the relation between these two that exhausts the analysis.

On the absence of head in Pāṇini. *Pradhāna* or *viśeṣya* or anything that may express any sort of the dominant-subordinate relation is absent in Pāṇini. Thus one may say that head is not defined in Pāṇini, neither the concept of head has been deployed in Pāṇini. So, Pāṇini is not bound to define head. The ensuing question would then be: why in Pāṇini there is no idea or concept of head.

The reasons behind the development of the concept of head in GB or X-bar theory would be cognitive. In the sense that idea behind getting an abstract, generalized phrase structure through X-bar emerged as there was a need to understand the economy and speed in the acquisition. Head-modifier abstraction in X-bar theory gives a general phrase structure for all the phrases, i.e. it successfully explains the common structure applicable to all the phrases. [15]

There are no such cognitive commitments in Pāṇini. Also, it seems that phrase has never been the concern of Pāṇini. In Paninian morphology it is the *pada*, which is at the center of analysis. It can also be said that no phrase-based analysis is carried out in Pāṇini, overtly or covertly. Though Paninian grammar may be claimed to be *vākya*-centric, it is only through the consistent *pada*-analysis that *vākya* may be analyzed in the totality of *pada*. Phrase, as something possessing a category as its head, is simply impossible in Pāṇini because categories have been fashioned differently in Pāṇini.

Though the terms *head* and *modifier* have strong affiliations to the generative grammars (which do not talk in terms of dependency, as dependency grammars do),[5] we, in this paper, use the terms headedness for *viśeṣyatā* and modification for *prakāratā*.

The analysis and parsing presented in this paper are distinct from the theories like dependency grammar or phrase structure grammar. Though we use terms like *kartṛ, karma* which may be identified in parallel with the functional categories, one important difference to be noted is that our analysis presents an intra-lexical analysis that holds suffix as the head (save verbs) and inter-lexical analysis as well.[6]

We certainly do not attempt to fuse dependency or phrase structure grammars. We use the terms modification and headedness as we find them convenient and suitable equivalents for *prakāratā* and *viśeṣyatā*. These terms have not been used to evoke any resemblance or solidarity towards the linguistic theories. One straightforward aim of this paper is to represent the system of sentence analysis and parsing found in the traditional knowledge system of *Nyāya* with some inputs from *Vyākaraṇa*.

1.2 Nyāya Terms Used in the Head-Modifier Analysis

Following Sanskrit terms (with abbreviations) are used for analyzing head-modifier relationship in Sanskrit sentences:

1. *niṣṭha* (नि)	Residing in.
2. *prakāratā* (प्र)	State of being a modifier/modification.
3. *nirūpita* (निरू)	In relation to.
4. *viśeṣyatā* (वि)	State of being a head/headedness.
5. *abhinna* (अ)	Same as.
6. *prakṛtyartha* (प्रकृ)	the root meaning
7. *pratyayārtha* (त्यर्थ)	meaning of the termination

This terminology is used in the texts of *Navya-Nyāya*. We have seen this terminology being used in several *Navya-Nyāya* texts, but never in the context

[5] See Kübler, et al (2009: 2) While the dependency structure represents head-dependent relations between *words*, classified by *functional* categories such as subject (SBJ) and object (OBJ), the phrase structure represents the grouping of words into *phrases*, classified by *structural* categories such as noun phrase (NP) and verb phrase (VP).

[6] See Kübler, et al (2009: 2) Other terms that are found in the literature are modifier or *child*, instead of *dependent*, and *governor, regent* or *parent*, instead of *head*. Note that, although we will not use the noun *modifier*, we will use the verb *modify* when convenient and say that a dependent *modifies* its head. Also, Kübler, et al (2009: 3).

It is also worth noting that many syntactic theories make use of hybrid representations, combining elements of dependency structure with elements of phrase structure. Hence, to describe dependency grammar and phrase structure grammar as two opposite and mutually exclusive approaches to natural language syntax is at best an over-simplification.

of parsing a sentence with special reference to head-modifier representation. We explain below the three main terms briefly:

1. *viśeṣya.* This word is in use since Pāṇini. It is a relative term and not absolute. It co-exists in relation with the state of being a modifier. Pāṇini in his grammar called *Aṣṭādhyāyī* uses it in relation with another term called *viśeṣaṇa.* The term *viśeṣya* literally means *something which is being distinguished* and the term *viśeṣaṇa* literally means *the instrument which distinguishes.* The term *viśeṣya* indicates the element which can be said to function as a 'head' in a particular structure in the scheme proposed in this paper. The suffix *tā* indicates *the state of being a head.*
2. *prakāra.* - This word is proposed to be used in this scheme in the sense of a *modifier.* This word came into circulation in this sense with the advent of *Navya-Nyāya.* The suffix *tā* indicates the *state of being a Modifier.*
3. *nirūpita.* - This word indicates the relation between the two concepts mentioned above. It means 'in relation to'. It is used as an intra-lexical as well as inter-lexical metalinguistic connector.

The state of being a head and *the state of being a modifier* are correlated, coexistent and are assumed *to rest in* the meaning of the word. This substratum is indicated here with the help of the word *niṣṭha.*

Thus using this terminology we can present in brief the head modifier relation for a representative word like *grāmaṃ* in the following way,
Analysis- *grāma-niṣṭha-prakāratā-nirūpita-karma-niṣṭha-viśeṣyatā* |[7]

2 *Prakāratā-Viśeṣyatā* Analysis

2.1 Sentence Level

Sanskrit scholars have proposed various *śābdabodha* theories. For grammarians meaning of the verb root is the most important element in the sentence. Therefore they have proposed *dhātvartha-mukhya-viśeṣyaka-śābdabodha.* According to them words of any given sentence can be classified as either *subanta* or *tiṅanta.* Since, the meaning of the verbal root is the most important component of the sentence, all *subanta* words will be connected to the *tiṅanta* word. Therefore at the sentence level, only *tiṅanta* becomes *viśeṣya.* All the *subantas* become *prakāras* of the *tiṅanta* word. Therefore, *tiṅanta* word in the sentence, *bālaḥ paṭhati* will be tagged as follows:

(bālaḥ) prakāratā-nirūpita (paṭhati) viśeṣyatā |

In some sentences, more than one *subanta* will have the same *vibhakti.* These *subantas* are grouped together as they have adjective-substantive relationship

[7] The English paraphrase of the same would be- the state of being a head, residing in 'karma', in relation to the state of being a modifier which resides in 'village'. This provides us with the information that it is 'village' which is the head and 'karma' is the modifier.

between them. The adjectives are linked to the substantive by "non-difference" (*abheda*). For example:

"aruṇayā ekahāyanyā piṅgākṣyā gavā somaṃ krīṇāti |"

In this sentence there are four *subantas* with the same *vibhakti aruṇayā, ekahāyanya piṅgakṣyā gavā*. Therefore they are grouped together. One *subanta* (here, *gavā*) is the substantive. The adjectives are shown to be linked with this substantive by the relation of non-difference (*abhinna*). This is shown below:

[(aruṇayā) a(ekahāyanyā) a(piṅgākṣyā)a gavā] somaṃ krīṇāti |

But in some cases, some *subantas* will be linked to the verbal root directly and some will be linked via these directly linked *subantas*. This we show in the following figure:

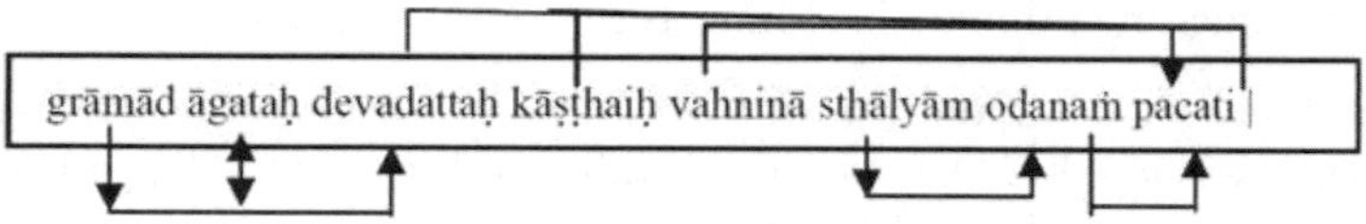

Fig. 1. Direct and indirect relation of *subantas* with the *tiṅanta*

Therefore, in this *prakāratā- viśeṣyatā* analysis, *subantas* that are linked to the *tiṅanta* directly, are related to the *tiṅanta* by the *prakāratā-nirūpita* tag. Whereas, the *subantas* that are linked to the *tiṅanta*-linked-*subantas* are grouped together. This group of *viśeṣaṇas* is placed before the *subanta* to which they are linked. These subantas will also be linked by the tag *prakāratā- nirūpita* but instead of *tiṅanta* they are linked to either *kartā* or *karma*. Therefore, the above mentioned sentence can be analyzed as:

Table 1. Direct and indirect relations of subantas with the verb

(odanaṃ) prakāratā- nirūpita pacati |
(sthālyām) prakāratā- nirūpita odanaṃ |
(vahninā) prakāratā- nirūpita pacati |
(kaāṣhaiḥ) prakāratā- nirūpita pacati |
(grāmāt) prakāratā- nirūpita agataḥ |
(āgataḥ)-(devadattaḥ) |
(devadattaḥ) prakāratā-nirūpita pacati |

Though all the *subantas* are ultimately linked to the *tiṅanta*, there is one exception. The genitive case which expresses relations[8] is not linked to the *tiṅanta*. The genitive case shows a relation with another noun in the sentence.

[8] P. II.3.50

Table 2. non-relatedness of the genitive case and verb

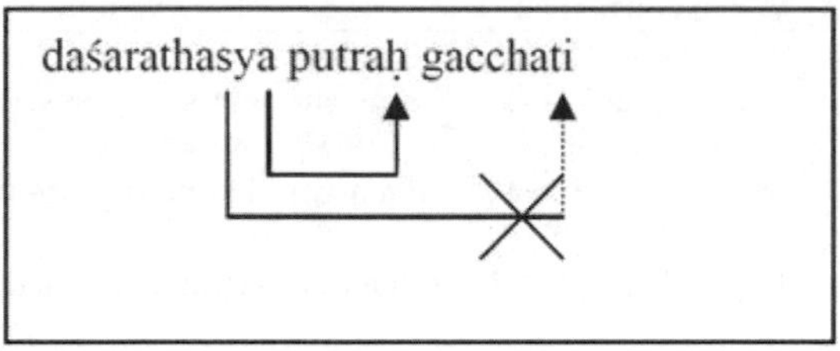

In this *prakāratā-viśeṣyatā* analysis, such genitive cases will be linked to the nouns to which they are related. Further, the genitive case will be tagged as *prakāratā-nirūpita* as it is the modifier of the noun. It is not the head of the noun to which it is related.

Thus the phrase *daśarathasya putraḥ* can be analyzed as (*daśarathasya*) *prakāratā- nirūpita putraḥ*. But the genitive word *daśarathasya* will not be linked to the verb *gacchati* in the sentence *daśarathasya putraḥ gacchati* |

2.2 Word Level *prakāratā-viśeṣyatā* Analysis Rules

Similar *prakāratā-viśeṣyatā* can be analyzed at the word level too. Each word in the Sanskrit sentence can be segmented into two parts: (a) *prakṛti* and (b) *pratyaya.*

***Prakāratā-viśeṣyatā* analysis within a *subanta* word:** Any *subanta* word in the sentence can be segmented into two parts: (1) *prakṛti* or *prātipadika* and (2) a suffix, *sup*. The meanings of these *sup pratyayas* are defined in the Paninian grammar. (Table.3)

Here the *pratyayārtha* is "head" so it is analyzed with *niṣṭha-viśeṣyatā*. The word *rāmam* will be analyzed as:

prakṛtyartha-rāma-niṣṭha- prakāratā- nirūpita-
pratyayārtha-karmaniṣṭha-viśeṣyatā.

-This (i.e. *karma-niṣṭha* etc.) particular tag is given when the particular kāraka is not expressed by the verbal suffix. Therefore (*kāraka*) is variable.

-But the nominative singular suffix does not express any *kāraka* relation and its meaning is *prātipadika* only therefore the particular *prātipadika* will be a part of the tag. In the case of indeclinables though *vibhakti* is deleted, the meaning of the suffix is *viśeṣya* and is treated like *prathamā* thus in the following case:

ca+su

ca+ 0

we observe following analysis:

prakṛtyartha-caniṣṭha- prakāratā- nirūpita - pratyayārtha - caniṣṭha- viśeṣyatā

Table 3. Meanings of *sup pratyayas*[9]

vibhakti	vibhaktyartha	Example
prathamā →	prātipadika →	rāma su (rāma -niṣṭha -viśeṣyatā)
dvitīyā →	karma →	rāma am (karma -niṣṭha- viśeṣyatā)
tṛtīyā →	karaṇa/kartṛ →	rāma ṭā (karaṇa/kartṛ- niṣṭha- viśeṣyatā)
caturthī →	saṁpradāna →	rāma ṅe (saṁpradāna - niṣṭha -viśeṣyatā)
pañcamī →	apādāna →	rāma ṅasi (apādāna- niṣṭha- viśeṣyatā)
ṣaṣṭhī →	saṁbandha/karma/kartṛ →	rāma ṅas (saṁbandha/karma/kartṛ -niṣṭha -viśeṣyatā)
saptamī →	adhikaraṇa →	rāma ṅi (adhikaraṇa - niṣṭha- viśeṣyatā)

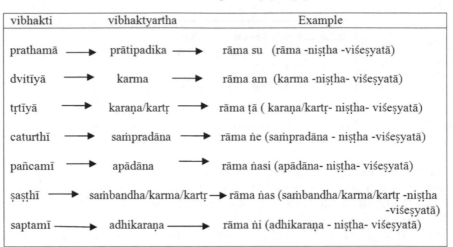

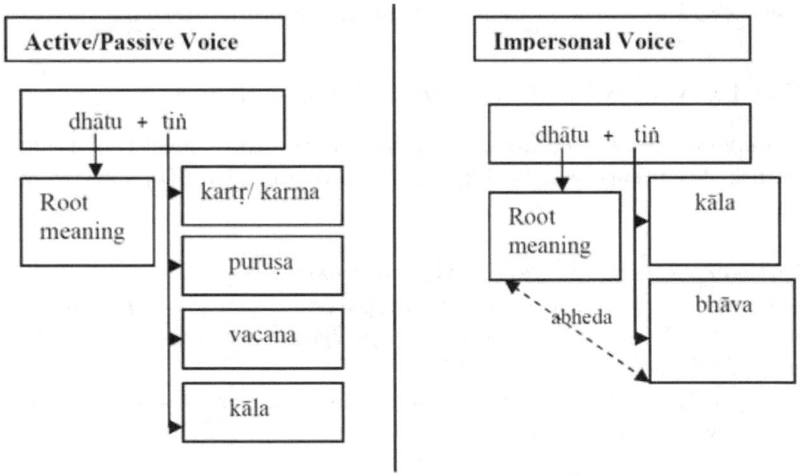

Fig. 2. Difference between the internal meaning structure of impersonal and active/passive verbs

***Prakāratā-viśeṣyatā* analysis within *tiṅanta* word:** In the case of verbs, root meaning is "head" and suffix meanings are modifiers. The *tiṅanta* suffix has four meanings (1) *kāraka*, (2) *puruṣa*, (3) *vacana* and (4) *kāla*. Tense or mood is related with the action and person and number are related with *kārakas* expressed by a *tiṅ* suffix. Overall two *kārakas* can be expressed by a *tiṅ* suffix and in the

[9] Even though more than one *kāraka* meaning is conveyed by one *vibhakti* we do not go into all the details of these mappings. We restrict ourselves for the present purpose to the most common mappings. Needless to say that we assume that the same technique would be equally applicable to other *kāraka* meanings as well.

absence of the expression of these *kārakas*; *bhāva* or action is expressed by the *tiṅ* suffix. When *bhāva* is expressed by a suffix, person and number are not related to it. Only *kāla* is. Here suffix expresses action in general and there is non-difference (*abheda*) between root meaning and suffix meaning. (See Fig. 2.)

3 From the above Discussion, We Formulate the Following Rules on Which We Base Our Analysis

3.1 Minimal Set of Rules on the Basis of Which the Headedness and Modification Will Be Determined at Different Levels

Rule no. 1. Suffix meaning is head[10] and root meaning is modifier.

Sub-rule a. In the verb,[11] root-meaning is the head and the suffix-meanings are the modifiers.

Rule no. 2. Modification of verbal suffix meanings-

2.a The number and person (which are the suffix meanings) are the modifiers of the *kāraka*.

2.b Time modifies the action[12] denoted by the verbal root.

Rule no. 3. The *kāraka* which is denoted by the *tiṅ* suffix matches with the nominative case in the sentence.[13]

Rule no. 4. All the *kārakas* modify the action denoted by the verbal root.

These rules can be said to be the rules for the head-modification within the *padārtha*

After applying these rules we get the following rule for headedness and modification on the sentence level.

Rule no. 5. The element which invariably has the tag of *niṣṭha viśeṣyatā* is the *mukhya viśeṣya* of the sentence.

After applying these rules we get the following general rule which is applicable on discourse level.

Rule no. 6. The head of the first sentence remains a head till it is modified by a sentence connective or gerund.

3.2 Complete *prakāratā- viśeṣyatā* Analysis

If both, sentence level and word level, *prakṛti* - *pratyaya* based analysis are combined, then the constituents of the meanings of the *subanta* words are linked as follows:

---

[10] *prakṛtipratyayārthayoḥ sahārthatve pratyayārthasya eva prādhānyam. paramalaghumañjūṣā dhātvarthaprakaraśam p. 29*

[11] *tiṅante kriyāyāḥ eva prādhānyaṁ śābdabodhe na pratyayārthasya iti bodhyam. kālaśca vyāpara viśeṣaṇam.* ibid p. 29

[12] ibid p.30

[13] *abhihite prathamā. vārtika on A. 2.3.2*

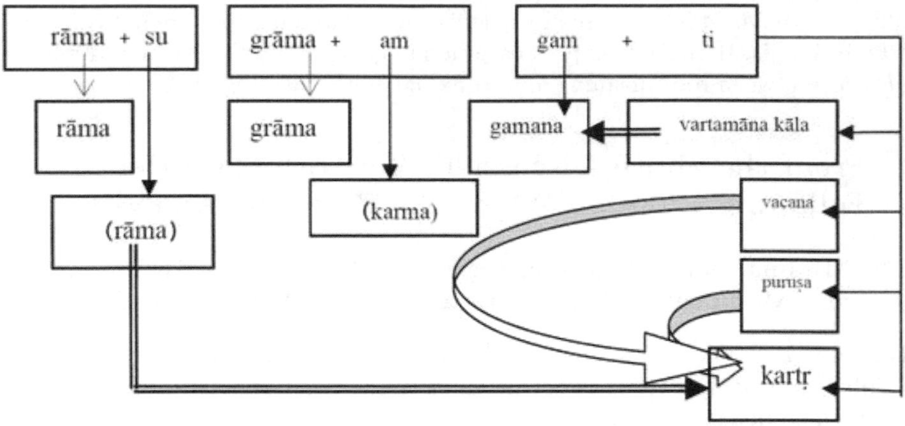

Fig. 3. Complete analysis of a sentence prior to prakāratā and viśeṣyatā tagging (Active Voice)

The *tiṅanta* suffix has four meanings (1) *kāraka*, (2) *puruṣa*, (3) *vacana* and (4) *kāla*. The *subanta* ending in the first case-ending is identified with the meaning *kāraka* of the *tiṅanta* suffix. Therefore, it is placed or moved in the *kāraka* meaning. Other two meanings of the *tiṅanta* suffix namely, *puruṣa*, and *vacana* become the *prakāra* of the *subanta* in the nominative case, linked with the *kāraka* meaning of the *tiṅanta* suffix. The complete sentence meaning after *prakāratā* and *viśeṣyatā* analysis will be as follows.(Table.4)

If the sentence is transformed to or is in the passive or impersonal form then we get slightly different sentences after the *prakāratā* and *viśeṣyatā* analysis. (Fig.4)

Since *karma* is expressed, (*abhihita*) in the passive voice, it is in the first case-ending. The *tiṅanta* suffix in the passive verb has the meaning *karma*. Therefore, as per our rule no. 3 this *subanta* which is in nominative case will be linked to the meaning *karma* of the *tiṅanta*.

The complete sentence after *prakāratā* and *viśeṣyatā* analysis of a passive voice will be as follows:

Table 4. *prakāratā-viśeṣyatā* analysis of the active voice sentence *ramaḥ grāmaṁ gacchati*

prakṛtyartha-grāma-niṣṭha-prakāratā-nirūpita-pratyayārtha-karma-niṣṭha-viśeṣyatā-niṣṭha-prakāratā-nirūpita-prathamapuruṣa-ekavacana-niṣṭha-prakāratā-nirūpita-prakṛtyartha-rāma-niṣṭha-prakāratā-nirūpita-pratyayārtha-rāma-niṣṭha-viśeṣyatā-abhinna-kartṛ-niṣṭha-viśeṣyatā-niṣṭha-prakāratā-nirūpita-vartamāna-niṣṭha-prakāratā-nirūpita-gamana-niṣṭha-viśeṣyatā-niṣṭha-viśeṣyatā-niṣṭha-viśeṣyatā

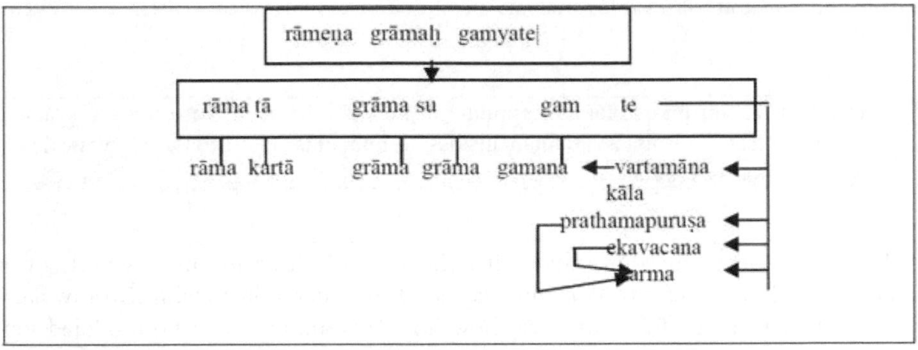

Fig. 4. Complete analysis of a sentence prior to *prakāratā* and *viśeṣyatā* tagging (Passive Voice)

Table 5. *prakāratā-viśeṣyatā* analysis of the passive voice sentence *rameṇa gramaḥ gamyate*

prakṛtyartha-rāmaniṣṭha-prakāratā-nirūpita-pratyayārtha-kartṛ-niṣṭha-viśeṣyatā-niṣṭha-prakāratā-nirūpita-prathamapuruṣa-ekavacana-niṣṭha-prakāratā-nirūpita-prakṛtyartha-grāmaniṣṭha-prakāratā-nirūpita-pratyayārtha-grāmaniṣṭha-viśeṣyatā-abhinna-karmaniṣṭha-viśeṣyatā-niṣṭha-prakāratā-nirūpita-vartamāna-niṣṭha-prakāratā-nirūpita-gamana -niṣṭha-viśeṣyatā-niṣṭha-viśeṣyatā-niṣṭha- viśeṣyatā |

Similarly, in the impersonal voice, the *kartṛ* is linked with *tiṅanta* but it is not expressed, *abhihita* by the *tiṅ* suffix. Therefore, *kartṛ* in such sentences is in instrumental case.

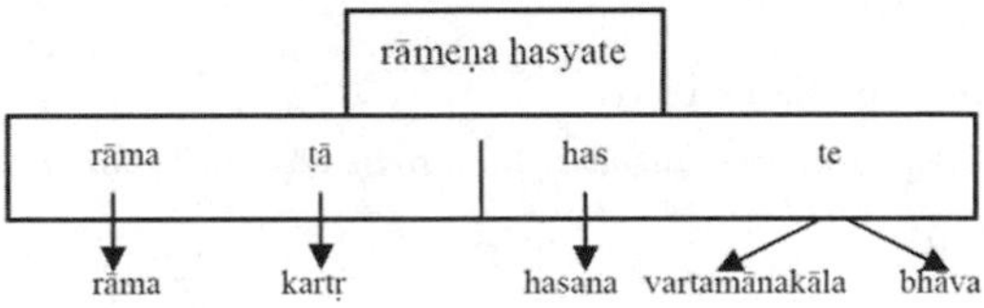

Fig. 5. Complete analysis of a sentence prior to *prakāratā* and *viśeṣyatā* tagging (impersonal voice)

Even though the suffix in the impersonal verb conveys two meanings (1) *kāla*, tense and (2) *bhāva*, the later meaning does not figure in the head-modifier analysis because the meaning *bhāva* actually means the action in general and therefore is connected to the meaning of the verbal root with non-difference (*abheda*). The head-modifier analysis of the sentence of the impersonal voice will be as shown below. (Table.6)

Table 6. *prakāratā-viśeṣyatā* analysis of the impersonal voice sentence : *rāmeṇa hasyate*

prakṛtyartha-rāmaniṣṭha-prakāratā-nirūpita- pratyayārtha -kartṛniṣṭha-viśeṣyatā-niṣṭha-prakāratā- nirūpita-vartamānaniṣṭha- prakāratā- nirūpita-hasananiṣṭha-viśeṣyatā-niṣṭha- viśeṣyatā

In this way, we can show that with the help of these minimal set of seven concepts and six rules, we can analyze head-modifier relationship in any Sanskrit sentence. In the following, we show how this analysis can be modeled formally and can be used along with the existing language processing tools for Sanskrit.

4 Automated or Semi-automated *prakāratā-viśeṣyatā* Analysis

This analysis can be modeled for the automated or semi-automated head and modifier relationship analysis of Sanskrit sentences. We propose a model using the rules mentioned in the above section. Any non-verse prose sentence can be analyzed by this model. Technical terminology can be shortened and analysis can be made much simpler if presented formally. We propose following short-forms for the formal representation of these rules (Table 7).[14]

For example, the head and modifier analysis for the active sentence *"rāmaḥ grāmaṃ gacchati"* will be as follows:

a) Word 1 or '*rāmaḥ*' is explained as-
prakṛtyartha-rāma-niṣṭha- prakāratā- nirūpita- pratyayārtha -rāmaniṣṭha-viśeṣyatā |
s1/1=)RM_प्रनि sTM_वि(
b) Word 2 or *grāmaṃ* is explained as-
prakṛtyarthagrāma-niṣṭha-prakāratā-nirūpita- pratyayārtha -karmaniṣṭha- viśeṣyatā |
S2/1(RM_प्रनि sTM_वि)

c) Word 3 or *gacchati*, the verb, can be explained in two parts as follows-
(3p/sing_प्रनि-कर्तृ-नि-वि)_प्रनि-(present tense_प्रनि-RM3(gamana)_वि)_वि।
or
(prathamapuruṣa-ekavacana-niṣṭha-prakāratā-nirūpita-kartṛniṣṭha-viśeṣyatā)-niṣṭha-prakāratā-nirūpita-(vartamāna-niṣṭha-prakāratā-nirūpita-gamana-niṣṭha-viśeṣyatā)-niṣṭha-viśeṣyatā |
t3/1= (tTM2,tTM3_प्रनि-tTM1_वि)_प्रनि(tTM4_प्रनि-RM_वि)_वि

[14] TM1=*kāraka*, (*kartṛ*, *karma*), TM2=*puruṣa*, TM3=*vacana*, TM4=*kāla* This is also mentioned in the discussion on fig.2.

Table 7. Abbreviations used in the formal representation of the head-modifier analysis in Sanskrit

देवनागरी short form	**Terms**	**English short form**
एव	*ekavacana*	Sig
द्विव	*dvivacana*	Du
बव	*bahuvacana*	Pl
प्रपु	*prathamapuruṣa*	3p
मपु	*madhyamapuruṣa*	2p
उपु	*uttamapuruṣa*	1p
प्रकृ	*prakṛtyartha*	RM
त्यर्थ	*pratyayārtha*	TM[14]
सु	*subanta*	s
ति	*tiṅanta*	t
	Short forms used for tagging head and modifier relations	
प्रनि	*prakāratā- nirūpita*	प्रनि
वि	*viśeṣyatā*	वि
[_] "Underscore"	*niṣṭha*	[_] "Underscore"

5 Hierarchical Bracket Parsing

The model that is proposed here takes primarily into consideration the brackets consisting of words and shows with the help of rules how the system can reduce

Table 8. Hierarchical bracket parsing of the active voice sentence *rāmaḥ grāmaṃ gacchati*

```
[
    (प्रकृ(grāma)_प्रनि-त्यर्थ-(karma)-वि)_प्रनि
    {
    (प्रपु-एव_प्रनि-प्रकृ(rāma)_प्रनि-त्यर्थ(rāma)_वि)-
    (कर्तृ)_वि)_प्रनि
    (vartamāna_प्रनि-(gamana _वि)
    }_वि
]
```

them one by one and ultimately how all these brackets representing words get reduced to a bracket of a sentence with the information of the *head* and *modifier* tagged to each bracket intact. We present below a brief description of the Hierarchical Bracket Parsing.

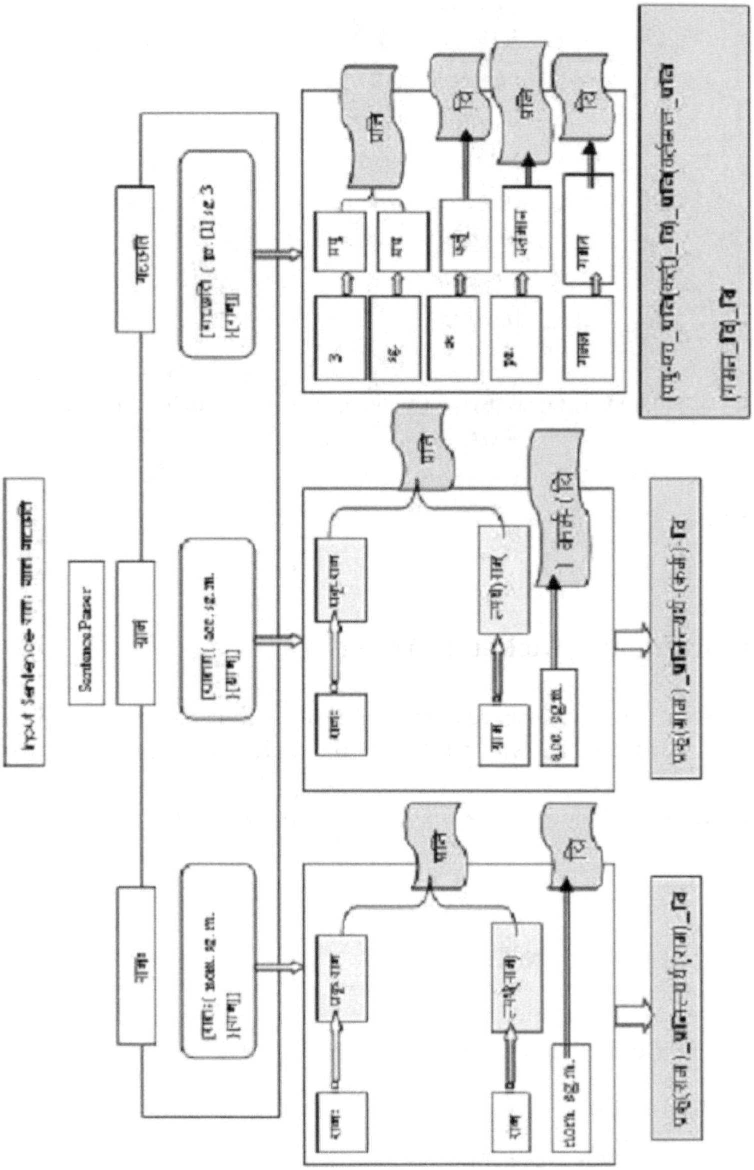

Table 9. Hierarchical Bracket Parsing of the passive voice sentence *rāmeṇa grāmaḥ gamyate*

[

(प्रकृ- rāma_प्रनि-त्यर्थ-(kartṛ)_वि)_प्रनि

{

(प्रपु-एव_प्रनि-प्रकृ- grāma _प्रनि-त्यर्थ- grāma _वि-(karma)_वि)_प्रनि

(vartamāna _प्रनि- gamana _वि)_वि

}_वि

]

Table 10. Hierarchical Bracket Parsing of the impersonal voice sentence *rāmeṇa hasyate*

[

(प्रकृ- rāma_प्रनि-त्यर्थ-(kartṛ)_वि)_प्रनि

{ vartamāna_प्रनि-hasana_वि

}_वि

]

This kind of analysis can be done not only on the simple sentences but also to the complex sentences from the literature. For example, we present the head and modifier analysis of the complex sentence, *prajāsu pālanīṃ vṛttiṃ veditum yaṃ ayuṅkta sa varṇiliṅgī viditaḥ vanecaraḥ dvaitavane yudhiṣṭhiraṃ samāyayau* |[15]

6 Application at Discourse Level

So far we presented analysis using the Navya-Nyāya terms of word-word meanings, and sentence-sentence meanings. We also presented this analysis in all the three voices and showed that the analysis does not get affected by the functional categories. Now we show how this technique can be used for discourse level. We present below the analysis of a group of three sentences. We will show how one element can be considered as head within all these three sentences and the rest as modifiers using this technique.

[15] Partial construal of *Kiratarjuniyam* 1.1

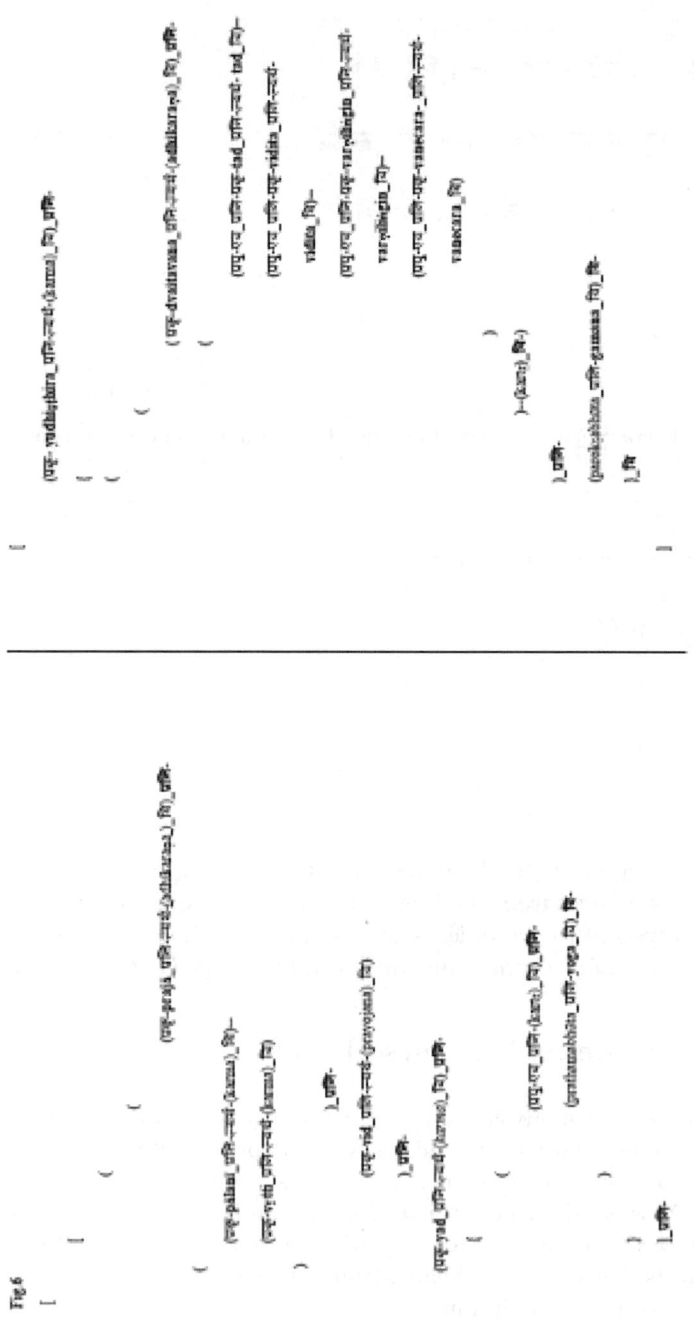

Fig. 6.

Table 11. prakāratā-viśeṣyatā analysis of multiple sentences

(S1) prakṛtyartha-mandira-niṣṭha-prakāratā-nirūpita-pratyayārtha-karma-niṣṭha-viśeṣyatā-niṣṭha-prakāratā-nirūpita-prathamapuruṣa-ekavacana-niṣṭha-prakāratā-nirūpita-prakṛtyartha-rāma-niṣṭha-prakāratā-nirūpita-pratyayārtha-rāma-niṣṭha-viśeṣyatā-abhinna-kartṛ-niṣṭha-viśeṣyatā-niṣṭha-prakāratā-nirūpita-vartamāna-niṣṭha-prakāratā- nirūpita-gamana-niṣṭha-viśeṣyatā-niṣṭha-viśeṣyatā-niṣṭha-viśeṣyatā

niṣṭha-prakāratā--nirūpita

(S2) -prakṛtyartha-tatas-niṣṭha-prakāratā-nirūpita-pratyayārtha-tatas-niṣṭha-viśeṣyatā-***niṣṭha-viśeṣyatā***-niṣṭha-prakāratā-nirūpita-prakṛtyartha-vidyālaya- niṣṭha-prakāratā-nirūpita- pratyayārtha-karma- niṣṭha-viśeṣyatā- niṣṭha-prakāratā-nirūpita-prathamapuruṣa-ekavacana-niṣṭha-prakāratā-nirūpita-kartṛ-niṣṭha- viśeṣyatā - niṣṭha-prakāratā-nirūpita-bhaviṣyat-niṣṭha-prakāratā-nirūpita-gamana-niṣṭha-viśeṣyatā-niṣṭha-viśeṣyatā-niṣṭha-viśeṣyatā - niṣṭha-viśeṣyatā

niṣṭha-prakāratā-nirūpita

(S3) -prakṛtyartha-anantara-niṣṭha-prakāratā-nirūpita-pratyayārtha-anantara-niṣṭha-viśeṣyatā- ***niṣṭha-viśeṣyatā***-niṣṭha-prakāratā-nirūpita-prakṛtyartha- gṛha- niṣṭha-prakāratā-nirūpita- pratyayārtha-karma- niṣṭha-viśeṣyatā- niṣṭha-prakāratā-nirūpita-prathamapuruṣa-ekavacana-niṣṭha-prakāratā-nirūpita-kartṛ-niṣṭha- viśeṣyatā - niṣṭha-prakāratā-nirūpita-bhaviṣyat-niṣṭha-prakāratā-nirūpita-gamana-niṣṭha-viśeṣyatā-***niṣṭha-viśeṣyatā-niṣṭha-viśeṣyatā*** - ***niṣṭha-viśeṣyatā***

Sentences:

S1. *rāmaḥ mandiram̩ gacchati*|
S2. *tataḥ vidyālayam̩ gamiṣyati*|
S3. *anantaram̩ gṛham̩ gamiṣyati*|(Table.11)

The last tag (*niṣṭha-viśeṣyatā*) in S3 indicates that the meaning of the verbal root in S3 remains a head in the context of (S1+S2). In other words, this tag shows the main meaning in the group of three sentences. This feature can be extended to a larger group of sentences.

7 Concluding Remarks and Future Work

Thus we showed that we can make optimal use of the existing theories within Sanskrit traditional *śāstras* for the purpose of parsing Sanskrit at the discourse level. Our approach combined insights from both *Nyāya* as well as *Vyākaraṇa* for parsing at various levels. For a sentential analysis, we used the *Vyākaraṇa* way of analysis and at the discourse level on the first step when we have to connect one previous sentence to the next one, we used the *Navya-Nyāya* way of analysis and then again we used *Vyākaraṇa* way of analysis.

Thus we have presented a model that sufficiently and minimally represents the head-modifier relationship between nominal and verbal elements. It is minimal in the sense that, it is the minimum number of terms employed from *Navya-Nyāya* that would sufficiently represent the analysis. We have also shown how it will be useful when we take into consideration the discourse level.

The analysis can be seen as an end in itself, such that it helps understand the sentence and words within it, in the scope of the proposed schema.

We propose to show in future how this model will capture the *viśeṣaṇa-viśeṣya* relation amongst nouns and also how it could prove useful for anaphora resolution in Sanskrit. We could also develop this model using *Mīmāṃsā* techniques to define sentence connectives that would prove useful for the purpose of discourse level analysis of Sanskrit. We presented an outline of a model that could very well be developed and tested in a full-fledged bracket parsing grammar under whose purview would fall issues like anaphora resolution, ellipsis and coordination etc.

Acknowledgements. We wish to thank all the reviewers for their invaluable comments that helped us revise the draft of this paper. We also wish to thank the editor for co-operation.

References

1. Bharati, A., Chaitanya, V., Sangal, R.: Natural Language Processing: A Paninian Perspective. Prentice Hall of India, New Delhi (1995)
2. Bhattacharya, G.: Viṣayatāvāda. Prasad, Varanasi (1943)
3. Carnie, A.: Syntax A Generative Introduction. Blackwell Publishing, Oxford (2006)
4. Corbett, G.G., Fraser, N.M., Scott, M. (eds.): Heads in Grammatical Theory. Cambridge University Press, Cambridge (1993)
5. Culicover, P.W.: Principles and Parameters: An Introduction to Syntactic theory. Oxford University Press, New York (2005)
6. Jha, G.N., Sobha, L., Diwakar, M., Surjit, K.S., Praveen, P.: Anaphors in Sanskrit. In: Johansson, C. (ed.) Proceedings of the Second Workshop on Anaphora Resolution, vol. 2. Cambridge Scholars Publishing, Cambridge (2008) ISSN 1736-6305
7. Jhalkikar, B.: Nyāyakośa. Bhandarakar Oriental Research Institute, Pune (1893)
8. Kübler, S., McDonald, R., Nivre, J.: Dependency Parsing. Morgan and Claypool Publishers, San Francisco (2009)
9. Mahalakshmi, G.S., Geetha, T.V., Kumar, A., Kumar, D., Manikandan, S.: Gautama - Ontology Editor Based on Nyaya Logic. In: Ramanujam, R., Sarukkai, S. (eds.) Logic and Its Applications. LNCS (LNAI), vol. 5378, pp. 232–242. Springer, Heidelberg (2009)
10. Mahalakshmi, G.S., Geetha, T.V.: Representing Knowledge Effectively Using Indian Logic. In: Sajja, A. (ed.) TMRF e-book, Advanced Knowledge based Systems, Model, Applications, Research, vol. I, pp. 12–28 (2010), `http://www.tmrfindia.org/eseries/ebookV1-C2.pdf`

11. Pedersen, M., Domenyk, E., Amin, S.K., Lakshmi P.: Relative clauses in Hindi and Arabic: a Paninian dependency grammar analysis. In: Coling 2004 workshop: Proceedings Recent Advances in Dependency Grammar. Prentice Hall of India Pvt. Ltd., Geneva (2004)
12. Potter, K.: Encyclopedia of Indian Philosophy. Motilal Banarasidas, Delhi (2000)
13. Sarma, V.V.S.: Survey of Indian Logic from the point of view of Computer Science. In: Sadhana, vol. 19 pp.196971–196983 (1994), `http://www.springerlink.com/content/lnv6656gg37022t0`
14. Shastri, G.: Patañjali's Vyākaraṇamahābhāṣya with Kaiyaṭa's Pradīpa and Bhaṭṭoji Dīkṣita's śabdakaustubha and Nāgeśabhaṭṭa's Udyota with commentary Rājalakṣmī. Rashtriya Sanskrit Sansthan, New Delhi (Reprint 2006)
15. Smith, N.: Chomsky Ideas and Ideals. Cambridge University Press, New York (2004)
16. Bhagavat, P.V.B.: (Tr.(in Marathi)): Paramalaghumañjuṣ. Paramarsha Prakashan, Pune Vidyapith Tattvajnana Vibhaga, Pune (1984)

Citation Matching in Sanskrit Corpora Using Local Alignment

Abhinandan S. Prasad and Shrisha Rao

International Institute of Information Technology, Bangalore
abhinandan.sp@iiitb.net, srao@iiitb.ac.in

Abstract. Citation matching is the problem of finding which citation occurs in a given textual corpus. Most existing citation matching work is done on scientific literature. The goal of this paper is to present methods for performing citation matching on Sanskrit texts. Exact matching and approximate matching are the two methods for performing citation matching. The exact matching method checks for exact occurrence of the citation with respect to the textual corpus. Approximate matching is a fuzzy string-matching method which computes a similarity score between an individual line of the textual corpus and the citation. The Smith-Waterman-Gotoh algorithm for local alignment, which is generally used in bioinformatics, is used here for calculating the similarity score. This similarity score is a measure of the closeness between the text and the citation. The exact- and approximate-matching methods are evaluated and compared. The methods presented can be easily applied to corpora in other Indic languages like Kannada, Tamil, etc. The approximate-matching method can in particular be used in the compilation of critical editions and plagiarism detection in a literary work.

Keywords: citation matching, local alignment, Smith-Waterman-Gotoh algorithm, Sanskrit, Mahābhārata, Mahābhārata-Tātparyanirṇaya.

1 Introduction

Citation matching in literature is the problem of finding which citations occur where in a given textual corpus. Citation matching is applied in various areas like authorship detection, content analysis, etc. Currently, citation matching is limited to scientific literature because most of the citation mapping work like autonomous citation matching, identity uncertainty, etc., are done on scientific literature. Autonomous citation matching identifies and groups variants of the same paper [5]. Identity uncertainty in the context of citation matching decides whether a set of citations corresponds to the same publication or not [9]. Citation matching is an unexplored area in Sanskrit literature due to various reasons like lack of encoded texts, complexity of the Sanskrit language compared to English, lack of Sanskrit knowledge among computer scientists, etc. As far as we know, no one has seriously attempted citation matching on Sanskrit literature.

This paper presents two methods: *exact matching* and *approximate matching*, to perform citation matching in Sanskrit texts. Exact matching is based on the

G.N. Jha (Ed.): Sanskrit Computational Linguistics, LNCS 6465, pp. 124–136, 2010.

idea of finding the precise character-by-character match between pattern and text. This method finds citations that are exactly the same as those in the corpus. The search space for a given text with citations is directly proportional to the size of the corpus from where citations are to be found. Sorting the corpus can help reduce the number of comparisons needed by obviating the need to make unnecessary comparisons. This method cannot find citations if there are occurrences of *pāṭhāntara* (variant readings) or *sandhi-viccheda* (splits of compounds). Such occurrences provide the motivation for the approximate matching method.

The *pāṭhāntara* and *sandhi-viccheda* problems can be easily handled if we consider similarity or closeness. Approximate matching is based on this idea. This method is widely applied in bioinformatics, where a common technique called *local alignment* is used. The Smith-Waterman algorithm [14] is one of the predominant algorithms in local alignment. The Smith-Waterman algorithm finds a pair of segments in the nucleotide or amino acid (protein) sequences such that there is no other segment with greater similarity [14]. The Smith-Waterman-Gotoh algorithm [4] is an extension of the Smith-Waterman algorithm which uses affine gap penalty to reduce the computational overhead of the basic Smith-Waterman algorithm.

Our approximate matching method computes the Smith-Waterman-Gotoh distance [4] between citation and text. The results are filtered based on the *similarity cutoff*. As our intuition would suggest, this method is computationally intensive compared to exact matching.

An experiment is conducted using the Mahābhārata-Tātparyanirṇaya as the base text where citations may exist, and the Mahābhārata [13] as the source text where to look for citations.[1] The Mahābhārata-Tātparyanirṇaya is a commentary on the Mahābhārata in digest form.[2] Its author Madhva (1238–1317 CE) indicates that a lot of its verses are taken from the Mahābhārata directly:

भारतेऽपि यथा प्रोक्तो निर्णयोऽयं क्रमेण तु ।
तथा प्रदर्शयिष्यामस्तद्वाक्यैरेव सर्वशः ॥ 2-53 ॥

However, a lot of the citations are yet to be traced. Madhva's citation of untraceable sources has in general been the subject of some controversy since the 17th century.[3]

---

[1] Mahoney [6] sorts some Sanskrit texts but as far as we know, no one has sorted the entire Mahābhārata. We have sorted the Mahābhārata to reduce the search size for exact matching. A previously unavailable sorted verse-index of the Mahābhārata has thus been created as part of this effort. See the URL `http://www.iiitb.ac.in/faculty/srao/other-information/` for a link to the same.

[2] See [12] for more information on the Mahābhārata-Tātparyanirṇaya and its author.

[3] In this regard, see [12] for some general discussion. Mesquita claims [7] that Madhva's citations are his own inventions, a charge countered by Rao and Sharma [10]. Mesquita recently has repeated [8] the charge in the specific context of Madhva's citations from the Mahābhārata.

Thus, beyond the general and computational aspects, we believe this work also has probative value in the context of this particular textual controversy.

The rest of the paper is structured as follows. Section 2 presents the similar work in other domains like bioinformatics. Section 3 describes the problem and the challenges. Section 4 presents the method to perform sorting of a Sanskrit corpus. Section 5 presents the exact matching method. Section 6 presents the approximate matching method. Section 7 presents some conclusions and suggestions for further work.

2 Related Work

Citation matching is one of the active areas of research. Citation matching is used for identifying authors of scientific papers [9] [5], knowledge discovery, etc. Citation matching is currently limited to scientific literature like research papers.

Sorting is performed to reduce the search space when we use the exact matching method. There are classical algorithms for performing sorting like heap sort, merge sort, etc.[4] There is an x86 assembly language program [16] which sorts Sanskrit texts. Mahoney [6] implements sorting using Perl. Neither [16] nor [6] specifies the maximum text size that these programs can handle. Also as far as our knowledge goes, nobody has sorted the entire Mahābhārata text. Gale and Church [3] describe a method for aligning sentences based on a sample statistical model of character length. This method is used by Csernel and Patte [2] for performing comparisons between Sanskrit manuscripts.

Local alignment is one of the sequence alignment techniques used in bioinformatics to find the similarity regions between two DNA, RNA or protein sequences of unequal length. The Smith-Waterman algorithm [14] is used widely for performing local alignment. The Smith-Waterman algorithm is more accurate but computationally intensive as compared to other methods like BLAST. The Smith-Waterman-Gotoh algorithm [4] is an extension of the Smith-Waterman algorithm.

Symmetric [17] is an open source similarity measurement library implemented in Java. Basic algorithms like edit distance and Smith-Waterman-Gotoh [4] are implemented in this library, which we have used.

3 Problem Definition

Citation matching is an unexplored area in natural language texts, especially in Sanskrit literature. Table 1 shows examples of sources and citations.

Citation matching in Sanskrit literature is complex compared to the same problem with scientific literature because of the following reasons.

[4] These classical algorithms can be used to sort the Sanskrit corpus by modifying the comparison function. There is no need for modification in English corpora but we cannot apply this method to Sanskrit corpora because the order of the syllables in the Sanskrit वर्णमाला is different from the Roman alphabet.

Table 1. Source and Citation Examples

Citation	Source
मानुषीं तनुमाश्रितं	मानुषीं तनुं आश्रितं
कालो वा कारणं राज्ञो	कालो वा कारणं राज्ञो
युद्धं तुमुलं लोमहर्षणं	युद्धं तुमुलं रोमहर्षणं
ब्रह्मचारी कौमारादपि पाण्डवः	ब्रह्मचारी कौमाराद् अपि पाण्डवः

(i) The Sanskrit alphabetical system is entirely different from that of the English language. In Sanskrit each letter is a combination of vowel and consonant like क् +अ = क, but this kind of combination is not present in English. Modern English does not support ligatures whereas Sanskrit supports ligatures like क् + क = क्क.
(ii) In many cases, the corpus has *pāṭhāntaras* with respect to the text, and there also are cases of differences due to *sandhi-viccheda* (e.g., तनुमाश्रितम् and तनुं आश्रितम्).
(iii) The size of the corpus may be very large (the ASCII text [13] of the Mahābhārata being about 8MB in size). It is challenging to handle these kinds of large data sets.
(iv) The conversion of Sanskrit texts into machine readable formats is prone to error as it involves human interaction. There is a high probability of error in the machine readable format which in turn affects the result significantly.

For both methods—exact matching and approximate matching—we use the Mahābhārata-Tātparyanirṇaya as the base text where citations may exist, and the Mahābhārata [13] as the source text or corpus where to look for citations. The Mahābhārata and Mahābhārata-Tātparyanirṇaya are both encoded in the ASCII-based Harvard-Kyoto format to evaluate these methods. The Harvard-Kyoto format is one of the common encoding schemes used to make Sanskrit texts machine readable.

4 Sorting

Exact matching compares each line of the source corpus with each citation. The number of comparisons is directly proportional to both corpus size and citations. If there are m lines of source corpus and n citations, then the total number of comparisons needed in exact matching is $m \times n$. Consider a corpus like the BORI Mahābhārata which has around 80000 verses or 160000 lines, and a text set with 100 citations; then the total number of comparisons are 16×10^6.

One approach to reduce the number of comparisons is to sort the source corpus. Sorting reduces the search space drastically. Consider a citation ककुदं तस्य चाभाति स्कन्धम् आपूर्य विष्ठितम्. It is enough to compare with the Mahābhārata hemistichs starting with क, and we can ignore the rest of the verses. In this way, sorting helps reduce the search space.

Table 2. Input and Output: Unix Sorting in Roman Order

Input	Output
कथं विराटनगरे मम पूर्वपितामहाः	अज्ञातवासं उषिता दुर्योधनभयार्दिताः
अज्ञातवासं उषिता दुर्योधनभयार्दिताः	गत्वाश्रमं ब्राह्मणेभ्य आचख्यौ
जनमेजय उवाच	जनमेजय उवाच
गत्वाश्रमं ब्राह्मणेभ्य आचख्यौ	कथं विराटनगरे मम पूर्वपितामहाः

Sorting is not straight-forward in Sanskrit texts compared to English. Even the classical sorting tools like the Unix `sort` fail to sort Sanskrit text. Table 2 shows the input and output of the `sort` command in Unix, which sorts in Roman, rather than Sanskrit, alphabetical order.

In Sanskrit each letter is a combination of vowel and consonant. So the comparison function should look at both vowel and consonant during comparison of words.

Consider an example of की and कि. कि should come before की because कि = क् + इ and की = क् + ई. Even though both have क् in common, they differ in their vowels, and इ comes before ई in the Sanskrit alphabetical system.

There are a large number of classical algorithms like quicksort, merge sort, heapsort, etc., to perform sorting [1]. Given that in our case the size of data involved is huge, quicksort, heapsort and merge sort are the candidate algorithms for performing sorting, because all these have log-linear time complexity. We prefer heapsort over the other two algorithms for the following reasons.

(i) The time complexity of heapsort in the worst case is $O(n \log n)$ whereas the time complexity of quicksort is $O(n^2)$.
(ii) The space complexity of heapsort is $\Theta(1)$ but the space complexity of merge sort is $\Theta(n)$.

Each character in the encoding scheme is assigned a positive integer corresponding to the position of the Sanskrit character it represents in the Sanskrit alphabetical system. In English, the ASCII values of the characters are implicitly used in the comparison, but this is not appropriate in Sanskrit. There should be a way of identifying which character appears before which other character. Hence we assign positive integers to the characters.

Let us consider an example. In the Harvard-Kyoto scheme, `A` represents आ and `u` represents उ. According to the Sanskrit alphabetical system, the positions of आ and उ in the वर्णमाला are 2 and 5 respectively.

The pseudocode for the modified comparison function of heap sort is as given in Algorithm 1. $CharVal(k)$ is the function which returns the positive number assigned to an encoded character k.

Let S_i, S_j be two strings with length i and j respectively.

In Algorithm 1, line 1 and line 2 check whether one of two strings is a substring of the other. Substrings appear before the base string in a lexicographic ordering. If two strings are not substrings, then the algorithm compares each character

input : Two strings S_i, S_j of size i and j respectively
output: An Integer

if `Substring(`S_i, S_j`)` **then** return 1;
if `Substring(`S_i, S_j`)` **then** return -1;
for $k \leftarrow 0$ **to** i **do**
  if $(S_i[k] = S_j[k])$ **then** continue;
  if (`charVal(`$S_j[k]$`)` $>$ `charVal(`$S_i[k]$`)` **then** return (`charVal(`$S_j[k]$`)` - `charVal(`$S_i[k]$`))`;
  else return(`charVal` $(S_i[k])$ - `charVal` $(S_j[k])$);
end
return 0 ;

Algorithm 1. Comparison Subroutine

of the strings at position k. The *for* loop in line 3 compares the character at position k in both the strings. If the characters are not equal, then the comparison function returns the difference of the values assigned to two characters.

Table 3 shows a partial list of the sorted Mahābhārata.

Table 3. Partial Listing of the Sorted Mahābhārata

Line	Reference
अकर्मोणो हि जीवन्ति स्थावरा नेतरे जनाः	(03.033.003)
अकल्ककस्य विप्रस्य भैक्षोत्करकृतात्मनः	(13.024.053)
अकल्कको निरारम्भो लघ्व् आहारो जितेन्द्रियः	(03.080.032)
अकल्कको ह्य् अतर्कश्च ब्राह्मणो भरतर्षभ	(13.024.030)
अकल्प्यन्त च मातणृगाः समनह्यन्त पत्तयः	(09.007.003)
अकल्मषः कल्मषाणां कर्ता क्रोधाश्रितस् तु सः	(03.211.020)

5 Exact Matching

Exact matching is based on the idea of finding a precise character-by-character match between pattern and text. This method finds the citations which are exactly the same as those in the corpus.

The notations used in the pseudocode are:

(i) T: Input Sanskrit corpus.
(ii) m: Size of the corpus T.
(iii) C: Set of input citations.
(iv) n: Cardinality of the citation set C.
(v) R: The set of citations present in T—clearly, $R \subseteq C$.

The pseudocode for this method is presented in Algorithm 2.

The complexity of this algorithm is $O(mn)$.

In Algorithm 2, each line of the source text is compared with each citation in the citation set. If both match exactly (all characters are identical), then the

```
input  : Sorted corpora T and set of citation C of size m and n
         respectively
Result: Set of citations R found in the corpus
R ← ∅ ;
for every line of corpus text T_i in T do
|   for every citation C_j in C do
|   |   if (T_i = C_j) then  R ← R ∪ C_j
|   end
end
```

Algorithm 2. Exact Matching

citation is included in the result set R; otherwise it is discarded. The source corpus is sorted using heap sort to reduce the search space.

The following table shows the partial result of the exact matching method.

Table 4. Partial Result Set for Exact Matching

Citation	Source
अस्मिन् युद्धे भीमसेन त्वयि (2.160)	अस्मिन् युद्धे भीमसेन त्वयि (05.075.018)
कालो वा कारणं राज्ञो (30.36)	कालो वा कारणं राज्ञो (05.130.015) (12.070.006)
ज्ञातुं द्रष्टुं च तत्त्वेन (1.108)	ज्ञातुं द्रष्टुं च तत्त्वेन (06.033.054)

This method finds those citations which are exactly the same in the text and in the source. Consider now the citation:

अवजानन्ति मां मूढा मानुषीं तनुमाश्रितं and the source text **अवजानन्ति मां मूढा मानुषीं तनुं आश्रितं**. Only the words **तनुमाश्रितं** and **तनुं आश्रितं** differ in this case. This is an example of *sandhi-viccheda*, occurring perhaps due to scribal or editorial choice. Exact matching does not capture these kinds of citations.

Exact matching is the first method that comes to our mind when we think of citation matching. Exact matching fails to consider similarity when citations are searched for. In Sanskrit literature, we have *pāṭhāntaras* of the same corpus occurring in different sources. Most of the times these corpora differ due to *sandhi-viccheda* also. Therefore, exact matching is not an ideal method when there are *pāṭhāntara* and *sandhi-viccheda*.

6 Approximate Matching

Sequence alignment is a technique from bioinformatics of arranging nucleotide or protein sequences to identify the similarity regions. This identification of these similar regions in an important problem in molecular sequence analysis [14].

Global alignment and *local alignment* are the two types of sequence alignment. Global alignment is applied on sequences of almost equal length. Local alignment is applied on sequences with unequal length.

The Smith-Waterman algorithm, based on dynamic programming, is used in local alignment.

The lengths of citation and source text are unequal in the present domain of citation matching in Sanskrit texts. Hence local alignment is more appropriate than global alignment. The approximate method used by us is based on the Smith-Waterman-Gotoh algorithm, with each line of source and citation is considered as a sequence. The comparison between text and citation can be reduced to finding the maximum similarity score between two sequences. The Smith-Waterman-Gotoh algorithm computes the similarity score of text and citation. Only the citations with scores which are equal or greater than some approximation cutoff are included in the result set.

6.1 Smith-Waterman-Gotoh Algorithm

Consider two sequences $A = a_1a_2 \ldots a_m$ and $B = b_1b_2 \ldots b_n$. The Smith-Waterman algorithm constructs a matrix H, where each cell h_{ij} in H is computed as follows [14]. It builds the matrix H based on the set of equations given below.

$$\forall \quad 1 \le i \le n, 1 \le j \le m$$

$$h_{ij} = \max \begin{cases} h_{i-1,j-1} + s(a_i, b_j) & \text{if } a_i \text{ and } b_j \text{ are associated} \\ \max_{k \ge 1}\{h_{i-k,j} - W_k\} & \text{if } a \text{ is at the end of deletion of length } k \\ \max_{l \ge 1}\{h_{i,j-l} - W_l\} & \text{if } b \text{ is at the end of deletion of length } l \end{cases}$$

$h_{k0} = h_{0l} = 0 \quad \forall 0 \le k \le n \quad and \quad 0 \le l \le m$

$s(a,b)$ = Similarity between a and b. W_k = Weight assigned to deletion of length k. $s(a,b)$ *and* W_k are chosen on an *a priori* statistical basis.

(i) $h_{i-1,j-1} + s(a_i, b_j)$ is the similarity, if a_i and b_j are associated.
(ii) $h_{i-k,j} - W_k$ is the similarity, if a_i is at the end of a deletion of length k.
(iii) $h_{i,j-l} - W_l$ is the similarity, if b_j is at the end of a deletion of length k.
(iv) Zero is included to prevent negative similarity. Occurrence of zero indicates absence of similarity between a_i and b_j.

The procedure to find the the optimal local alignment is:

(i) Locate the highest similarity score in the matrix H.
(ii) Go backwards from this cell until a cell with zero value is encountered.
(iii) The path from the cell with maximum score to one with zero value gives the maximum sequence.

Consider the biological sequences TACTAG and ACCTAG with $s(\text{match}) = 2$, $s(\text{mismatch}) = W_k = W_l = -1$.

The corresponding similarity matrix is

$$\begin{array}{c} \\ - \\ T \\ A \\ C \\ T \\ A \\ G \end{array}\begin{array}{c} \begin{array}{ccccccc} - & A & C & C & T & A & G \end{array} \\ \begin{pmatrix} 0 & 0 & 0 & 0 & 0 & 0 & 0 \\ 0 & 1 & 1 & 0 & 2 & 1 & 0 \\ 0 & 2 & 1 & 0 & 1 & 4 & 3 \\ 0 & 1 & 4 & 3 & 2 & 3 & 2 \\ 0 & 0 & 3 & 2 & 5 & 4 & 3 \\ 0 & 2 & 2 & 2 & 4 & 7 & 6 \\ 0 & 1 & 1 & 1 & 3 & 6 & 9 \end{pmatrix} \end{array}$$

The optimal local alignment in the above example is -CTAG.

However, in this paper we are not concerned about the optimal local alignment. Our interest is only in the maximum similarity score.

The Smith-Waterman algorithm is computationally intensive compared to other methods like BLAST. The Smith-Waterman-Gotoh algorithm is an extension of the Smith-Waterman algorithm which uses affine gap penalty [4] to reduce the computational overhead.

We have redefined the parameters *costfunction*, *windowsize* and the limits of *gapfunction* of the Smith-Waterman-Gotoh algorithm for performing citation matching. The new *windowsize* and *gapfunction* are as follows. We use $windowsize \leftarrow 100$ and $0 \leq gapfunction \leq 5$. Table 5 shows the scores associated with different matches.

There are seven groups of characters. Let $char1$ and $char2$ be two characters in $String_1$ and $String_2$ at position k respectively. $char1$ and $char2$ are considered as being approximate match if $char1$ and $char2$ are any two characters in the same group. Table 6 shows these seven groups.

Table 5. Scores Associated With Different Types of Matches

Matching Type	Score
Exact	+5
Approximate	+3
Mismatch	-3

Table 6. Groups and Characters Present in the Groups

Group	Characters
I	d,t
II	g,j
III	l,r
IV	m,n
V	b,p,v
VI	a,e,i,o,u
VII	, and .

The $Smith - Waterman - Gotoh - Distance(string_1, string_2)$ function takes two strings $string_1$ and $string_2$ as arguments. It builds the similarity matrix between $string_1$ and $string_2$ similar to the above example with redefined parameters, and finally returns the maximum similarity score in the matrix.

6.2 Algorithm

Algorithm 3 uses the notation *cutoff* in addition to the notations used by Algorithm 2. *cutoff* is defined as the similarity score cutoff.

The pseudocode for the approximate matching is presented below.

```
input : Corpus text T and set of citation C of size m and n respectively.
        cutoff score cutoff
Result: Set of citations R which are greater than cutoff found in the
        corpus

R ← ∅;
for every line of corpus T_i in T do
    for every citation C_j in C do
        if Smith-Waterman-Gotoh-Distance(T_i, C_j) > cutoff then
        R ← R ∪ C_j
    end
end
```

Algorithm 3. Approximate Matching

In Algorithm 3, Line 4 computes the Smith-Waterman-Gotoh distance between a citation and a line of text in the source corpus. The citation is included in the result set only if the Smith-Waterman-Gotoh distance is greater than the *cutoff*. This step is repeated for every line of text in the corpos.

Table 7 shows the partial result set of this method.

Table 7. Partial Result Set for Approximate Matching

Citation	Source
समस्ता हतप्रवीराः सहसा (22.285)	समन्ताद् धतप्रवीराः सहसा (03.152.019)
भीमो दुर्योधनं हन्ता (21.350)	अहं दुर्योधनं हन्ता (02.068.026)
युद्धं तुमुलं लोमहर्षणं (27.66)	युद्धं तुमुलं रोमहर्षणं (06.043.011)
ब्रह्मचारी कौमारादपि पाण्डवः (28.162)	ब्रह्मचारी कौमाराद् अपि पाण्डवः (05.057.001)
बलवान् स जातः शूरस्तपस्वी (22.283)	बलवान् स जातः शूरस् तरस्वी (03.152.017)

The exact matching method does not include **शूरस् तरस्वी** and **कौमाराद् अपि** in the result set because of *pāṭhāntara* and *sandhi-viccheda* respectively. This shows that approximate method can handle *pāṭhāntara* and *sandhi-viccheda*.

If *cutoff* ← 1 then approximate matching behaves exactly as the exact matching method. This of course is tantamount to saying that if the similarity score cutoff is 1, then only exact matches are considered similar, so the approximate matching results are identical to the exact matching results.

This method takes closeness or similarity between text and citation into account when searching the citations. This is the strength of this method. However this method is computationally intensive as compared to exact matching and usually also requires manual review of results.

We performed the experiment with different values of *cutoff*. Initially we started with *cutoff* = *0.9*. In this case the number of valid results is 87 and the rest of the results are false positives or very small matches like उमा उवाच. We repeated the same experiment with *cutoff* = *0.8*. In this case the number of valid results are 103 and there was also an increase in the count of false positives.

The similarity score is a measure of how one string can be transformed into another given string. If one of the strings is a substring of the other, then naturally the similarity score is high. This is the main reason for the rise in false positives when *cutoff* is lowered. The number of false positives is inversely proportional to *cutoff*.

7 Conclusions

Citation matching is one the active areas in research but unfortunately it is limited to scientific literature. We have presented approximate matching based on local alignment in this paper. The known limitations of exact matching provide the inspiration for approximate matching.

We can easily see that the results obtained with approximate matching are far superior to those with exact matching. The exact matching method is unable to identify even some well-known verses like अवजानन्ति मां मूढा मानुषीं तनुं आश्रितम् because of *sandhi-viccheda*, but the approximate matching method can.

Heap sort is the best sorting method, preferable to other methods like quick sort, merge sort, etc., for sorting Sanskrit corpora, due to its worst-case log-linear time complexity and constant space complexity.

Approximate matching can be extended to other Indic languages like Kannada. We can use this method as long as the encoding schemes used by both source and citation is consistent.

Citation matching is used in content analysis. We can extend this to Sanskrit literature using the approximate matching method. It is noteworthy that many classical authors and commentators in Sanskrit quote earlier texts implicitly, without indicating their sources or indicating that the words/sentences/verses used are not their own. To discover these hidden citations, given the prevalence of *pāṭhāntaras* and *sandhi-vicchedas*, our approach can be used.

The approximate matching method can be easily extended to detect plagiarism in literature because the Smith-Waterman algorithm is used elsewhere [15] to detect plagiarism. However, in our opinion, plagiarism is a modern concept and not likely to apply significantly in the case of classical Sanskrit texts. However, critical editions of many texts like the Mahābhārata need to be prepared and revised, and in such cases variant readings recorded by long-ago authors and commentators may be utilized in coming up with a text that, while not certain to be the same as a putative *ur*-text, is nonetheless "the text that best explains all the extant documents" [11]. Approximate matching has a role in this, and local alignment is thus an important tool.

Acknowledgements

The work of S. Rao was supported in part by the Centre for Artificial Intelligence and Robotics, DRDO, under contract CAIR/CARS-06/CAIR-23/0910187/09-10/170. His work was also partially supported by a UGSI Travel and Research Grant from Unisys.

References

1. Cormen, T.H., Leiserson, C.E., Rivest, R.L., Stein, C.: Introduction to Algorithms, 2nd edn. MIT Press/McGraw-Hill (2001)
2. Csernel, M., Patte, F.: Critical edition of Sanskrit texts. In: Huet, G., Kulkarni, A., Scharf, P. (eds.) Sanskrit Computational Linguistics 2007/2008. LNCS (LNAI), vol. 5402, pp. 358–379. Springer, Heidelberg (2009)
3. Gale, W.A., Church, K.W.: A program for aligning sentences in bilingual corpora. In: Meeting of the Association for Computational Linguistics, pp. 177–184 (1991)
4. Gotoh, O.: An improved algorithm for matching biological sequences. Journal of Molecular Biology 162(3), 705–708 (1982)
5. Lawrence, S., Bollacker, K., Giles, L.C.: Autonomous citation matching. In: Etzioni, O. (ed.) Proceedings of the Third International Conference on Autonomous Agents. ACM Press, New York (1999)
6. Mahoney, R.: Arbitrary lexicographic sorting: Sort UTF-8 encoded Romanised Sanskrit, `http://www.indica-et-buddhica.org/sections/repositorium-preview/materials/software/sort-utf8-sanskrit`
7. Mesquita, R.: Madhva's Unknown Literary Sources: Some Observations. Aditya Prakashan, New Delhi (2000)
8. Mesquita, R.: Madhva's Quotes from the Puranas and the Mahabharata: An Analytical Compilation of Untraceable Source-Quotations in Madhva's Works along with Footnotes. Aditya Prakashan, New Delhi (January 2008)
9. Pasula, H., Marthi, B., Milch, B., Russell, S., Shpitser, I.: Identity uncertainty and citation matching (2002)
10. Rao, S., Sharma, B.N.K.: Madhva's unknown sources: A review. Asiatische StudienÉtudes Asiatiques LVII 1, 181–194 (2003)
11. Robinson, P.: The one text and the many texts. Literary and Linguistic Computing 15(1), 5–14 (2000)
12. Sharma, B.N.K.: History of the Dvaita School of Vedanta and its Literature, 3rd edn. Motilal Banarsidass, Delhi (2000)
13. Smith, J.: The Mahabharata (2009), `http://bombay.indology.info/mahabharata/statement.html`
14. Smith, T.F., Waterman, M.S.: Identification of common molecular subsequences. Journal of Molecular Biology 147(1), 195–197 (1981)

15. Su, Z., Ahn, B.R., Eom, K.Y., Kang, M.K., Kim, J.P., Kim, M.K.: Plagiarism detection using the Levenshtein Distance and Smith-Waterman algorithm. In: International Conference on Innovative Computing, Information and Control. IEEE Computer Society, Los Alamitos (2008)
16. Dudenskt, http://www.sanskritweb.net/koko/dudenskt.pdf
17. Chapman, S.: SimMetrics - open source Similarity Measure Library, http://www.dcs.shef.ac.uk/~sam/simmetrics.html

RDBMS Based Lexical Resource for Indian Heritage: The Case of Mahābhārata

Diwakar Mani

Centre of Development of Advance Computing, Pune, India
diwakarmani@gmail.com
http://sanskrit.jnu.ac.in/mb/index.jsp

Abstract. The paper describes a lexical resource in the form of a relational database based indexing system for Sanskrit documents Mahābhārata (MBh) as an example. The system is available online on http://sanskrit.jnu.ac.in/mb with input and output in Devanāgarī Unicode, using technologies such as RDBMS and Java Servlet. The system works as an interactive and multi-dimensional indexing system with search facility for MBh and has potentials for use as a generic system for all Sanskrit texts of similar structure. Currently, the system allows three types of searching facilities- 'Direct Search', 'Alphabetical Search' and 'Search by Classes'. The input triggers an indexing process by which a temporary index is created for the search string, and then clicking on any indexed word displays the details for that word and also a facility to search that word in some other online lexical resources.

Keywords: Mahābhārata, Indexing, Mahābhārata Search Engine, Online Indexing, Mahābhārata Indexing System, Mahābhārata Indexer.

1 Introduction

MBh, the great epic of India ascribed to Veda Vyāsa, can be un-hesitatingly given the honour of being the cultural encyclopaedia of India. It is the story of a great war that ended one age and began another. The story has been passed down to us in a classical canon of Sanskrit verses - some 90,192 stanzas (including additional *Harivaṁśa*) long, or some 1.8 million words in total (among the longest epic poems worldwide) divided in 18 parvas and 97 upaparvas which are again divided into 1995 chapters.[1] The MBh is about 12 times the length of the *Bible* and 8 times longer than the *Iliad* and the *Odyssey*. The work therefore holds a significant place in the cultural history of India. An overview of structure of the MBh is given below (Table 1).

2 Why Is Indexing System Necessary for MBh?

The MBh extols its greatness itself in the following words: *yadihāsti tadanyatra yannehāsti na tat kvacit.* The saying *vyāsocchiṣṭaṃ jagatsarvam* also stresses

[1] The structure of Mahābhārata is based on BORI's Critical Edition of Mahābhārata.

G.N. Jha (Ed.): Sanskrit Computational Linguistics, LNCS 6465, pp. 137–149, 2010.

Table 1. Structure of Mahābhārata (According to the BORI's critical edition)

Parvan	Sub-parvan	Ākhyāna	Adhyāya	Ślokas
Ādiparva	19	09	225	07197
Sabhāparva	09	00	072	02390
Āraṇyakaparva	16	00	299	10338
Virāṭaparva	04	00	067	01824
Udyogaparva	12	05	197	06063
Bhīṣmaparva	04	00	117	05406
Droṇaparva	08	00	173	08192
Karṇaparva	01	00	069	03871
Śalyaparva	04	00	064	03315
Sauptikaparva	02	00	018	00772
Strīparva	03	00	027	00730
Śāntiparva	03	00	353	12902
Anuśāsanaparva	02	22	154	06439
Aśvamedhikaparva	02	00	096	02743
Āśramavāsikaparva	03	00	047	01062
Mausalaparva	01	00	009	00273
Mahāprasthānikaparva	01	00	003	00106
Svargārohaṇaparva	01	00	005	00194
Total (in 18 parvans)	**95**	**36**	**1995**	**73,817**
Harivaṃśa (Khilaparva)	02	—	—	16,375
Total (including khila)	**97**	**36**	**1995**	**90,192**

this point. MBh is neither a history in the modern sense of the term, nor a chronicle. But it stands in incomparable isolation, defying all definitions. It is a veritable encyclopaedia comprising heterogeneous material from all branches of knowledge. Taking the core-story of the feud between two branches of a royal family and the circumstances leading to a catastrophic war, several branches of knowledge including philosophy, law, ethics, statecraft, warfare, history and ethnology are embodied in its structure. The MBh is a comment on the human condition with all its richness, complexity and subtlety. The MBh is the source of many compositions like- *abhijñāanaśākuntalam* of Kālidāsa, *naiṣadhīya-caritam* of śrīharṣa etc. It is the text that is most sought for in order to enrich cultural, social and any other type of knowledge about Indian civilization.

While on the one hand, the text is very important, on the other, it is so huge that it becomes virtually very difficult and time consuming for someone to search a specific keyword in it manually. The searching process combines the benefit of search and extraction. The indices thus prepared will constitute a separate text in itself due to the size of the MBh and will be of tremendous use to the researchers and users.

2.1 Uses of the Indexing System

The indexing system of Sanskrit documents can be used in various NLP applications like building Sanskrit WordNet, dictionaries, Sanskrit- Indian Language

Machine Translation System (MTS) etc. Unique words are a basic need of a WordNet and a dictionary. Automatic indexing and sorting is a good tool to extract unique words from a given text. Though, there is not a direct use of indexing system in machine translation but can be helpful in generating a context lexicon (a lexicon with the entries including the uncommon words frequently occurring in the context of use of that particular word, which, later assigning the sense may be used as a lexicon for word sense disambiguation). This work, besides being an essential resource in NL system of Sanskrit, may also be useful for authentic and referential knowledge about Indian heritage. The system can also be very useful for the researches of historical, socio-political and geographical researches by providing the facts from the huge text which cannot be easily read.

3 Previous Work

The history of textual indexing is very rich in India. Śaunaka, a great scholar of Vedas, made a Vedic index named *sarvānukramaṇī*. "A Word Concordance of Mahābhārata" is a research tool supplying well alphabetized and grammatically analyzed record of each and every word-unit occurring in the text of different Parvans of MBh. It is a gigantic collaboration project at Sanskrit evam Prācya Vidyāsaṃsthāna, Kurukshetra University. Six parts of the same have already been published. Seventh part is in press and press copies of next two parts are ready for publication and the tenth part is near completion. This work is based on BORI's critical edition of MBh.

A western scholar S. Sörensen created "An Index to the Names in the Mahābhārata" with short explanations and a concordance to the Bombay and Calcutta editions and P. C. Roy's translation which was published from Williams and Norgate, London (1904-1925) and reprinted from Motilal Banarasidass, Delhi (1963). Here, all proper names are given with a complete listing of the places of their occurrence. It also contains, under the names of the 100 sub-parvans, very brief chapter by chapter summaries of the contents of the sub-parvans.

For the emerging R'&'D area of Sanskrit informatics, it is necessary to make indices available online. Unfortunately, the task of electronic indices for Indian heritage has not attracted required attention of computational linguists. Some efforts made in this area, directly and indirectly, are listed here:

1. The important work in the area of *Online Indexing of Indian heritage* has been done in University of Göttingen, Germany. It includes a *word indexing* of complete MBh in Roman transliteration. It is accessible at `http://www.sub.uni-goettingen.de/ebene_1/fiindolo/gretil/1_sanskr/2_epic/mbh/sas/mahabharata.html`
2. The Indology Department, University of Wuerzburg, Germany has created "Multimedia Database to Sanskrit drama" which is mainly focused on *word indexing* of Bhāsa's drama but also includes *mudrārākṣasa* of Viṣākhadatta. It is available online at `http://www.indologie.uni-wuerzburg.de/bhasa/index.html`

3. Linguistics Research Centre of The University of Texas has an online *ṛgveda* in Romanized transliteration format. This online text is based on the Barend A. Van Nooten and Gary B. Holland's electronic version. This can be seen at- `http://www.utexas.edu/cola/centers/lrc/RV/`
4. A project on the complete text of the Critical Edition of Mahābhārata begun by J. A. B. Van Buitenen and now continued under the chief editorship of James Fitzgerald of Brown University. Van Buitenen (translated the first three volumes, comprising the first five major parvas of the epic) and Fitzgerald (translated volume seven, comprising *strīparva* and the first half of *śāntiparva*) have translated the BORI version of MBh text into modern English prose. At present, ten volumes are projected, and four volumes have been published by the University of Chicago Press.
5. Barend A. Van Nooten, an emeritus Professor of Sanskrit at the University of California at Berkeley, has written "*The Mahābhārata Attributed to Kṛṣṇa Dvaipāyana Vyāsa*" which was publish in 1971 by Twayne Publishers, New York. This work provides a well informed but non-technical overview of the MBh. It includes a detailed, book-by-book summary of the story, discussions of the religious, philosophical, and ethical components of the text, and an outline of the MBh's influence since the eighteenth century.
6. "*The Sanskrit Epics*" - a comprehensive guide to the Mahābhārata and the Rāmāyaṇa is written by Professor John Brockington (Leiden: Brill, 1998), former Head of the Department of Sanskrit at the University of Edinburgh and General Secretary of the World Sanskrit Association, gives an up to date, general survey of the history of the two Sanskrit epics and the scholarship upon them.
7. Paradigmatic index on *rāmopākhyāna* of MBh is the work of Peter Scharf and Malcolm Hyman, which gives the Devanāgarī text, Roman transliteration, analysis of sandhi, inflection, glossary, prose paraphrases, syntactic and cultural notes, and English translation. It was described by Scharf (2001) at the international conference on Mahābhārata conference in Montreal (18-20 May 2001) and published in "*The Mahābhārata: What is not here is nowhere else*" by Munshiram Manoharlal, Delhi, 2005 (T.S. Rukmani (ed.)). It is available at `http://sanskritlibrary.org`
8. Electronic text of the Critical Edition of the MBh is available in downloadable text format of several commonly-used encodings, such as- Unicode Devanagri, Unicode Roman, ISCII, ASCII, Norman etc. at the home page of Professor John Smith, Cambridge University. Smith's revision is based on Prof. Muneo Tokunaga's first digital and searchable version of the MBh text. It is available at
`http://bombay.indology.info/mahabharata/statement.html`
9. Maharshi University of Management has created PDF files of all 18 Parvas of Mahābhārata. The site is
`http://is1.mum.edu/vedicreserve//itihas.htm`

10. Kishori Mohan Ganguli has translated the MBh in English which is available at- http://www.sacred-texts.com/hin/maha/
11. Aryabharati International Society for Hindu Ved Vignan & Atomic Research has created pages for complete MBh (in Unicode Devanagari & Roman transliteration format) on http://www.aryabharati.org/mahabharat/index.asp. This text has been cross-referenced with Kishori Mohan Ganguli's English translation on a book-by-book basis.
12. http://sanskritdocuments.org/ has created online versions of Sanskrit documents including Vedas, MBh and many more in ITX, HTML, PS, XD-VNG, GIF and PDF format.
13. A project to translate the full epic into English prose began to appear in 2005 from the Clay Sanskrit Library, co-published by New York University Press and JJC Foundation. The translation is based not on the Critical Edition but on the version known to the commentator Nīlakaṇṭha. Currently *sabhā, āraṇyaka, virāṭa, udyoga, bhīṣma, droṇa, karṇa, śalya, sauptika, strī*, and *śāntiparva* are available. It is accessible at- http://www.claysanskritlibrary.org/

3.1 Distinction of Present System over Previous Ones

"Mahabharata Online" at University of Goettingen provides two services, book and chapter (parva and adhyāya)-wise browsing, and alphabetically sorted index, each entry linking to the part of the text where it occurs. Our system is a dynamic search indexer which provides three kinds of search facility - string input search, search by listing of words by first letter, and browsing the word by parva>upaparva>adhyāya and ākhyāna. The search result displays information of the entry with detailed extracted information of the verse of its occurrence, including number and name of parva, upaparva, adhyāya, śloka number and ākhyāna (if available). One more difference between the two is in their output format. The previous displays the text in Roman IAST scheme and later uses Devanagari Unicode for input and output.

4 Methodology for the MBh Indexer

The MBh, an encyclopedia of Indian civilization, has always remained attractive not only to Indian scholars, but also to the western. The MBh being a popular epic has several versions. The edition selected for this work is the critical edition of Mahābhārata, critically edited by V.S. Sukthankar and others and published by Bhandarkar Oriental Research Institute (from 1919 to 1966). It has been digitized by Prof. John Smith based on the work of Muneo Tokunaga in Unicode Devanāgarī Text format.

To provide more comprehensive search, the text is segmented according to Pāṇinian sandhi rules. The tokens so obtained are basically padas (word-forms). After this, the text has been adapted to the database system. The original and segmented text has been stored in database tables. The other information of the structure of the text has been stored in different tables and those are connected with each other. The connections of the table complete the reference of the searched query and connect the entire data with other relations. The database has five tables having the information of parva, upaparva, adhyāya, ākhyāna and the ślokas respectively.

The structure for database storage is as follows:

Table 2. The 'shloka' table

Shloka_Id	ParvaId	UpaParvaId	AdhyayaId	AkhyanaId	ShlokaId	ShlokaSamhita	ShlokaPada
01	01	01	01	00	00	नारायणं नमस्कृत्य नरं चैव नरोत्तमम् । देवीं सरस्वतीं चैव ततो जयमुदीरयेत् ॥	नारायणम् नमः कृत्य नरम् च एव नरोत्तमम् । देवीम् सरस्वतीम् च एव ततः जयम् उदीरयेत् ॥
02	01	01	01	00	01	लोमहर्षणपुत्र उग्रश्रवाः सूतः पौराणिको । नैमिषारण्ये शौनकस्य कुलपतेर्द्वादशवार्षिके सत्रे ॥	लोमहर्षणपुत्र उग्रश्रवाः सूतः पौराणिकः । नैमिष अरण्ये शौनकस्य कुलपतेः द्वादश वार्षिके सत्रे ॥
03	01	01	01	00	02	समासीनानभ्यगच्छद्ब्रह्मर्षीन् संशितव्रतान् । विनयावनतो भूत्वा कदाचित् सूतनन्दनः ॥	समासीनान् अभ्यगच्छद् ब्रह्मर्षीन् संशितव्रतान् । विनय अवनतः भूत्वा कदाचित् सूतनन्दनः ॥
---	---	---	---	---	---	---	---
---	---	---	---	---	---	---	---
2603	01	07	69	01	43	तामेवमुक्त्वा राजर्षिर्दुःषन्तो महिषीं प्रियाम् । वासोभिरन्नपानैश्च पूजयामास भारत ॥	ताम् एवम् उक्त्वा राजर्षिः दुःषन्तः महिषीम् प्रियाम् । वासोभिः अन्नपानैः च पूजयामास भारत ॥

Table 3. The 'adhyaya' table

Id	ParvaId	UpaParvaId	ActAdhId	Adh_Nm
01	01	01	00	प्रथमोऽध्यायः
02	01	02	00	द्वितीयोऽध्यायः
03	01	03	00	तृतीयोऽध्यायः
---	---	---	---	---
---	---	---	---	---
62	01	07	01	द्विषष्ठितमोऽध्यायः
63	01	07	01	त्रिषष्ठितमोऽध्यायः

Table 4. The 'akhyana' table

Id	ParvaId	UpaParvaId	AdhyayaId	Akhyana
01	01	07	62	शकुन्तलोपाख्यान
02	01	07	70	ययात्युपाख्यान
03	01	07	81	उत्तरयायातम्
04	01	11	160	तापत्योपाख्यान
05	01	11	164	वाशिष्ठोपाख्यान

Table 5. The 'upaparva' table

Id	Parva_Id	Upaparva
01	01	अनुक्रमणीपर्व
02	01	पर्वसंग्रहपर्व
03	01	पौष्यपर्व
04	01	पौलोमपर्व

Table 6. The 'parva' table

Id	Parva
01	आदिपर्व
02	सभापर्व
03	वनपर्व (आरण्यकपर्व)
04	विराटपर्व
05	उद्योगपर्व

5 Development of the MBh Indexer

A dynamic search engine-cum-indexer has been developed which is built in the front-end of Apache Tomcat Web server using JSP and Java servlets. It has its data in MS-SQL Server 2005 with Unicode. For connecting the front-end to the database server the MS-JDBC connectivity has been used. The following model describes the multi-tiered architecture of the MBh indexing system:

5.1 Process Flow of the System

There are three ways to give input to the system e.g. Direct Search, Alphabet search and Search by the structure of the text in Devanāgarī UTF-8 format.

Step I: Preprocessing. Preprocessing a text mainly consists of transformation of raw data required to facilitate further cartographic processing. For example - preprocessor will remove any non Devanāgarī characters, punctuations that may have been inadvertently introduced by the user like "@es" in "महा@भारesत" and other similar cases.

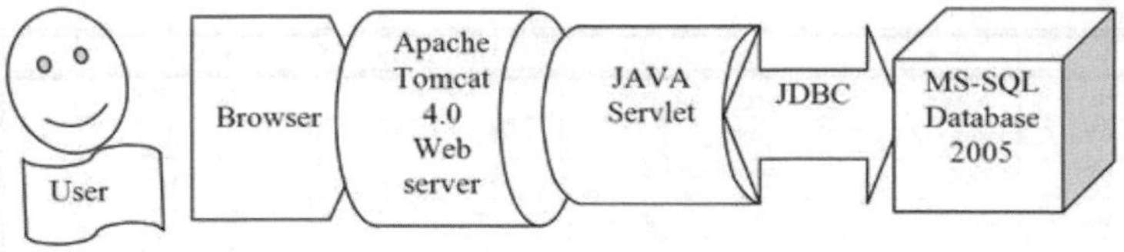

Fig. 1. Multi-tiered architecture of the MBh indexing system

Step II: MBh Indexer and Database. At this step, the indexer makes an indexed list of exact and partially matching words. Getting the query as an input, the indexer, after a light preprocessing, sends it to the database. If the word has its occurrence in the database, the system gives the output.

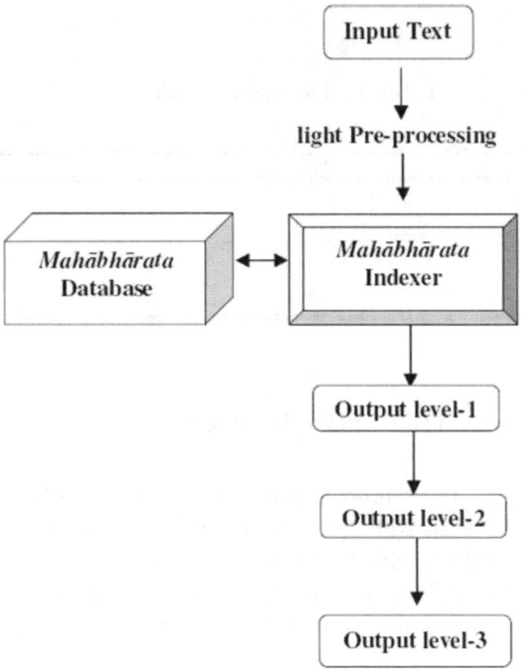

Fig. 2. Process flow of the system

Step III: Output level-1. At this level, the indexer gives all the occurrences of the searched query with its numerical reference in a hyperlinked mode.

Step IV: Output level -2. Clicking any hyperlinked word, system shows its original place in the śloka and also gives its full reference in the text. It also asks for further information from other online lexical resources.

Step V: Output - final level. Here, the indexer gives a list of online lexical resources and also gives the facility to do morphological analysis of the query with the help of POS Tagger[2] and Subanta Analyzer.[3]

5.2 Front-End of the MBh Indexer

The front-end of the system is developed in utf-8 enabled Java Server Pages (JSP) and HTML. The front-end of the software enables the user to interact with the indexing system with the help of Apache Tomcat web-server. The JSP technology helps to create web based applications by combining Java code with HTML. The web server runs the Java code and displays the results as HTML. For this system, there are two JSP pages, one is the main search page and the other is the cross-referential search page which searches the searched query in different online lexical and linguistic resources. The snapshots of the indexing system are as follow:

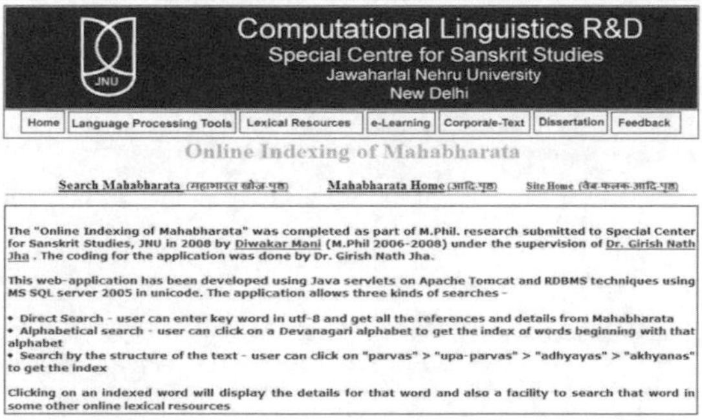

Fig. 3. Homepage of the MBh Indexing System

Fig. 4. Search page of the MBh indexing system

[2] http://sanskrit.jnu.ac.in/post/post.jsp

[3] http://sanskrit.jnu.ac.in/subanta/rsubanta.jsp

Fig. 5. Results of the searched query in hyperlinked mode with their numerical references

Fig. 6. Referential page for the searched query

Fig. 7. Cross-referential page where the user can click any one link to know further linguistic, grammatical or cultural knowledge

5.3 The Back-End of the System

The back-end of the indexing system consists of RDBMS, which contains co-relative data tables. This Tomcat server based program connects to MS-SQL Server 2005 RDBMS through JDBC connectivity. The lexical resources are stored as Devanāgarī utf-8. There are five tables namely; *'shloka', 'adhyaya', 'akhyana', 'upaparva'* and *'parva'*.

A design of the indexing system of MBh database is given below:

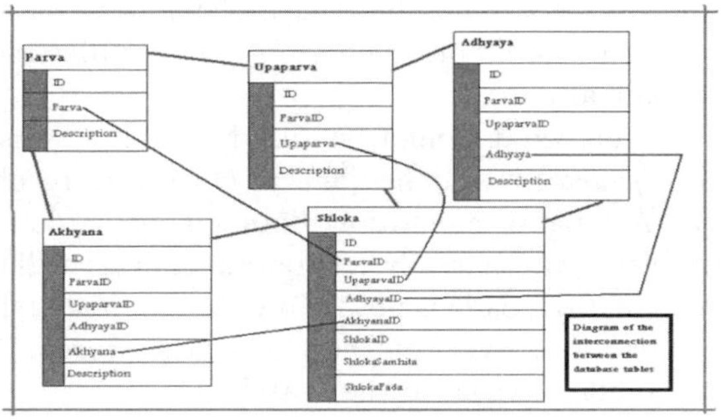

Fig. 8. Diagram of the interconnection between the database tables

5.4 Database Connectivity

The database connectivity is done through the JDBC driver software. JDBC Application Programming Interface (API) is the industry standard for database independent connectivity for Java and a wide range of database-SQL databases. JDBC technology allows to use the Java programming language to develop 'Write once, run anywhere' capabilities for applications that require access to large scale data. JDBC works as bridge between Java program and Database. SQL server 2005 and JDBC support input and output in Unicode, so this system accepts Unicode Devanāgarī text as well as prints result in the same format.[4]

6 Limitations of the System

1. The system has fixed input and output mechanism. One can search his query in Unicode Devanāgarī only and the output will be in the same format. The work on transcoding of input and output in another format is being developed.

[4] `http://java.sun.com/javase/technologies/database/`

2. At present, the search facility is accessible only for previously displayed links. For enhancement of the system, we are planning to add other online lexical resources like MW Dictionary of Cologne Digital Sanskrit Lexicon (`http://www.sanskrit-lexicon.uni-koeln.de/monier/indexcaller.php`), The Sanskrit Library project at Brown University and Hindi WordNet (`http://www.cfilt.iitb.ac.in/wordnet/webhwn/wn.php`) of IIT, Mumbai. The reason to link with Hindi WordNet is because of a preponderance of Sanskrit words (*tatsama*) approximately 60-70% words in Hindi.[5]
3. At present, the system is unable to give the translation in other language.
4. In this version, it may fail to search a word which is in sandhi form. While sandhi-split version of text has been stored for only *ādiparva*, efforts are being made to do this for all other parvans. The sandhi splitting of the text has been done manually.
5. If a base word is searched it cannot be found in all its forms. For example, if "*brahman*" (*prātipadika*) is searched, it will not return results for "*brahmā*", "*brahmaṇi*" etc. A subanta generation module is being developed separately to solve this problem. Alternatively, a subanta analyzer will enable the search to look for the stem if the pada is not found. At this point, the system directs the user to a subanta analyzer at the end of the search. The next version will make this module internal to the search.
6. It has only a string search facility so it cannot search synonymous words. For example, if *'kṛṣṇa'* is searched, it cannot return *'nārāyaṇa'*, *'vāsudeva'*, *'hari'*, *'muralīdhara'*, *'yogeśvara'* etc. A separate module of *Amarakośa* has been developed. In near future, this module will be used to handle this issue.

References

1. Bhise, U.R.: Nāradīyā Śikṣā with the Commentary of Bhaṭṭa Śobhākara. Bhandarkar Oriental Research Institute, Poona (1986)
2. von Böhtlingk, O.: Pāṇini's Grammatik (Primary source text for our database). Olms, Hildesheim (1887)
3. Bronkhorst, J.: Greater Magadha Studies in the Culture of Early India. Brill, Leiden (2007)
4. Dhūpakara, A.Ś.: Śrīpiṅgalanāga-viracitaṃ Chandaḥśāstram. Parimal Publications, Delhi (1985)
5. Deshpande, M.M.: Semantics of Kārakas in Pāṇini: An Exploration of Philosophical and Linguistical Issues. In: Matilal, B.K., Bilimoria, P. (eds.) Sanskrit and Related Studies: Contemporary Researches and Reflections, pp. 33–57. Sri Satguru Publications, Delhi (1990)
6. Gladigow, B.: Sequenzierung von Riten und die Ordnung der Rituale. In: Stausberg, M. (ed.) Zorostrian Rituals in Context, Numen Book Series, Studies in the History of Religions, vol. CII, pp. 57–76. Brill, Boston (1999)
7. Joshi, S.D., Roodbergen, J.A.F.: On siddha, asiddha and sthānivat, vol. LXVIII, pp. 541–549. Annals of the Bhandarkar Oriental Research Institute, Poona (1987)

[5] `http://www.cfilt.iitb.ac.in/gwc2010/pdfs/67_Sanskrit_Wordnet__Kulkarni.pdf`

8. Joshi, S.D., Roodbergen, J.A.F.: The Aṣṭādhyāyī of Pāṇini, with Translation and Explanatory Notes, vol. II. Sahitya Akademi, New Delhi (1993)
9. Houben, J.E.M.: 'Meaning statements' in Pāṇini's grammar: on the purpose and context of the Aṣṭādhyāyī. Studien zur Indologie und Iranistik 22, 23–54 (1999-2001)
10. Houben, J.E.M.: Memetics of Vedic Ritual, Morphology of the Agniṣṭoma. In: Griffiths, A., Houben, J.E.M. (eds.) The Vedas: texts, language & ritual, pp. 23–47. Forsten, Groningen (2004)
11. Katre, S.M.: Aṣṭādhyāyī of Pāṇini. Motilal Banarsidass, Delhi (1989)
12. Lüders, H.: Die Vyâsa-Çikshâ besonders in ihrem Verhältnis zum Taittirîya-Prâtiçâkhya, Göttingen (1894)
13. Michaels, A.: 'Le rituel pour le rituel' oder wie sinnlos sind Rituale? In: Caduff, C., Pfaff-Czarnecka, J. (eds.) Rituale heute: Theorien - Kontroversen - Entwürfe, pp. 23–47. Reimer, Berlin (1999)
14. Michaels, A.: The Grammar of Rituals. In: Michaels, A., Mishra, A. (eds.) Grammar and Morphology of Ritual, Ritual Dynamics and the Science of Ritual, vol. I, pp. 15–36. Harrassowitz Verlag, Wiesbaden (2010)
15. Mishra, A.: Simulating the Pāṇinian System of Sanskrit Grammar. In: Huet, G., Kulkarni, A., Scharf, P. (eds.) Sanskrit Computational Linguistics 2007/2008. LNCS (LNAI), vol. 5402, pp. 127–138. Springer, Heidelberg (2009)
16. Mishra, A.: Modelling the Grammatical Circle of the Pāṇinian System. In: Huet, G., Kulkarni, A., Scharf, P. (eds.) Sanskrit Computational Linguistics 2007/2008. LNCS (LNAI), vol. 5402, pp. 40–55. Springer, Heidelberg (2009)
17. Mishra, A.: On the Possibilities of a Pāṇinian Paradigm for a Rule-based Description of Rituals. In: Michaels, A., Mishra, A. (eds.) Grammar and Morphology of Ritual, Ritual Dynamics and the Science of Ritual, vol. I, pp. 15–36. Harrassowitz Verlag, Wiesbaden (2010)
18. Sarup, L.: The Nighaṇṭu and The Nirukta. Motilal Banarsidass, Delhi (1967)
19. Oppitz, M.: Montageplan von Ritualen. In: Caduff, C., Pfaff-Czarnecka, J. (eds.) Rituale heute: Theorien - Kontroversen - Entwürfe, pp. 23–47. Reimer, Berlin (1999)
20. Sarup, L.: The Nighaṇṭu and the Nirukta. Motilal Banarasidass, Delhi (1967)
21. Shastri, K.: āgeśabhaṭṭa-kṛta Vaiyākaraṇa-siddhānta-parama-laghu-mañjūṣā. Kurukshetra University Press, Kurukshetra (1975)
22. Shastri, M.D.: The Ṛgveda-Prātiśākhya with the commentary of Uvaṭa. Vaidika Svadhyaya Mandir, Varanasi (1959)
23. Scharf, P.M.: Modeling Pāṇinian Grammar. In: Huet, G., Kulkarni, A., Scharf, P. (eds.) Sanskrit Computational Linguistics 2007/2008. LNCS (LNAI), vol. 5402, pp. 95–126. Springer, Heidelberg (2009)
24. Scharfe, H.: A new perspective on Pāṇini. In: Indologica Taurinensia, vol. XII, pp. 1–273 (2009)
25. Staal, F.: Rules without meaning: ritual, mantras and the human sciences. Lang, New York (1989)
26. Thieme, P.: Meaning and form of the 'grammar' of Pāṇini. Studien zur Indologie und Iranistik 8/9, 12–22 (1982)
27. Varma, V.K. (ed.): Vājasaneyi-Prātiśākhyam. Chaukhamba Sanskrit Pratishthan, Delhi (1987)
28. Varma, V.K.: Ṛgveda-Prātiśākhyam. Banaras Hindu University Sanskrit Series, Varanasi (1972)
29. Wezler, A.: On the quadruple division of the Yogaśāstra, the caturvyūhatva of the Cikitsāśāstra and the 'four noble truths' of the Buddha. In: Indologica Taurinensia, vol. XII, pp. 289–337 (1984)

Evaluating Tagsets for Sanskrit

Madhav Gopal[1], Diwakar Mishra[2], and Devi Priyanka Singh[3]

[1] Centre for Linguistics, SLL & CS
[2] Special Center for Sanskrit Studies
[3] Centre for Indian Languages, SLL & CS
Jawaharlal Nehru University, New Delhi
{mgopalt,diwakarmishra,devisi.translation}@gmail.com

Abstract. In this paper we present an evaluation of available Part Of Speech (POS) tagsets designed for tagging Sanskrit and Indian languages which are developed in India. The tagsets evaluated are - JNU-Sanskrit tagset (JPOS), Sanskrit consortium tagset (CPOS), MSRI-Sanskrit tagset (IL-POST), IIIT Hyderabad tagset (ILMT POS) and CIIL Mysore tagset for the Linguistic Data Consortium for Indian Languages (LDCIL) project (LDCPOS). The main goal behind this enterprise is to check the suitability of existing tagsets for Sanskrit from various Natural Language Processing (NLP) points of view.

Keywords: Aṣṭādhyāyī, POS tagging, POS tagger, tagset, morphology, Pāṇini, WSD, machine learning.

1 Introduction

POS tagging assigns a suitable part of speech label for each word in a sentence of a language. In other words, POS tagging is the process of identifying lexical/grammatical category of a word according to their function in the sentence. For NLP tasks, annotated corpora of a language have a great importance. Annotated corpora serve as an important resource for such well-known NLP tasks as POS-Tagging, chunking, parsing, structural transfer, word sense disambiguation (WSD) etc. POS tagging is also referred to as morpho-syntactic tagging or annotation. A POS Tagger (POST) is a piece of software that reads text in some language and assigns parts of speech to each word (and other token), such as noun, verb, adjective, etc., although generally computational applications use more fine-grained POS tags like 'noun-masculine' etc.

The quality of the POS annotation in a corpus is crucial for the development of POST. The design of the annotated Sanskrit corpus is rapidly going on in India. Natural languages are intrinsically very complicated and Sanskrit is not an exception to this. Sanskrit is morphologically and lexically very rich language. It has a variety of words, lexemes, morphemes, and a rich productive mechanism of forming new words. Due to identical inflections for various cases, many of its words are ambiguous and their disambiguation for NLP tasks is a must. Moreover, there are many factors involved in causing ambiguity in this

G.N. Jha (Ed.): Sanskrit Computational Linguistics, LNCS 6465, pp. 150–161, 2010.

language. A tagged Sanskrit corpus could be used for a wide variety of research like developing POS Taggers, chunkers, parser, WSD etc.

A tagged text corpus is useful in many ways. More abstract levels of analysis benefit from reliable low-level information, e.g. parts of speech, so a good tagger can serve as a preprocessor. It is also useful for linguistic research for example to help find instances or frequencies of particular constructions in large corpora. Apart from this it is good for stemming in information retrieval (IR), since knowing a word's part of speech can help tell us which morphological affixes it can host. Automatic POS taggers can help in building automatic word-sense disambiguating algorithms; POS taggers are also used in advanced ASR language models such as class-based N-grams. POS tagging is also useful for text to speech synthesis and alignment of parallel corpora.

2 Characteristics of the Sanskrit Language

Sanskrit is one of the most studied languages of the world. It has been analyzed through many schools of grammar in India and abroad as well. Traditionally, the number of grammatical categories of Sanskrit varies from one to five. According to Indra school of *v*yakaraṇa there is only one category of words (*arthaḥ padam*). Pāṇini recognizes two categories: nominal and verbal (*suptiṅantam padam* - Pāṇini 1.4.14). The Nyaya philosopher Jagadisha in his *śabdaśaktiprakaśika* accepts only three categories of words: base, suffix and particles (*prakṛti*, *pratyaya* and *nipata*). The most popular categorization is of Yaska who describes the language in four categories: nominal base, verb root, prefix and indeclinable (*catvāri padajātāni nāmākhyate copasarganipātāśca* – Nirukta 1.1). The new grammarians (navya vaiyākaraṇa) add one more category to the list of Yaska, that is, *karmapravacaniya* and accept five categories of words. The different views on number of word categories are given in Durga's commentary on *Nirukta* [19]. If we take Yaska's grammatical categories for POS tagging, it will be very coarse grained tagging. But it will help us give high accuracy result as less number of tags lead to efficient machine learning. Moreover, two of these categories (prefix and indeclinable) belong to the closed class and the remaining two categories have totally different set of inflectional suffixes which are relatively easily identifiable. Apart from this, the set of roots of one category (verb) is also a closed list, that is, *dhātupāṭha* (list of verb roots listed by Pāṇini).With so shallow information this accuracy will not be of much help in many kinds of NLP tasks like in WSD which requires very fine-grained analysis.

Sanskrit is a relatively free word order language. In Sanskrit, a syntactic unit is called *pada*. Cardona [3] posits the formula for Sanskrit sentence (N-En)p... (V-Ev)p. A *pada* can be a nominal (*subanta*) or verbal (*tiṅanta*) expression. *Padas* with *sup* (nominal) inflections constitute the NPs (*subanta padas*), and those with *tiṅ* (verbal) can be called constituting the VPs (*tiṅanta pada* along with object *subantas*). In the former, the bases are called *prātipadikas* which undergo *sup* suffixations under specifically formulated conditions of case, gender, number, and also the end-characters of the bases to yield nominalsyntactic words.

The rules for *subanta padas* are found scattered in *Aṣṭādhyāyī* mostly in chapters 7-1, 7-2, 7-3, 6-1, 6-4. However, these rules have been treated in the subanta chapter of *Siddhānta Kaumudā* from rule number 177 to 446 [10].

The derivational morphology in Sanskrit studies primary forms (*kṛdanta*) and secondary forms (*taddhita*), compounds (*samāsa*), feminine forms (*strīpratyayānta*) etc. These can be inflected for 21 case (7 cases x 3 number) affixes to generate 21 inflected forms.

The verb morphology (*tiṅanta*) is equally complex. Sanskrit has approximately 2014 verb roots including *kaṇvādi* according to Pāṇinian *dhātupāha* classified in 10 gaṇas to undergo peculiar operations; it can also be subclassified in 12 derivational suffixes. A verb root conjugates for tense, mood, number and person information. Further, these can have *ātmanepadī* and *parasmaipadī* forms in 10 *lakāras* and 3x3 person and number combinations. There are12 secondary suffixes added to verb roots to create new verb roots. A verb root may have approximately 2190 (tense, aspect, number etc.) morphological forms. Mishra et al. [17] have done a rough calculation of all potential verb forms in Sanskrit to be more than 1029,60,000.

3 Evaluation Discussion of Tagsets

This section presents a survey of tagsets designed specifically for Sanskrit or for all Indian languages and then their evaluation with respect to Sanskrit. The tagsets are given in tables in their respective sub-sections.

3.1 ILMT Tagset (IIIT-Hyderabad)

NLP researchers at IIIT-Hyderabad have designed a standard POS tagset called ILMT tagset, guidelines for POS tagging and Chunking for Indian Languages and also brought out a collection of Application Program Interfaces (APIs) and tools for NLP called *Sanchay*, which also includes syntactic annotation interface. Their tagset is a flat one derived from the Penn tagset for English. They have made certain changes in the original tagset to make it suitable for Indian languages. Wherever the Penn tags were found to be inadequate for Indian languages, either new tags were introduced or existing tags were modified. It has less number of tags with the assumption that having less tags will facilitate machine learning and will give higher accuracy and coverage. This tagset covers mainly lexical categories and does not consider syntactic function of the word in the sentence, citing a reason that this reduces confusion involved in manual tagging. The machine is able to establish a word-tag relation which leads to efficient machine learning. The following table (Table. 1) contains all the tags listed in this tagset.

Thus, this tagset allows only coarse linguistic analysis and avoids fine-grained distinctions among linguistic items. At the same time, the designers of the tagset believe that "too coarse an analysis is not of much use". They try to seek a balance between fineness and coarseness. According to them, analysis should not be so fine as to hamper machine learning and also should not be so coarse as

Table 1. ILMT Tagset

NN	Noun	CC	Conjunct
NNP	Proper Noun	QW	Question Words
PRP	Pronoun	QF	Quantifier
VFM	Verb Finite Main	QFNUM	Number Quantifiers
VAUX	Verb Auxiliary	INTF	Intensifier
VJJ	Verb NonFinite Adjectival	NEG	Negative
VRB	Verb NonFinite Adverbial	NNC	Compound Common Nouns
UNN	Verb NonFinite Nominal	NNPC	Compound Proper Nouns
JJ	Adjective	NVB	Noun in kriya mula
RB	Adverb	JVB	Adjective in kriya mula
NLOC	Noun Location	RBVB	Adv in kriya mula
PREP	Postposition	UH	Intrjection Words
RP	Particle	SYM	Special

to lose important information. This tagset does not include attributes of lexical categories such as gender, number, person etc. Apart from this the tags for incorporated adjective and adverb are not applicable in Sanskrit. In Sanskrit, like noun and verb, Particles also require more tags as they have different and valuable function in the language, so in order to suit Sanskrit this tagset will have to be modified significantly. Keeping in mind the complex morphology of Sanskrit this flat tagset will hardly be able to do WSD.

3.2 JNU-Sanskrit Tagset (JPOS)

R. Chandrashekar's doctoral thesis titled *Part-of Speech Tagging for Sanskrit* [4] deals with the POS tagging problem for Sanskrit for the first time. In the linguistic analysis, JPOS follows the Pāṇinian grammatical description very strictly. His classification of Sanskrit linguistic items is very exhaustive and it provides a very unique tag applying the Sanskrit grammatical terms as much as possible. In the following table (Table. 2) the tag description of main categories is given.

In this tagset, there are 65 word class tags, 43 feature sub-tags, and 25 punctuation tags and one tag UN to annotate unknown words – a total of 134 tags. A single full tag is a combination of word class tag and feature sub-tags (indeclinable and punctuation tags do not have sub-tags). This is helpful in ambiguity resolution as person and number sub-tags have an agreement between the nouns and verbs in Sanskrit. The complete table for verb tags is not given here; I have given only two tags for it- *Parasmai Pada* Verb and *Atmane Pada* Verb. Verb takes *lakāra*, and person-number tags. Adjectives are included with their degrees, that is, they have 3 tags.

Table 2. JNU-Sanskrit Tagset

Common Noun (*nāmapada*)	N
Proper noun	NA
Proper Noun Country (*nāma abhidhāna deśa*)	NAD
Patronymic Noun	NAP
Metronymic Noun	NAS
Proper Noun Nationality (*nāma abhidhāna tadrāja*)	NAT
Noun Desiderative	NS
Coordinative Enumerative Compound (*dvandva itaretara*)	NCDI
Coordinative Collective Compound (*dvandva saṃhara*)	NCDS
Determinative Compound Accusative	NCT2
Determinative Compound Instrumental	NCT3
Determinative Compound Dative	NCT4
Determinative Compound Ablative	NCT5
Determinative Compound Genitive	NCT6
Determinative Compound Locative	NCT7
aluk Compound	NCAI
Negative Compound	NCNT
karmadhāraya Compound	NCK
dvigu Compound	NCD
bahuvrīhi Compound	NCB

Adverbial Compound	NCA
Pronoun	SN
Pronoun First Person	SNU
Pronoun Second Person	SNM
Pronoun Reflexive	SNA
Pronoun Demonstrative	SNN
Pronoun Interrogative	SNP
Pronoun Relative	SNS
Parasmai Pada Verb	AV
Atmane Pada Verb	PV
Cardinal Number	SAM
Ordinal Number	SAMY
Past Active Participle	KV1
Past Middle/Active Participle	KV2
Past Passive Participle	KB1
Past Active Participle	KB2
Future Active Participle	KAa
Future Passive Participle	KAb
kṛdanta vidhyarthaka 1	KVI1
kṛdanta vidhyarthaka 2	KVI2
kṛdanta vidhyarthaka 3	KVI3
Particles	AV
Negative Particle	AVN
Conjunctive Particle	AVC
Disjunctive Particle	AVD
Interrogative Particle	AVP
Infinitive Particle	AVT
avyaya ktvānta	AVK
avyaya lyabanta	AVL
Adverb Particle	AVKV
Interjection Particle	UD

This is a flat and fine-grained tagset capturing linguistic features according to the traditional Sanskrit grammar. Its coverage is very high. Almost every kind of linguistic item was treated discretely (especially *taddhita* and *kṛidanta* constructions). This tagset is very informative but we assume that its specificity will suffer while tagging the data. For instance, among noun tags there are 8 tags and the basis of counting them separately seems to be the special morphology for these items like Patronymic Noun (e.g. *rāghavaḥ, jānakī*), Metronymic Noun (e.g. *pārthaḥ, saumitriḥ*), (both tags could be treated as Proper Noun), Agent Noun (e.g. *kartuḥ, adhyāpakaḥ*) (could be treated as Common Noun). Likewise under Compound category there are 14 types. Many of which could be clubbed

Table 3. LDCIL Tagset

Pronoun	PRP	Cardinal	CRD
Demonstrative	DEM	Ordinal	ORD
Verb Finite	VM	Classifier	CL
Verb Auxiliary	VX	Intensifier	INTF
Adjective	ADJ	Interjection	INJ
Adverb	ADV	Negation	NEG
Postposition	PSP	Quotative	UT
Particles	PRT	Symbol	SYM
Conjuncts	CNJ	Reduplicative	RDP
Question Words	WQ	Echo word	ECH
Quantifiers	QTF	Unknown	UNK
Noun	NN	Proper Noun	NNP
NLOC	NST	Compound (any)	XC

under other categories as well. For example, the words like *rājapurusaḥ* and *devendraḥ* have been assigned Determinative Compound (genitive) which could be very well given Common noun and Proper noun respectively. The division of participle tags is also very lengthy; *kṛidanta vartamāna* has two separate tags which does not seem serving any purpose and the same is with *kṛidanta vidhyarthaka* which has 3 tags.

Apart from this, the human annotator will have enormous cognitive load as this tagset requires remembering this huge number of tags along with exhaustive knowledge of *Pāṇinian* grammar in order to apply them correctly. Only a sophisticated pundit of *Pāṇinian* grammar can handle this task. At experiment level, however, no data have been tagged to test its efficiency, so the above is our assumption only. Moreover, its adaptation for other Indian languages will invite enormous changes in the tagset. So the tagged multilingual corpora will hardly be compatible with each other.

3.3 LDCIL Tagset

Linguistic Data Consortium for Indian Languages(LDC-IL) is a consortium for the development of language technology in India and was set up by a collective effort of the Central Institute of Indian Languages, Mysore and several other like-minded institutions working on Indian Languages technology. LDCIL tagset is available in the website of Linguistic Data Consortium for Indian Languages: http://www.ldcil.org/POSTagSet.html. This tagset was formulated by the Standardization Committee on August 6, 2007 at IIIT, Hyderabad and is a further extension of ILMT tagset with some additional tags to achieve higher level of granularity. It is a standard tagset for annotating the corpora of Indian languages. It is a flat tagset designed in the pattern of ILMT tagset and has a total of 26 tags. The table below (Table. 3) contains all the tags. This tagset is similar to ILMT tagset, so the same evaluation goes to this also.

Table 4. IL-POSTS Sanskrit tagset categories and their Types

Category	Type
Noun (N)	Common (NC)
	Proper (NP)
Verb (V)	
Pronoun (P)	Pronominal (PPR)
	Reflexive (PRF)
	Reciprocal (PRC)
	Relative (PRL)
	Wh (PWH)
Nominal Modifier (J)	Adjective (JJ)
	Quantifier (JQ)
Demonstrative (D)	Absolutive (DAB)
	Relative (DRL)
	Wh- (DWH)
Kridant (KD)	Participle (KDP)
	Gerundive (KDG)
Particle (C)	Coordinating (CCD)
	Subordinating (CSB)
	Gerundive (CGD)
	Interjection (CIN)
	Negative (CNG)
	Emphatic (CEM)
	Interrogative (CNT)
	Adverb (CAD)
	Ambiposition (upapada) (CPP)
	Quotative (CQT)
	Comparative (CCM)
	Reduplicative (CRD)
	Other (CX)
Punctuation (PU)	
Residual (RD)	Foreign word (RDF)
	Symbol (RDS)
	Others (RDX)

Table 5. IL-POSTS Sanskrit tagset attributes and their values

Category	Type
Noun (N)	Common (NC)
	Proper (NP)
Verb (V)	
Pronoun (P)	Pronominal (PPR)
	Reflexive (PRF)
	Reciprocal (PRC)
	Relative (PRL)
	Wh (PWH)
Nominal Modifier (J)	Adjective (JJ)
	Quantifier (JQ)
Demonstrative (D)	Absolutive (DAB)
	Relative (DRL)
	Wh- (DWH)
Kridant (KD)	Participle (KDP)
	Gerundive (KDG)
Particle (C)	Coordinating (CCD)
	Subordinating (CSB)
	Gerundive (CGD)
	Interjection (CIN)
	Negative (CNG)
	Emphatic (CEM)
	Interrogative (CNT)
	Adverb (CAD)
	Ambiposition (upapada) (CPP)
	Quotative (CQT)
	Comparative (CCM)
	Reduplicative (CRD)
	Other (CX)
Punctuation (PU)	
Residual (RD)	Foreign word (RDF)
	Symbol (RDS)
	Others (RDX)

3.4 IL-POSTS Sanskrit Tagset

This tagset has been derived from IL-POSTS, a standard framework for Indian languages developed by MSRI. This is a hierarchical tagset which is based on guidelines similar to EAGLES. This tagset encodes information at three levels- Categories, Types, and Attributes. Till date (as far as we know) four language specific tagsets have been derived from this framework; Hindi, Bangla, Tamil and Sanskrit. The tagset for Sanskrit is given here in these two tables (Table. 4, 5)- one having categories and their respective types and the other containing attributes and their respective values:

The standard which has been followed in this tagset takes care of the linguistic richness of Indian languages including Sanskrit. It allows the "selective inclusion and removal of features for a specific language/project, thereby keeping the framework a common standard across languages/projects" [1]. This tagset is supposed to provide cross-linguistic compatibility, reusability, and interchangeability. Keeping in mind its higher accuracy results, it could prove a boon to Indian NLP which still needs to achieve a good position in the world as Indian languages are lacking adequate resources in terms of data and tools (see [1] and [11]).

This framework is very sophisticated in covering the linguistic features of Indian languages. The tags used are extremely fine-grained, and incorporate a great deal of information about case, gender, number and so on. The IL-POST is able to accommodate all desired linguistic features of Sanskrit and the tagged corpus remains compatible with other languages tagged by a brethren tagset. Apart from this, as this tagset follows the standard of European languages it could enable us to access the European language data for Indian NLP and vice versa.

MSRI has developed an annotation tool called *MSRI Part-of-Speech Annotation Interface* which is a GUI for assigning the appropriate POS label for each word in a sentence. This is designed for annotating within the framework of a hierarchical tagset designed at MSRI. The tool also provides the facility to provide the morphological attributes with their values. The interface supports various operations for viewing and editing the POS labels and morphological information associated with the word. This tool reduces the cognitive load of the human annotator and enables one to speed up the tagging task.

3.5 Sanskrit Consortium Tagset (CPOS)

This tagset is designed for tagging sandhi-free Sanskrit corpus. It is based on traditional Sanskrit grammatical categorization. All the tags seem to be the extensions of Yāska's four grammatical categories: nāma, ākhyāta, upasarga and nipāta. This tagset has 28 tags. All the tags are given in the following table (Table. 6):

This is apparently a flat tagset and allows the annotation of major categories only. Most of the categories appear to have been adapted either from the JPOS or the ILMT tagset. For morphological analysis it will take help from Morphological Analyzer, so morpho-syntactic features are not included in the tagset. It

Table 6. Sanskrit Consortium tagset

NP	nāmapada (common noun)
NS	nāmasanjnā (proper noun)
NV	nāmaviśesan (adjective)
KN	kridantanāma (participles)
KNV	kridantanāma viśeṣaṇa (participle adjective)
SN	sarvanām (pronoun)
SNV	sarvanām-viśeṣaṇa (demonstrative)
SSN	sāmbandhika sarvanām (relative pronoun)
SMK	saṃkhyavācaka (number)
SMKY	saṃkhyeyavācaka (cardinal)
SMKP	saṃkhyāpūraṇavācaka (ordinal)
KP	kriyāpada (verb)
A	avyaya (indeclinable)
UNK	Unknown

AN	niṣedhārthaka (negation)
AP	praśnārthaka (interrogative)
AUP	upapada (ambiposition)
ASMB	saṃbodhanaṃ (vocative)
AUD	uddharaṇa (quotative)
KKS	Gerundive
KV	kriyāviśesan (adverb)
ABS	(udgāra) interjection
APY	pada yojaka (connective)
AVK	vakya yojaka (sentential connective)
DV	dwirukti (reduplication)
ASD	sādṛśyādi (comparative)
AB	other language
PUNC	Punctuation

is supposed to give higher accuracy result by having less number of tags. A tool called *Sanchay* is being used to facilitate manual tagging and thus it reduces the cognitive load of annotator. For Sanskrit tagging, it can be useful as long as we know what will be the use of tagged data in this tagset. At an experimental level two taggers were developed using this tagset - a rule based and a statistical and have reported 79-81% results. The rule based tagger can be accessed using this link http://sanskrit.jnu.ac.in/cpost/post.jsp. As this tagset does not follow any standard framework the tagged data will hardly allow any sharing and reusability. Due to its coarse-grained tagging and unavailability of Indian language morph analyzers, it will be interesting to see how this tagset can be put to actual use.

4 Conclusion and Further Research

From the above evaluation of the tagsets it appears that the most useful tagged corpus is the one that ensures the maximal use and sharing of it, and it will only be possible when it is tagged by a well-established standard tagset or paradigm. The tagsets that we have discussed in this article are of both kinds- those following global standards/guidelines and those which have evolved from particular language specific requirements. The tagsets like ILPOST, ILMT and LDCIL are designed to take care of all Indian languages, and this could provide resource sharing and reusability of the linguistic resources. The flat tagsets may be easier to process but they cannot capture higher level of granularity without an extremely large list of independent tags. They are hard to adapt too. So far, ILPOSTS has been used for four languages Hindi, Tamil, Bangla and Sanskrit, and an adapted version is reportedly being used in the LDC-IL project at CIIL Mysore for many more languages. So, in order to enrich Indian NLP resources, ILPOSTS seems to be a better option. However, further research is needed to substantiate this claim.

For further research a certain amount of data could be manually tagged with the help of these tagsets by some (1 or 2) human annotator. And on the basis of the experience of these annotators certain facts can be discovered, for example, which tagset is easy to use for Sanskrit and which tagset gives better results and which is less confusing and so on. Thereafter, rule based or statistical taggers can be tested for the efficiency of these tagsets and also their compatibility across languages.

References

1. Baskaran, S., Bali, K., Bhattacharya, T., Bhattacharyya, P., Choudhury, M., Jha, G.N., Rajendran, S., Saravanan, K., Sobha, L., Subbarao, K.V.S.: A Common Parts-of-Speech Tagset Framework for Indian Languages. In: LREC, Marrakech, Morocco (2008)
2. Baskaran, S., et al.: Framework for a Common Parts-of-Speech Tagset for IndicLanguages (2007), `http://research.microsoft.com/~baskaran/POSTagset`
3. Cardona, G.: Pāṇini: His work and its traditions. Motilal Banarasidass, Delhi (1988)
4. Chandrashekar, R.: Parts-of-Speech Tagging For Sanskrit. Ph.D. thesis submitted to JNU, New Delhi (2007)
5. Greene, B.B., Rubin, G.M.: Automatic grammatical tagging of English. Department of Linguistics, Brown University, Providence, R.I. (1981)
6. Hardie, A.: The Computational Analysis of Morphosyntactic Categories in Urdu. PhD Thesis submitted to Lancaster University (2004)
7. Hellwig, O.: SANSKRITTAGGER, A Stochastic Lexical and POS Tagger for Sanskrit. In: Huet, G., Kulkarni, A. (eds.) Sanskrit Computational Linguistics 2007. LNCS (LNAI), vol. 5402. Springer, Heidelberg (2009)
8. Huet, G.: The Sanskrit Heritage Site, `http://sanskrit.inria.fr/`
9. IIIT-Tagset. A Parts-of-Speech tagset for Indian Languages, `http://shiva.iiit.ac.in/SPSAL2007/iiit_tagset_guidelines.pdf`

10. Jha, G.N.: Generating nominal inflectional morphology in Sanskrit. In: SIMPLE 2004, IIT-Kharagpur Lecture Compendium, Shyama Printing Works, Kharagpur (2004)
11. Jha, G.N., Gopal, M., Mishra, D.: Annotating Sanskrit Corpus: adapting IL-POSTS. In: Vetulani, Z. (ed.) Proceedings of the 4th Language and Technology Conference: Human Language Technologies as a Challenge for Computer Science and Linguistics, pp. 467–471 (2009)
12. Jha, G.N., Mishra, S.: Semantic processing in Panini's karaka system. In: Huet, G., Kulkarni, A., Scharf, P. (eds.) Sanskrit Computational Linguistics 2007/2008. LNCS (LNAI), vol. 5402. Springer, Heidelberg (2009)
13. Kale, M.R.: A Higher Sanskrit Grammar. MLBD Publishers, New Delhi (1995)
14. Leech, G., Wilson, A.: Recommendations for the Morphosyntactic Annotation of Corpora. EAGLES Report EAG-TCWG-MAC/R (1996)
15. Leech, G., Wilson, A.: Standards for Tag-sets. In: van Halteren, H. (ed.) Syntactic Word class Tagging. Kluwer Academic, Dordrecht (1999)
16. Leech, G.: Grammatical Tagging. In: Garsire, Leech, McEnery (eds.) Corpus Annotation: Linguistic Information for Computer Text Corpora. Longman, London (1997)
17. Mishra, S., Jha, G.N.: Identifying verb inflections in Sanskrit morphology. In: Proceedings of SIMPLE 2004, IIT Kharagpur (2005)
18. Ramkrishnamacharyulu, K.V.: Annotating Sanskrit Texts Based on Sabdabodha Systems. In: Kulkarni, A., Huet, G. (eds.) Sanskrit Computational Linguistics. LNCS (LNAI), vol. 5406, pp. 26–39. Springer, Heidelberg (2009)
19. Rishi, U.S.S. (ed.): Yaska-pranitam niruktam, vol. I. Chowkhamba Vidyabhawan, Varanasi (2005)
20. Santorini, B.: Part-of-speech tagging guidelines for the Penn Treebank Project. Technical report MS-CIS-90-47, Dept. Of Computer and Information Science, University of Pennsylvania (1990)

Appendices

Appendix 1 Sample of Tagged data by Sanskrit Consortium tagset

दमनकः\NS आह\KP -\PUNC "\PUNC यद्यपि\AVK न\AN किंचित्\SN देवपादानाम्\NP अस्माभिः\SN प्रयोजनम्\NP ,\PUNC तथापि\AVK भवताम्\SN प्राप्तकालम्\NV वक्तव्यम्\KN ,\PUNC यत्\SSN उत्तममध्यमाधमैः\NV सर्वैः\SN अपि\ASD राज्ञाम्\NP प्रयोजनम्\NP ।\PUNC उक्तम्\KN च\AVK -\PUNC

दन्तस्य\NP निष्कोषणकेन\NP नित्यम्\KV कर्णस्य\NP कण्डूयनकेन\NP वा\AVK अपि\A ।\PUNC
तृणेन\NP कार्यम्\NP भवति\KP ईश्वराणाम्\NP किमङ्ग\ABS वाग्धस्तवता\NV नरेण\NP ॥\PUNC 77\PUNC ॥\PUNC

तथा\AVK वयम्\SN देवपादानाम्\NP अन्वयागताः\NV भृत्याः\NP आपत्सु\NP अपि\ASD पृष्ठगामिनः\NV यद्यपि\AVK स्वम्\NV अधिकारम्\NP न\AN लभामहे\KP ,\PUNC तथापि\AVK देवपादानाम्\NP एतत्\SN युक्तम्\NV न\AN भवति\KP ।\PUNC

Appendix 2 Sample of Tagged data by IL-POST tagset

दमनकः\NP.mas.sg.nom.i आह\V.ppd.sg.3.prs.n -\PU "\PU यद्यपि\CSB न\CNG किंचित्\CX देवपादानाम्\NC.mas.pl.gen.vi अस्माभिः\PPR.0.pl.1.ins.iii.n.n.prx प्रयोजनम्\NC.neu.sg.0.i ,\PU तथापि\CSB भवताम्\PPR.mas.pl.3.gen.vi.n.n.prx प्राप्तकालम्\JJ.neu.sg.0.i.n.n वक्तव्यम्\KDG.neu.sg.0.i ,\PU यत्\CSB उत्तममध्यमाधमैः\JJ.neu.pl.ins.iii.n.n सर्वैः\PPR.neu.pl.3.ins.iii.n.n.0 अपि\CEM राज्ञाम्\NC.mas.pl.gen.vi प्रयोजनम्\NC.neu.sg.0.i ।\PU उक्तम्\KDP.neu.sg.0.i च\CCD -\PU

दन्तस्य\NC.mas.sg.gen.vi निष्कोषणकेन\NC.mas.sg.ins.iii नित्यम्\CAD कर्णस्य\NC.mas.sg.gen.vi कण्डूयनकेन\NC.mas.sg.ins.iii वा\CCD अपि\CEM ।\PU

तृणेन\NC.neu.sg.ins.iii कार्यम्\KDG.neu.sg.0.i भवति\V.ppd.sg.3.prs.n ईश्वराणाम्\NC.mas.pl.gen.vi किमङ्ग\CIN वाग्धस्तवता\JJ.mas.sg.ins.iii.n.n नरेण\NC.mas.sg.ins.iii ॥\PU 77\RDS ॥\PU

तथा\CCM वयम्\PPR.0.pl.1.nom.i.n.n.prx देवपादानाम्\NC.mas.pl.gen.vi अन्वयागताः\JJ.mas.pl.0.i.n.n भृत्याः\NC.mas.pl.0.i आपत्सु\NC.mas.pl.loc.vii अपि\CEM पृष्ठगामिनः\JJ.mas.pl.0.i.n.n यद्यपि\CSB स्वम्\PRF.mas.sg.acc.ii अधिकारम्\NC.mas.sg.acc.ii न\CNG लभामहे\V.apd.pl.1.prs.n ,\PU तथापि\CSB देवपादानाम्\NC.mas.pl.gen.vi एतत्\PPR.neu.sg.3.nom.i.n.n.prx युक्तम्\JJ.neu.sg.0.i.n.n न\CNG भवति\V.ppd.sg.3.prs.n ।\PU

Performance of a Lexical and POS Tagger for Sanskrit

Oliver Hellwig

SAI, Universität Heidelberg, Germany
hellwig7@gmx.de

Abstract. Due to the phonetic, morphological, and lexical complexity of Sanskrit, the automatic analysis of this language is a real challenge in the area of natural language processing. The paper describes a series of tests that were performed to assess the accuracy of the tagging program `SanskritTagger`. To our knowlegde, it offers the first reliable benchmark data for evaluating the quality of taggers for Sanskrit using an unrestricted dictionary and texts from different domains. Based on a detailed analysis of the test results, the paper points out possible directions for future improvements of statistical tagging procedures for Sanskrit.

1 Overview and Previous Research

The quality of automatic language analysis methods is one of the central issues in natural language processing (NLP). Most research in this area concentrates on English and the voluminous corpora available in this language. For example, the tagsets and the precision of the POS tagging applied to the Penn treebank were described in [6]. The authors used a small POS tagset that consisted of 48 categories and included tags such as "existential *there*" and "comma". Applying "a cascade of stochastic and rule-based taggers" to the treebank, they obtained an error rate of 2-6% using this tagset. A similar approach is found in [5] where `CLAWS4`, a tagger for the British National Corpus, is described. The authors used a multi-level model for analysis and reported error rates of 1-5% for POS tagging. In more recent publications, however, it has been pointed out that many of the assumptions on which the popular NLP systems for English are based do not apply to other languages. For tagging Hungarian, for instance, a completely different tagset was used in [2]. This tagset consisted of 744 tags, thus even exceeding the size of the tagset applied by `SanskritTagger`. Using a morphological analyzer and a maximum entropy model, the authors reported error rates of 1-3% for POS assignment in a Hungarian corpus.

Obviously, the precautions brought forward in [2] apply to the computational analysis of Sanskrit to an even higher degree. Like in Hungarian, the linguistic complexity of Sanskrit is partly caused by its inflectional nature. In addition, the automatic processing of Sanskrit faces a number of problems that occur on different levels of language analysis and are not well studied in NLP literature dealing with European and especially Germanic languages:

G.N. Jha (Ed.): Sanskrit Computational Linguistics, LNCS 6465, pp. 162–172, 2010.

1. Segmenting a Sanskrit phrase into its inflected components is the first and perhaps most demanding task in the analysis of Sanskrit texts. This situation is mainly caused by the phonetic rules called Sandhi, and it is aggravated by the almost complete absence of typographic conventions such as punctuation or the lower/upper case opposition, which is so useful in analyzing German texts. Similar problems are, for example, known from the automatic processing of Chinese, where word segmentation is a well-studied, though not satisfactorily solved problem (cmp. [8] for an overview of this area).
2. On a higher level of linguistic analysis, the large number of homonyms and the comparatively free word order complicate the automatic processing of Sanskrit. Studies performed for languages whose word order is less regimented than in English suggest that tagging strategies that are based on positional information are less successful in processing such languages (see, for example, [7, 280]). The situation is further complicated by the structure of digital dictionaries of Sanskrit. Many of these dictionaries are not designed for the needs of NLP, but are created from scanned versions of printed dictionaries. Due to the exigencies of philological work, many printed dictionaries contain collocations or infinite verbal forms that may be interesting from a philological perspective, but deteriorate the quality of automatic processing.

This means that the stages of processing Sanskrit differ strongly from what is known for European languages and especially English. To analyze Sanskrit phrases, a multi-level strategy had to be designed whose single components are much more entangled than in the processing of European languages (cmp. [4]). Due to these differences in the analysis process and thus in the test design, POS error rates found for European languages are hardly comparable with those for Sanskrit.

The following section first sketches the basic ideas of the tagging process applied in `SanskritTagger` and then describes the testing procedure and its results.

2 Testing the `SanskritTagger`

2.1 The Test Design

In the statistical model used by `SanskritTagger`, the result of the analysis is influenced by two factors: The language data from which statistical key values are estimated and the procedures that make use of these statistical values to analyze a given piece of text. Since the procedures were outlined in [4], we will concentrate on the question of how the lexical data and the amount of data used for training influences the quality of the analysis.

To begin with, let us shortly reconsider some details of the analysis process to make clear on which kind of data the analysis is based. In the following, W denotes an uninterrupted string, this means a string of letters that does not contain break markers such as spaces or *daṇḍa*s. If W is formed correctly, it can be split into at least one substring w_i, which is a grammatically correct Sanskrit form. `SanskritTagger` divides the analysis of W roughly into three stages:

1. Initially, a list of candidate solutions is generated for each substring w_i that may be contained in W. The number of candidate solutions depends on the well-known features that complicate the computational (and human!) analysis of Sanskrit: the number of possible Sandhi breakpoints and the morphological and lexical complexity of w_i.
2. In the second step, the program searches for a path through the elements of the candidate lists that is optimal from the perspective of phraseology or lexicography. The path is calculated using a first-order Markov chain whose parameters (i.e., relative frequencies of single lexemes and transition probabilities between bigrams of lexemes) are the statistical key values mentioned above (cmp. [4, 273/74]). These values are estimated from the texts stored in the database of `SanskritTagger`. As a first hypothesis, we may assume that the size of this text database strongly influences the quality of the estimated parameters and, therefore, also the accuracy of the analysis. At this point, it should be taken into consideration that previous research has shown that the correlation between the amount of training data and the analysis quality is not linear (see, e.g., [9, 363ff.] for an early treatment of this problem). Therefore, we will variate the number of texts used to calculate these parameters for reaching a realistic picture of how the accuracy of the tagging is influenced by the size of the text database (cmp. page 168).
3. After the candidate lists have been filtered by lexical criteria, the third stage of the analysis process uses similar methods to find an optimal POS path through the remaining candidates. The role of the text database and the parameter estimation corresponds with the scenario described in the second step.

For performing the tests, we used the standard leave-one-out strategy with consistent train-test splits. First, the set T of all texts t that contain more than 10.000 words was retrieved from the corpus.[1] Next, for each $t \in T$, the statistical data was rebuilt without including t, and t was analyzed using this data and the built-in routines of `SanskritTagger`. To evaluate the test result, the new analysis a_{new} of each separable string was compared with the analysis of the string a_{DB} that is stored in the program database. This step seems trivial at first view. However, consider the following case:

String: *kṛṣṇapādasevanatatparaḥ*
a_{DB}: *kṛṣṇa-pāda-sevana-tatparaḥ*
a_{new}: *kṛṣṇa-pādasevana-tatparaḥ*

If the new analysis a_{new} were marked as completely wrong, the error rate would be overestimated because 2 of 4 lexemes (*kṛṣṇa* and *tatpara*) were identified correctly. Therefore, we had to devise an evaluation procedure that was able to cope with such imperfect matches.

A simple solution of this problem is offered by alignment techniques (see [3] for a survey of techniques for pairwise alignment and [1, 368ff] for an Indological

[1] The only large text that was skipped is the MAHĀBHĀRATA, whose processing time would have amounted to several hours.

application of such algorithms). The basic idea of these techniques can be sketched as follows. Given two vectors of symbols v_1 and v_2, a matrix is created that has the dimensions $(|v_1|+1)\cdot(|v_2|+1)$. The first row and the first column of the matrix are filled with ascending numbers, i.e., $c_{[i,1]} = i - 1$ for the first row and $c_{[1,i]} = i - 1$ for the first column. Now, the comparison algorithm fills the remaining $|v_1| \cdot |v_2|$ cells of the matrix. The value $c_{[x,y]}$ of each cell $[x, y]$ is calculated by inspecting the three preceeding cell values $c_{[x,y-1]}$, $c_{[x-1,y]}$ and $c_{[x-1,y-1]}$:

$$c_{[x,y]} = \min \begin{cases} c_{[x,y-1]} + \gamma \\ c_{[x-1,y]} + \gamma \\ c_{[x-1,y-1]} + \texttt{compare}(v_{1x}, v_{2y}) \end{cases}$$

γ is the gap penalty with which the non-alignment of two symbols is penalized. $\texttt{compare}(v_{1x}, v_{2y})$ is a function that compares the symbols v_{1x} and v_{2y}, which are represented by the analysis of a substring stored in the database (v_{1x}) and its new analysis (v_{2y}) in our case. The value of `compare` ranges from 1 (= v_{1x} and v_{2y} are derived from different lexemes) to 0 (= v_{1x} and v_{2y} are identical). It is calculated by inspecting to which degree the analyses of v_{1x} and v_{2y} are different. A difference in the lexical analysis is the most severe kind of error because it invalidates the following morphological and POS analysis. `compare` returns 1 in this case. This kind of error is caused by two scenarios. In the first scenario, the string is segmented differently in the database and by the new analysis. The string *kṛṣṇapādasevanena* may, for instance, be analyzed as *kṛṣṇa*[comp.]-*pāda*[comp.]-*sevanena*[instr.] in the database, but as *kṛṣṇa*[comp.]-*pādasevanena*[instr.] by the new analysis. In many cases, these errors are caused by redundancies found in the Sanskrit dictionaries (cmp. page 163). In the second scenario, both procedures have split the string in the same way, but have assigned different lexemes to at least one substring. In our example, this may result in analyses such as "the worship of the feet of Kṛṣṇa" in the database and "the worship of black feet" by the new analysis due to the homonymy of the noun and the adjective *kṛṣṇa*. A more detailed analysis of this class of errors is given in section 2.2. If the lexical analysis has succeeded, errors may occur in the morphological and POS analysis of the words (third step of the analysis). The most frequent types of errors are caused by *bahuvrīhi* composites whose differing gender is not recognized correctly, and by morphologically ambiguous forms such as *vanam*, which may be a nominative, an accusative, or a vocative. In these cases, `compare` returns a positive value larger than zero that represents the number of correct decisions the algorithm has made. Errors in the analysis of a noun, for example, can occur during the four stages of lexical analysis, this means in case, number, and gender. If the lexical analysis, case, and number were recognized correctly, `compare` returns $\frac{3}{4} = 0.75$.

After the algorithm has traversed the whole matrix, the cell $c_{[|v_1|+1,|v_2|+1]}$ contains the accumulated costs of the optimal alignment of v_1 and v_2. This value may be used to calculate a measure of quality e for the analysis process:

$$e(v_1, v_2) = 1 - \frac{c_{[|v_1|+1,|v_2|+1]}}{\gamma \cdot \max(|v_1|, |v_2|)}.$$

Using $\gamma \cdot \max(|v_1|, |v_2|)$ in the divisor is motivated by the fact that the accumulated costs amout to $\max(|v_1|, |v_2|) \cdot \gamma$ if two sequences contain no pairwise identical symbols. Therefore, $e(v_1, v_2)$ is in the range $[0, 1]$ with $e(v_1, v_2) = 1$ for a perfect match. To extract the details about the errors made by the program, the matrix is traversed in backward direction by using pointers set during the forward pass. The error information gained in this step is examined in detail in the next section.

2.2 Results and Evaluation

All texts in the SanskritTagger database that contain more than 10.000 words were analyzed using the testing scenario described in the preceeding section. The test data will be evaluated in two steps. First, we will give an overview of general trends in the performance of the tagger, which includes questions such as the influence of the text genre and the style on the analysis accuracy (page 166). Second, we will have a closer look at the lexical errors, which constitute the most frequent class of errors (page 168).

The general performance. The raw test results are listed in table 1, which contains some types of information that need further explanation:

1. **Lexical richness** is the ratio of the number of different lexemes divided by the number of all words contained in a text (n_T). It ranges from $\frac{1}{n_T}$ (each word of the text is derived from the same lexeme, something like *śiva śiva śiva*) to $\frac{n_T}{n_T} = 1$ (each word is derived from another lexeme).
2. Six kinds of **errors** are distinguished in the tests. Apart from a lexical error e_{LEX} (cmp. page 165 and page 168 for a detailed analysis of this kind of error), a wrong POS tag can be assigned to a correctly identified lexeme (e_{POS1-5}). e_{POS1} occurs only when an infinite verbal form is confused with a finite one.
3. The **hit rate** H records the overall hit rate of SanskritTagger. If C is the number of words that were analyzed correctly and $\omega = \sum_{i=1}^{5} e_{POSi}$, it is defined as

$$H = \frac{C}{C + e_{LEX} + \omega}.$$

The lexical hit rate H_{LEX} ignores POS errors because the lexeme has been recognized correctly in such cases:

$$H_{LEX} = \frac{C + \omega}{C + e_{LEX} + \omega}.$$

To compare the accuracy of SanskritTagger with that of other POS taggers, the POS error rate can be calculated as follows (values not displayed in table 1):

$$E_{POS} = \frac{\omega}{C + \omega}.$$

Table 1. Results of the leave-one-out tests; refer to page 166 for the exact meanings of the columns

Text	Lexical richness	Avg. phrase length	Analyzed correctly (C)	Error: lexical analysis (e_{LEX})	Error: diff. word classes (e_{POS1})	Error: case, number (e_{POS2})	Error: gender (e_{POS3})	Error: tense, mode (e_{POS4})	More than one non-lexical error (e_{POS5})	Hit rate	Lex. hit rate
Rasaratnasamuccayaṭīkā	0.18	9.8	9986	800	8	357	299	1	234	0.855	0.932
Rasamañjarī	0.22	7.2	10550	694	5	576	244	0	281	0.854	0.944
Bhāvaprakāśa	0.25	7.0	11096	878	2	314	280	0	161	0.872	0.931
Laṅkāvatārasūtra	0.16	11.6	9305	1730	17	347	254	3	388	0.773	0.856
Rasendracintāmaṇi	0.25	7.7	10845	922	4	555	337	0	244	0.840	0.929
Bodhicaryāvatāra	0.21	7.4	11257	1219	10	520	288	5	198	0.834	0.910
Yājñavalkyasmṛti	0.27	7.0	11764	1059	19	557	234	1	150	0.853	0.923
Gokarṇapurāṇasāraḥ	0.19	6.6	12497	1100	18	431	220	4	144	0.867	0.924
Śārṅgadharasaṃhitādīpikā	0.17	8.1	12449	833	2	532	372	1	305	0.859	0.943
Daśakumāracarita	0.28	16.9	12081	1283	19	439	357	5	252	0.837	0.911
Rasaprakāśasudhākara	0.19	7.4	13263	988	2	608	288	2	219	0.863	0.936
Skandapurāṇa	0.20	6.8	13827	1242	15	559	317	6	157	0.858	0.923
Rasendracūḍāmaṇi	0.21	7.2	15417	888	9	553	345	0	335	0.879	0.949
Viṣṇusmṛti	0.24	5.7	15071	1378	29	809	371	4	257	0.841	0.923
Buddhacarita	0.20	9.1	15748	1485	22	723	536	5	237	0.840	0.921
Mṛgendraṭīkā	0.16	12.4	17443	1902	23	705	569	5	478	0.826	0.910
Sātvatatantra	0.30	7.1	6240	1259	4	309	143	0	1923	0.632	0.873
Arthaśāstra	0.29	9.3	8877	962	5	450	318	0	248	0.817	0.911
Hitopadeśa	0.23	7.1	9822	900	12	414	167	3	77	0.862	0.921
Rasaratnasamuccaya	0.18	7.2	23311	1656	5	1175	672	1	546	0.852	0.939
Mugdhāvabodhinī	0.13	12.7	24613	2087	20	1268	797	0	443	0.842	0.929
Rasārṇava	0.13	6.8	26198	1895	10	1427	1031	1	711	0.838	0.939
Manusmṛti	0.17	7.0	31177	2728	57	1650	890	9	342	0.846	0.926
Āyurvedadīpikā	0.12	12.8	31355	2834	51	1259	902	10	545	0.848	0.923
Kūrmapurāṇa	0.14	6.6	33308	2856	26	1244	715	17	391	0.864	0.926
Bhāgavatapurāṇa	0.17	7.1	33840	4076	76	1958	1062	22	625	0.812	0.902
Bṛhatkathāślokasaṃgraha	0.12	6.8	49372	5482	114	1997	1401	28	864	0.833	0.907
Ānandakanda	0.11	6.9	68767	5896	17	3720	1731	17	1791	0.839	0.928
Carakasaṃhitā	0.11	10.5	73239	6830	157	3582	2189	79	1763	0.834	0.922
Skandapurāṇa (2)	0.08	6.8	97348	7645	112	3951	1627	44	1250	0.869	0.932
Matsyapurāṇa	0.10	6.8	101860	9805	141	3996	2518	25	1517	0.850	0.918
Liṅgapurāṇa	0.08	6.9	105185	10477	63	4137	3447	23	1557	0.842	0.916
Suśrutasaṃhitā	0.09	8.9	113237	11904	209	5769	3809	30	2274	0.825	0.913
Rāmāyaṇa	0.05	6.8	220398	18643	387	8292	4437	64	2619	0.865	0.927

Figure 1 displays the boxplots of H, H_{LEX} and E_{POS}. Since the analysis of Sanskrit differs strongly from that of other languages, H and H_{LEX} are hardly comparable with rates recorded in previous research. Obviously, however, both H and H_{LEX} need significant improvements in future versions of the program. If we take into account the complex considerations that the linguistic phenomena of composite creation and *bahuvrīhi* make necessary for POS tagging of Sanskrit, the POS accuracy of `SanskritTagger` compares acceptably with POS error rates reported for less complex languages ($\mu_{e\,POS} = 0.090$, $\sigma_{e\,POS} = 0.035$).

To go deeper into the details of table 1, we should try to determine the influence of literary features on the analysis quality. If lexical richness and the average phrase length are used to get an estimation of the complexity of a literary style, we get the four plots displayed in figure 2. Astonishingly, the correlations between these two features and the (lexical) hit rates are almost unnoticeable. This result is supported by inspecting linear regressions performed with these data none of whose parameters (intercept, slope) are significant at the 10% level.

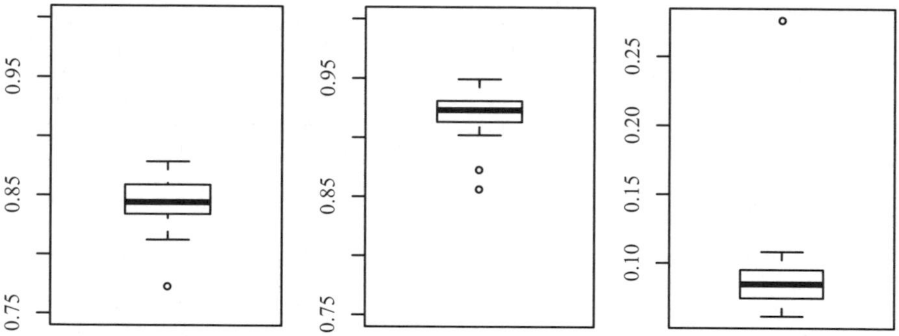

Fig. 1. Boxplots of the hit rates recorded in table 1; left: hit rate H, middle: lexical hit rate H_{LEX}, right: POS error rate

In the introduction, we mentioned the theory that the amount of training data influences the analysis accuracy. To assess this theory, the analysis of the RASAMAÑJARĪ was repeated using statistical key values that were calculated from randomly selected subsets of the text corpus. Figure 3 displays graphically the influence between the amount of training data and the analysis quality. Although the analysis accuracy obviously increases with the amount of training data, we achieve comparatively high rates of accuracy even when only 20% of the corpus are used for training. Therefore, it seems that substantial increases in the accuracy of the tagger cannot be achieved by increasing the size of the corpus, but only by improving the analysis procedures.

Lexical errors. Errors in the lexical analysis constitute the major part of the errors made by `SanskritTagger`. Since a reduction of this class of errors would strongly increase the accuracy of the tagger, we should detail the subclasses of this kind of error:

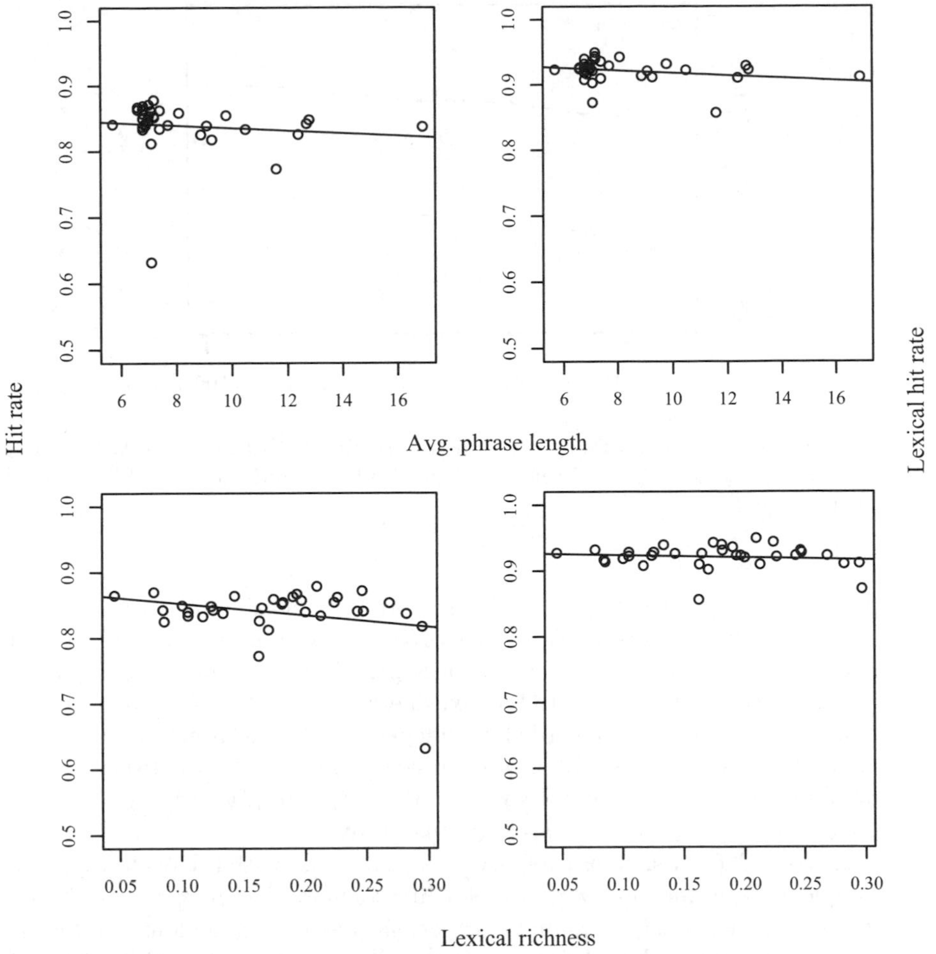

Fig. 2. Correlation between the average phrase length (upper row)/lexical richness (lower row) and the hit rate (left column)/lexical hit rate (right column)

Confusion of (partly) homonymous words. In most cases, homonyms belonging to different declensional classes (e.g., adjectives and nouns) such as *pṛthu* = "broad" and *pṛthu* = "name of a man" are confused in this subclass. One of the most productive pairs of words that belongs to this error class is *tad* = "this" and the suffix *tā*, which denotes abstract nouns. Since the nominative and accusative plural of *tā* are identical with the nominative/accusative plural fem. of the pronoun, the program selects the much more frequent pronoun for the configuration *any word - tāḥ* as long as no overwhelming testimony for the pair *any word - tā* is recorded in the database. Less frequent are confusions of two or more homonymous nouns that are entered as different lexemes in the dictionary. Among these cases, we

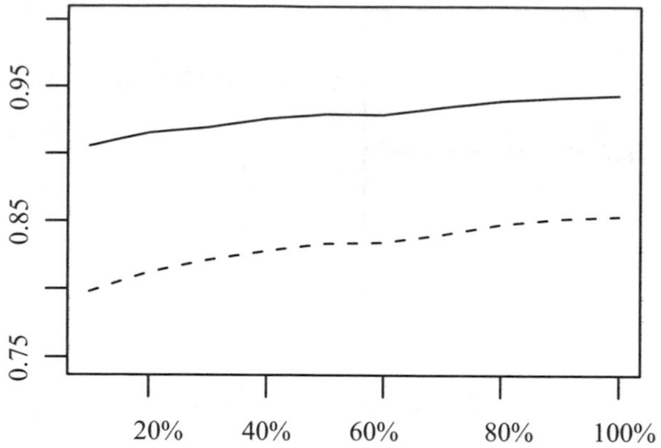

Fig. 3. Accuracy of `SanskritTagger` (y-axis) correlated with the size of the training corpus (x-axis); the solid line indicates the lexical hit rate and the dashed line the hit rate (cmp. page 2)

find, for example, the triple *śiva* (m.) = "the god Śiva", *śiva* (n.) = "bliss", and *śiva* (adj.) = "auspicious", some of whose inflected forms are identical. Another subtype comprised in this class pertains to comparatively few lexemes, but contributes much to the lexical error rate: When inflected forms of nouns or adjectives are recorded as independent lexical items in the dictionary, the lexical analysis often has problems to distinguish between these – finally identical – instances. Examples are *prāya* (m.)/*prāyeṇa* (adv.), *pratyaha* (adj.)/*pratyaham* or *āśu* (adj.)/*āśu* (adv.).

Infinite verbal forms. This category partly coincides with the preceeding one. Frequently, gerundives or participles are included as independent lexical items in the dictionary. Examples are *hāsya* (adj.) = "ridiculous" and *hāsya* as the gerundive of the root *has* or *kartā* = nom. sg. of *kartṛ* (adj.) = "doing" or as the periphrastic future of the verb *kṛ*. The Monier-Williams, from which the lexical database of `SanskritTagger` is built, contains a large number of such lemmatized participles. If the meaning of such lemmata can be derived directly from the meaning of the root (e.g., *saṃjihāna*), these lemmata are removed from the digital dictionary. Although this – still ongoing – manual cleanup is time consuming, it reduces the number of alternatives and thus the possibilities of lexical errors. Some few lemmatized participles have developed independent meanings and are therefore kept in the dictionary (e.g., *vidagdha*= "clever" or *samuddhata* = "arrogant"). In spite of these improvements, there remains a group of high frequency lemmatized participles such as *sat* = "good" or *kṛta* that are responsible for the majority of errors in this class.

Segmentation errors. In this class, two types of errors can be distinguished. The first one is caused by redundancies found in the Monier-Williams. If the dictionary contains both a composite noun and the components constituting it, a substring w_i corresponding to the composite noun can be analyzed in two ways. An example is the term *suhṛdvākya* for which the Monier-Williams gives the meaning "the advice of a friend" and whose components *suhṛd* and *vākya* are also recorded in the dictionary. Obviously, the lemma *suhṛdvākya* is redundant because its meaning can be derived directly from the meaning of its components when standard rules of compound analysis are applied. Similar to most of the lemmatized participles mentioned above, such lemmata should therefore be removed from the digital dictionary. The second class of errors are real missegmentations, which are comparatively rare.

3 Conclusions and Future Developments

To our knowledge, the tests described in this paper have yielded the first reliable estimations of error rates that occur during the automatic processing of Sanskrit using an unrestricted dictionary and texts from diverse knowledge domains. The largest group of errors consists of lexical misinterpretations, which are mostly caused by high frequency homonyms and redundancies in the digital dictionary. While the redundancies will be reduced by an ongoing revision of the dictionary, the homonyms pose a serious challenge to the computational analysis of Sanskrit. Considering the weak correlation between the amount of training data and the accuracy of the tagger, we cannot expect these and similar errors will disappear with an increasing size of the text corpus. Manually designed decision rules or a more detailed context analysis may offer a way out of this problem. When compared with results reported for other languages, the POS error rate of `SanskritTagger` is still too high. A combination of morphological and syntactic analysis steps may help to decrease the frequency of this error; however, the basic steps necessary for such an analysis have first to be implemented in `SanskritTagger`. In conclusion, the results reported in this paper indicate that the three steps of analysis should be merged into one step in future versions of `SanskritTagger`.

References

1. Csernel, M., Patte, F.: Critical edition of Sanskrit texts. In: Huet, G., Kulkarni, A., Scharf, P. (eds.) Sanskrit Computational Linguistics 2007/2008. LNCS (LNAI), vol. 5402, pp. 358–379. Springer, Heidelberg (2009)
2. Halácsy, P., Kornai, A., Oravecz, C., Trón, V., Varga, D.: Using a morphological analyzer in high precision POS tagging of Hungarian. In: Proceedings of 5th Conference on Language Resources and Evaluation (LREC), Citeseer, pp. 2245–2248 (2006)
3. Haque, W., Aravind, A., Reddy, B.: Pairwise sequence alignment algorithms: a survey. In: Proceedings of the 2009 conference on Information Science, Technology and Applications, pp. 96–103 (2009)
4. Hellwig, O.: SanskritTagger, a stochastic lexical and POS tagger for Sanskrit. In: Huet, G., Kulkarni, A., Scharf, P. (eds.) Sanskrit Computational Linguistics 2007/2008. LNCS (LNAI), vol. 5402, pp. 266–277. Springer, Heidelberg (2009)

5. Leech, G., Garside, R., Bryant, M.: CLAWS4: The tagging of the British National Corpus. In: Proceedings of the 15th International Conference on Computational Linguistics, Kyoto pp. 622–628 (1994)
6. Marcus, M.P., Marcinkiewicz, M.A., Santorini, B.: Building a large annotated corpus of English: The Penn Treebank. Computational Linguistics 19(2), 313–330 (1993)
7. Megyesi, B.: Improving Brill's POS tagger for an agglutinative language, pp. 275–284.
8. Sproat, R., Shih, C.: Corpus-based methods in Chinese morphology and phonology. In: Proceedings of the 19th International Conference on Computational Linguistics (2002)
9. Weischedel, R., Schwartz, R., Palmucci, J., Meteer, M., Ramshaw, L.: Coping with ambiguity and unknown words through probabilistic models. Computational Linguistics 19(2), 360–382 (1993)

The Knowledge Structure in Amarakośa

Sivaja S. Nair and Amba Kulkarni

Department of Sanskrit Studies,
University of Hyderabad,
Hyderabad
sivaja.s.nair@gmail.com, apksh@uohyd.ernet.in

Abstract. *Amarakośa* is the most celebrated and authoritative ancient thesaurus of Sanskrit. It is one of the books which an Indian child learning through Indian traditional educational system memorizes as early as his first year of formal learning. Though it appears as a linear list of words, close inspection of it shows a rich organisation of words expressing various relations a word bears with other words. Thus when a child studies Amarakośa further, the linear list of words unfolds into a knowledge web. In this paper we describe our effort to make the implicit knowledge in Amarakośa explicit. A model for storing such structure is discussed and a web tool is described that answers the queries by reconstructing the links among words from the structured tables dynamically.

Keywords: Amarakośa, Synset, Polysemy, Semantic relations, Ontology.

1 Introduction

The Indian tradition of transmitting knowledge orally is on the verge of extinction. As the oral transmission demands, Indian traditional educational culture was organised to be *formal and intensive* as opposed to the modern culture which is more *informal and extensive* (Wood, 1985). In traditional circumstances, a child would receive his education largely by oral transmission, mainly through rote-learning. The method employed was through recitation and remembering. A child is taught the alphabet (varṇamālā), he would memorise a few verses, subhāṣitas, and then start reciting a dictionary of synonymous words – the Amarakośa – till it is memorised. It typically would take anywhere between 6 months to a year to memorise a list of approximately 10,000 Sanskrit words arranged as a list of synonyms. The close inspection of the structure of the Amarakośa gives much more insight into the way the words are organised. When a student memorises it, though in the beginning it appears as a linear list of words, as he starts understanding the meaning of the words, reads the commentaries on this text and starts using these words, the linear structure unfolds into a knowledge web with various links.

The Amarakośa printed in the form of a book just shows the linear order, and the index at the end of the book point to various words for easy references. But there is much more to it than just a linear order. The knowledge a student acquires through various commentaries and also its practical use in his own field of

G.N. Jha (Ed.): Sanskrit Computational Linguistics, LNCS 6465, pp. 173–189, 2010.

expertise – be it *Āyurveda*, *Vyākaraṇa* or *Sāhitya*, is in the form of various links. With the modern education culture that is dominated by the use of computers as a tool, which relies more on the secondary memories such as books, computers, and the World Wide Web, than the human memory, it is necessary to make the *implicit* knowledge in Amarakośa *explicit*. The computers have an advantage over the printed books. Computers can represent multi-dimensional objects, and thus one can navigate through the whole structure and at the same time with the powerful search facilities can search complex queries. In this paper, we illustrate with examples various kinds of links one can 'visualise' in Amarakośa, and provide a database model to store these links in order to facilitate automatic extraction of these links as an answer to a search query.

2 Amarakośa

Amarakośa primarily named *Nāmaliṅgānuśāsana* (a work that deals with instructions related to the gender of nouns) is authored by Amarasiṃha - 4th century A.D. (Oka, 1981) - and is the most celebrated and authoritative ancient thesaurus of Sanskrit with around 60 commentaries and translations into modern Indian as well as foreign languages such as Chinese, Tibetan, French, etc. (Patkar, 1981). It is considered as an essential requisite for a Sanskrit scholar and as such a child is asked to memorise it even before he starts his studies formally. It consists of 1608 verses composed in anuṣṭup meter[1] and are divided into 3 chapters called Kāṇḍas.[2]

Classification. Each of the three *kāṇḍas* is further subdivided into various *vargas*. The classification of three *kāṇḍas* into 25 vargas is as given below.

- ***prathamakāṇḍam*:**
  svargavargaḥ (heaven)
  vyomavargaḥ (sky)
  digvargaḥ (direction)
  kālavargaḥ (time)
  dhīvargaḥ (cognition)
  śabdādivargaḥ (sound)
  nāṭyavargaḥ (drama)
  pātālabhogivargaḥ (nether world)
  narakavargaḥ (hell)
  vārivargaḥ (water)

- ***dvitīyakāṇḍam*:**
  bhūmivargaḥ (earth)
  puravargaḥ (towns or cities)

[1] śloke ṣaṣṭaṃ guruṃ jñeyaṃ sarvatra laghu pañcamam |
dvicatuḥpādayorhrasvaṃ saptamaṃ dīrghamanyayoḥ ||
[2] And as such is known as *Trikāṇḍī*.

śailavargaḥ (mountains)
vanauṣadhivargaḥ (forests and medicines)
siṃhādivargaḥ (lions and other animals)
manuṣyavargaḥ (mankind)
brahmavargaḥ (priest tribe)
kṣatriyavargaḥ (military tribe)
vaiśyavargaḥ (business tribe)
śūdravargaḥ (mixed classes)

– ***tṛtīyakāṇḍam:***
viśeṣyanighnavargaḥ (adjective)
saṁkīrṇavargaḥ (miscellaneous)
nānārthavargaḥ (polysemous)
avyayavargaḥ (indeclinables)
liṅgādisaṅgrahavargaḥ (gender)

Amarakośa contains 11,580 content words (tokens). Some of the tokens are repeated either within a *kāṇḍa* or across the *kāṇḍas* leading to only 9,031 types. The *kāṇḍa*-wise distribution of the tokens and types is shown in Table 1.

Table 1. Tokens and types in each *kāṇḍas*

kāṇḍa	tokens	types
prathamakāṇḍam	2465	2300
dvitīyakāṇḍam	5827	5282
tṛtīyakāṇḍam	3288	2271

Synset. A set of synonymous words is termed as a synset. Each synonym may span over one or more verses. The following verse, e.g., provides a synonym for the word *jaṃbuka*.

striyām śivā bhūrimāyagomāyumṛgadhūrtakāḥ |
sṛgālavañcakakroṣṭupherupheravajambukāḥ ||*2.5.5* ||

Polysemy. Amarakośa has 4,017 synsets. Some of the words fall under more than one synset, and thus are ambiguous. Most of these polysemous words belong to the *nānārthavarga* of the third kāṇḍa which lists the polysemous words alphabetically according to their endings. The polysemy distribution in the Amarakośa is summarised in Table 2. There is only one word *hari* in Amarakośa which has as many as 14 senses, the word *antara* belongs to 13 synsets, and the word *go* has 12 synsets. We note that almost 65% words (7459 words) belong to a single synset and thus are not ambiguous.

Gender. A few verses in the beginning of the Amarakośa describe the metalanguage and the techniques employed to indicate the gender of various words. The word *striyām*, for example, in (2.5.5) above is not a token but a word

Table 2. Polysemy distribution

No. of meanings	No. of words	Words
14	1	*hari*
13	1	*antara*
12	1	*go*
10	2	*kriyā, kūṭa*
9	2	*rasa, vṛṣa*
8	8	*dhātu, dharma, vasu, ariṣṭa...*
7	9	
6	18	
5	49	
4	136	
3	330	
2	1015	
1	7459	

from the meta-language indicating the gender of the following word *śivā* to be feminine. In addition to these general guidelines, in the *liṅgādisaṅgrahavarga* Amarasiṃha gives certain grammatical and phonological clues for deciding the gender of a word. In the event of absence of any rule, the gender of the remaining words in 2.5.5, constituting two compound words "*bhūrimāyagomāyumṛgadhūrta kāḥ*" and "*sṛgālavañcakakroṣṭupherupheravajambukāḥ*" is inferred to be masculine from their compounding-forms.

The avyayavarga lists synsets consisting of indeclinables.

3 Organisation of Synsets within a *Varga*

Except the polysemous words (nānārthavarga), all other synsets in a *varga* show some semantic relation to the *varga* it belongs to and sometimes even to the preceding or following synsets. These semantic relations indicate various kinds of relations. They may be classified as hierarchical or associative. The hypernym indicating a more general term or the hyponym showing a more specific term are the examples of hierarchical relation. Similarly the holonym-meronym relation marking the whole-part relation is also a hierarchical relation. In addition various other relations are indicated by the adjacency of the synsets. These may be termed as associative relations, which indicate some kind of association of one synset with the other. This association may be the association among human beings, or the association of certain objects with certain other objects. We illustrate below some such relations with examples.

3.1 Example 1: Viṣṇuḥ

The verses from 1.1.18 to 1.1.29 describe various synsets representing *Viṣṇu*, and objects related to/associated with *Viṣṇu*. The relations, as is evident from the following description, are kinship relations such as father, brother, son, grandson, wife, and also associated objects such as conch, discus, sword, vehicle, etc. (See Figure 1).

Viṣṇuḥ (1.1.18 - 1.1.22)[3]
 Kṛṣṇa's father (1.1.22)
 Kṛṣṇa's elder brother (1.1.23 - 1.1.24)
 kāmadevaḥ (1.1.25 - 1.1.26)
  floral arrows of kāmadevaḥ (1.1.26)
  physical arrows of kāmadevaḥ (1.1.26)
  son of kāmadevaḥ - aniruddhaḥ (1.1.27)
 wife of Viṣṇuḥ - lakṣmī (1.1.27)
 Special devices/equipments of Viṣṇuḥ(1.1.28)
 (conch, discus, sword, jewel, bow, horse, mark,etc.)
 Kṛṣṇa's charioteer, minister (1.1.28)
 Kṛṣṇa's younger brother (1.1.28)
 Viṣṇu's vehicle - garuḍaḥ (1.1.29)

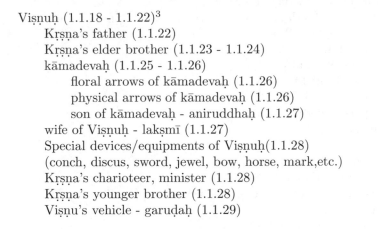

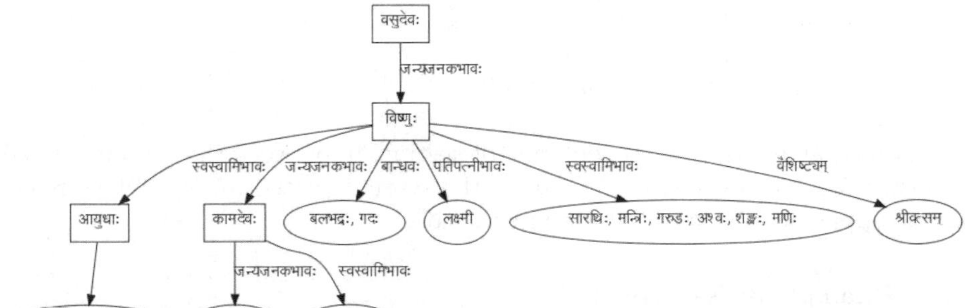

Fig. 1. Relations of Viṣṇu

3.2 Example 2: Samayaḥ

The verses from 1.4.1 to 1.4.9 deal with words related to time, units of measurement, special names of special days, etc.

Time (1.4.1)
 Lunar day (1.4.1)
  First lunar day (1.4.1)
 {Day (1.4.2)
  Morning (1.4.2 - 1.4.3)

[3] The English translations of the subheadings, which are given here and in the following examples, describing the *ślokas* are taken from Colebrooke's commentary on Amarakośa (Colebrooke, 1808).

Twilight (1.4.3)
Evening (1.4.3)
First four hours of a day (1.4.3)
Second four hours of a day (1.4.3)
Third four hours of a day (1.4.3)
Period of the day (1.4.3)
Night (1.4.3 - 1.4.4)
A dark night (1.4.5)
A moonlight night (1.4.5)
A night and two days (1.4.5)
First part of night (1.4.6)
Midnight (1.4.6)
Sequence of nights (1.4.6)
Space of three hours (1.4.6) }
Last day of the half month (1.4.7)
Precise moment of the full or the new moon (1.4.7)
Full moon day (1.4.7)
Full moon whole day(1.4.8)
Full Moon with a little gibbous on part of a day (1.4.8)
No moon day (1.4.8)
wanning crescent (1.4.9)
No moon whole day (1.4.9)

In this example we also see violation of nesting. In between the synsets related to *lunar day* and *last day of the month*, the synsets related to *day* (which refers to the apparent solar motion) are intervened.

3.3 Example 3: Kṣatriyaḥ

Here is a group of verses from 2.8.1 to 2.8.10 belonging to the *kṣatriyavarga*. The words here refer to the king, military, ministers, various category of people engaged in the services of kings, etc.

Man of the military tribe (2.8.1)
King (2.8.1)
Universal monarch (2.8.2)
An emperor (2.8.2)
King over a country (2.8.2)
Paramount sovereign (2.8.3)
Multitude of kings (2.8.3)
Multitude of military tribe (2.8.4)
Minister (2.8.4)
Deputy minister (2.8.4)
Priest (2.8.5)
Judge (2.8.5)
King's companions (2.8.5)

Body guards of a king (2.8.6)
Warder (2.8.6)
Superintendent (2.8.6)
    Village Superintendent (2.8.7)
    Superintendent of many villages (2.8.7)
    Superintendent of Gold (2.8.7)
    Superintendent of Silver (2.8.7)
    Superintendent of the womens' appartments (2.8.8)
        Outside guard of the womens' appartment (2.8.8)
    attendant of a king (2.8.9)
        eunuch (2.8.9)
    Prince whose territories lie on the frontiers of those of the enemy (2.8.9)
    Neighboring prince (2.8.9)
    Prince whose territories lie beyond those of the friend (2.8.10)
    Enemy in the rear (2.8.10)

3.4 Implicit Relations

These were three samples from three distinct topics involving totally different kind of relations. All these relations are semantic in nature. A more detailed study of such examples showed that following relations occur more frequently.

- *avayavāvayavī* (part-whole relation)
- *parāparājāti* (is a kind of relation)
- *janyajanaka* (child-parent relation)
- *patipatnī* (husband-wife relation)
- *svasvāmi* (master-possession relation)
- *ājīvikā* (livelihood)

There are a few other relations such as kinship relations, *ādhāra-ādheya, vaṃśa-vaṃśīya* etc. The extraction of such relations and marking is still ongoing. But the instances of such relations were found to be rare.

4 Amarakośa-jñāna-jāla

In the recent past there have been notable efforts by Sanskrit computational linguists with focus on Amarakośa. Jha et. al. (2010, Online Multilingual Amarakosha) have developed a searchable web interface to Amarakośa which provides the Indian language equivalents of the Amarakośa words in addition to the original sanskrit text. Bharati et. al. (2008) and Nair et. al. (2009) did comparative study of the Amarakośa with the existing Hindi WordNet in order to find the usefulness of Hindi WordNet in augmenting the Amarakośa with relational information. There have been efforts by Kulkarni et. al. (2008, 2010) that describe the development of Sanskrit WordNet. The present effort is altogether an innovative effort that helps reveal the internal structure of the Amarakośa.

The Amarakośa-jñāna-jāla is developed as a web application. The application provides a search result of a query dynamically generated using the structured lexicon of the Amarakośa and the supplementary tables marking the relations.

The structured lexicon as well as the supplementary tables showing the explicit relations are simple ASCII text files. Sanskrit words are stored in a roman transliterated scheme (WX notation).[4] There are two advantages of storing the text in WX notation. The first advantage is, it facilitates a lexicographer to use simple unix tools such as grep, sed, etc. for her day-to-day work of updating the knowledge-base. Unicode for Devanagari mixes the phonemes with the syllables, making it un-natural to write the search expressions. The second advantage, of course, is the size. The size of the tables in UTF-8 for Devanagari is more than 2 times the corresponding files in roman transliteration such as WX notation.

4.1 Structured Lexicon

The main structured lexicon consists of synsets stored in the form of a set of records. Each record corresponds to a word in the Amarakośa (excluding the meta-language words). It consists of 5 fields as described below.

Stem. Amarakośa lists words in nominative cases. However, we decided to go for the nominal stem instead of the nominative case word form. In case of feminine words, this field contains the feminine stem, i.e. the stem after adding the feminine suffix. In case of *nānārthavarga* (part of the Amarakośa dealing with polysemous words), the polysemous word is entered in this field.

The reason for choosing nominal stem over the nominative case form is the ease in linking the Amarakośa words with the existing computational resources such as morphological analysers and generators and various e-lexicons, which typically expect a *prātipadikam* and not a *prathamānta* (ending in nominative case).

Amarakośa index. This field contains a reference to an entry in the Amarakośa, as a 5 tuple of numbers, separated by dots. The 5 numbers in the 5 tuple refer to the *kāṇḍa, varga, śloka, pāda* and the word number respectively. Table 3 shows a sample entry corresponding to the following śloka,

> *svaravyayaṃ svarganākatridivatridaśālayāḥ* |
> *suraloko dyodivau dve striyāṁ klībe triviṣṭapam* || 1.1.6 ||

***Liṅgam* (gender).** This field contains the gender of the stem. The gender of a word in a *śloka* is decided with the help of meta-language employed by Amarakośa. These are further cross checked with Devadatta Tiwari's *Devakośa arthāt Amarakośa* (Tiwari, 1989) and Colebrooke's commentary on Amarakośa (Colebrooke, 1808) when in doubt.

Sanskrit has 3 values for gender viz. masculine, feminine and neuter. Thus there are 8 possible combinations (an indeclinable is assigned no gender, and

[4] http://sanskrit.uohyd.ernet.in/~anusaaraka/sanskrit/samsaadhanii/wx.html

Table 3. Words and references of the *svarga*-synset

Word	**Reference**
svar	1.1.6.1.1
svarga	1.1.6.1.2
nāka	1.1.6.1.3
tridiva	1.1.6.1.4
tridaśālaya	1.1.6.1.5
suraloka	1.1.6.2.1
dyo	1.1.6.2.2
div	1.1.6.2.3
triviṣṭapa	1.1.6.2.4

the adjectives are the ones which take all the three genders). In addition, Amarakośa also provides information about words that are always plural or dual by nature. Following combinations of gender, number information were found in the Amarakośa.

- Indeclinable - (*avya.*)
- Feminine - (*strī.*)
- Masculine - (*puṃ.*)
- Neuter - (*napuṃ.*)
- Masculine and Feminine - (*strī-puṃ.*) [aśani[5]]
- Feminine and Neuter - (*strī-napuṃ.*) [uḍu[6]]
- Feminine dual - (*strī-dvi.*) [dyāvāpṛthvyau[7]]
- Feminine plural - (*strī-bahu.*) [apsaras[8]]
- Masculine and Neuter - (*puṃ-napuṃ.*) [daivatāni[9]]
- Masculine dual - (*puṃ-dvi.*) [nāsatyau[10]]
- Masculine plural - (*puṃ-bahu.*) [gṛhāḥ[11]]
- Neuter and indeclinable - (*napuṃ-avya.*) [apadiśaṃ[12]]
- Adjective - (*vi.*)

Vargaḥ. This field contains the name of the *varga*, as given in the commentaries to which the entry belongs.

Head Word. The first four fields cover all the explicit information that can be easily extracted automatically. The important feature of Amarakośa is that it provides synonymous words. The marking of synonymous words is obvious only

[5] aśanirdvayoḥ (1.1.47)
[6] tārakāpyuḍu vā striyāṃ (1.3.21)
[7] dyāvāpṛthvyau (2.1.19)
[8] striyāṃ bahuṣvapsarasaḥ (1.1.52)
[9] daivatāni puṃsi vā (1.1.9)
[10] nāsatyāvaśvinau dasrāvāśvineyau ca tāvubhau (1.1.51)
[11] gṛhāḥ puṃsi ca bhūmnyeva (2.2.5)
[12] klībāvyayaṃ tvapadiśam (1.3.5)

through the world knowledge or through the commentaries. To provide a handle to each set of synonymous words – called as synset, we created a field termed as *Head Word* which provides a name to each synset. Thus these Head Words are unique and act as a reference ID for a synset. The total number of Head Words give us the total number of synsets in the Amarakośa. We denote the synset corresponding to a Head Word *W* by *Syn(W)*.

The choice of Head-Words is mainly guided by the Bhānuji Dīkṣitā's *Sudhā* commentary on *Amarakośa* (Pandit, 1915). When a better choice was available in the Malayalam commentary *Triveṇī* (Moosath, 1956) or *Pārameśvarī* (Moosath, 1914) or the Hindi commentary *Prabhā*, it was chosen. Table 4 shows an example of a śloka 2.5.5 converted to a structured table, and figure 2 shows the search result of the Amarakośa-jñāna-jāla for the word śṛgāla.

4.2 Tables Marking Various Relations

The relations are among various Head Words and are marked as records. Each record corresponds to one synset ID. The first field of each record consists of the synset ID, and remaining six fields correspond to the Head Words that bear a relation of *is_a_part_of (avayavāvayavi)*, *is_a_kind_of(parāparājāti)*, *janya-janaka-bhāva*, *pati-patnī-bhāva*, *sva-svāmi-bhāva*, *ājīvikā* with the synset ID in the first field.

Table 4. Example of Head-Word

Token	Reference	Gender	Varga-name	Head-Word
śivā	2.5.5.1.1	strī	siṃhādivargaḥ	jaṃbhūkaḥ
būrimāya	2.5.5.1.2	puṃ	siṃhādivargaḥ	jaṃbhūkaḥ
gomāyu	2.5.5.1.3	puṃ	siṃhādivargaḥ	jaṃbhūkaḥ
mṛgadūrtaka	2.5.5.1.4	puṃ	siṃhādivargaḥ	jaṃbhūkaḥ
śṛgāla	2.5.5.2.1	puṃ	siṃhādivargaḥ	jaṃbhūkaḥ
vañjaka	2.5.5.2.2	puṃ	siṃhādivargaḥ	jaṃbhūkaḥ
kroṣṭu	2.5.5.2.3	puṃ	siṃhādivargaḥ	jaṃbhūkaḥ
pheru	2.5.5.2.4	puṃ	siṃhādivargaḥ	jaṃbhūkaḥ
pherava	2.5.5.2.5	puṃ	siṃhādivargaḥ	jaṃbhūkaḥ
jaṃbuka	2.5.5.2.6	puṃ	siṃhādivargaḥ	jaṃbhūkaḥ

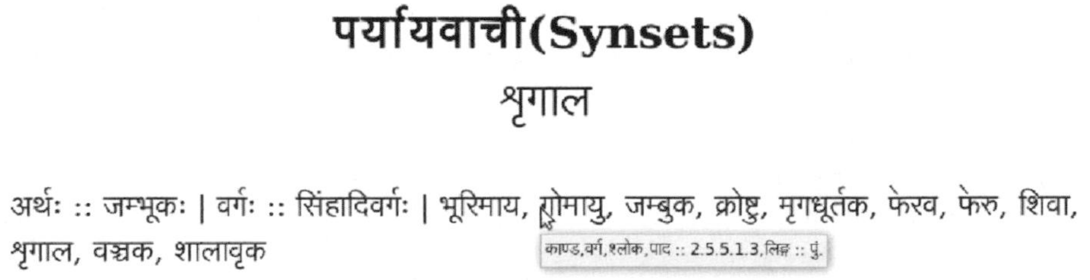

Fig. 2. Example of a synset

1. Is a part of (*avayavāvayavi*)
   This field marks *is a part of* relation. Let *W* be the synset-ID. Then this field will have an entry *W'* if the member of *SynW* is a part of member of *SynW'* (See Table 5).

   For example,
   Syn(*rātriḥ*)=*śarvarī, kṣaṇadā, kṣapā, niśā, niśīthinī, rajanī, rātri, vibhāvarī, tamasvinī, tamī, triyāmā, yāminī, naktam, doṣā, vasati, śyāmā.*
   and
   Syn(*rātrimadhyaḥ*) = *ardharātra, niśītha.*

   Now, *ardharātra, niśītha* are part of *niśā, rajanī, rātri*, etc.. Hence *rātrimadhyaḥ* is marked to be is_a_part_of (*avayava* of) *rātriḥ*
   Similarly *pradoṣa, rajanīmukha* (∈ *Syn*(*rātriprāraṃbhaḥ*)) are also part of *niśā, rajanī, rātri*, etc.. Hence *rātriprāraṃbhaḥ*, where Syn (*rātriprāraṃbhaḥ*)= *pradoṣa, rajanīmukha*
   also bears a part_of relation with *rātriḥ*.

   Table 5. Example of is-a-part relation

Head-Word W	**part (*avayava*)-of W**
rātrimadhyaḥ	rātriḥ
rātriprāraṃbhaḥ	rātriḥ

2. Is a kind of (*parāparājāti*)
   This field marks *is a kind of* relation. The entry contains the Head Word W' such that synset ID W bears a relation of *is a kind of* with W'. The hypernymy and hyponymy relation can be extracted using this field. Here are some entries: (see Table 6.)

   Table 6. Example of is_a_kind_of relations

Head-Word W	**kind (*parājātiḥ*) of W**
gaṅgā	nadī
yamunā	nadī
narmadā	nadī

3. *Janya-janaka-bhāva* (parent-child relation)
   This field marks the relation of parent-child (*janya-janaka-bhāva*). (see Table 7.) Where Syn (*jayantaḥ*) = *pākaśāsani, jayanta.*
   and
   Syn (*indraḥ*) = *indra, biḍaujas, maghavan, marutvat, pākaśāsana, sunāsīra, vṛddhaśravas, purandara, puruhūta, jiṣṇu, lekharṣabha, śakra, śatamanyu, divaspati, vṛṣan, vṛtrahan, gotrabhid, sutrāman, vāsava, vajrin, balārāti,*

śacīpati, surapati, vāstoṣpati, harihaya, jambhabhedin, namucisūdana, svarāj, meghavāhana, saṅkrandana, turāṣā, duścyavana, ākhaṇḍala, ṛbhukṣin, sahasrākṣa, kauṣika, ghanāghana, parjanya, hari.

Syn (*sanatkumāraḥ*) = *sanatkumāra, vaidhātra*
and
Syn (*brahmā*) = *ātmabhū, brahman, caturānana, hiraṇyagarbha, lokeśa, parameṣṭhin, pitāmaha, surajyeṣṭha, svayambhū, abjayoni, aṇḍaja, haṁsavāhana, kamalāsana, kamalodbhava, nābhijanman, nidhana, prajāpati, pūrva, rajomūrtin, satyaka, sadānanda, svaṣṭṛ, vedhas, viriñci, viśvasṛj, vidhātṛ, vidhi, dhātṛ, druhiṇa, ka, ātman, śambhu.*

Table 7. Example of *Janya-janaka* relation

Head-Word W	**Child (*janya*) of W**
indraḥ	jayantaḥ
brahmā	sanatkumāraḥ
śivaḥ	gaṇeśaḥ

4. *Pati-patnī-bhāva* (husband-wife relation)
This field marks the husband-wife relation, as shown below. (see Table 8.) Where Syn(*lakṣmī*) = *bhārgavī, haripriyā, indiraā, kamalā, kṣīrasāgarakanyakā, kṣīrodatanayā, lakṣmī, lokajananī, lokamātṛ, mā, padmā, padmālayā, ramā, śrī, vṛṣākapāyī.*
and
Syn(*viṣṇuḥ*) = *hṛṣīkeśa, keśava, kṛṣṇa, mādhava, nārāyaṇa, svabhū, vaikuṇḍha, viṣnu, viṣṭraśravas, dāmodara, acyuta, garuḍadhvaja, govinda, janārdana, pītāmbara, puṇḍarīkākṣa, śārṅgin, viṣvaksena, daityāri, cakrapāṇi, caturbhuja, indrāvaraja, madhuripu, padmanābha, upendra, vāsudeva, trivikrama, adhokṣaja, balidhvaṁsin, kaṁsārāti, puruṣottama, śaurī, śrīpati, vanamālin, xevakīnandana, jalaśāyin, kaiṭabhajit, mukunda, muramardana, narakāntaka, purāṇapuruṣa, śrīvatsalāñchana, viśvambhara, viśvarūpa, vidhu, yajñapuruṣa, lakṣmīpati, murāri, aja, ajita, avyakta, vṛṣākapi, babhru, hari, vedhas.*

Table 8. Example of *Pati-patnī* relation

Head-Word W	**Husband (*pati*) of W**
lakṣmī	viṣṇuḥ
pārvatī	śivaḥ
lopāmudrā	agastyaḥ

Table 9. Example of *Sva-svāmi* relation

Head-Word W	**master (*svāmi*) of W**
viṣṇoḥ mantriḥ	viṣṇuḥ
viṣṇoḥ sārathiḥ	viṣṇuḥ
garuḍaḥ	viṣṇuḥ

5. *Sva-svāmi-bhāva* (master-possession relation)
   This field marks the master-possession or sva-svāmi-bhāva relation as shown below: (see Table 9.)
6. Ājīvikā (livelihood)
   This field marks the livelihood relation between two syn-sets. For example, the synset with Head Word *matsya* is (*aṇḍaja*, *jhaṣa*, *matsya*, *mīna*, *pṛthuroman*, *śakulī*, *vaisāriṇa*, *visāra*, *animiṣa*) denotes objects which act as a livelihood for the objects expressed through the concept of *dhīvara*, and hence the livelihood for the objects belonging to the synset *dhīvara* is marked as a *matsya*. (see Table 10.)

Table 10. Example of *Ājīvikā* relation

Head-Word W	**Livelihood (Ājīvikā) of W**
dhīvaraḥ	matsyaḥ
nartakī	nṛtyam
nāvikaḥ	naukā
sevakaḥ	sevā

4.3 Quantitative Analysis

For every headword, one or more of the relations as specified above are marked. As was expected, the hierarchical relations viz. is_a_kind_of and is_a_part_of appear prominently than the associative relations. The occurrence of various relations in terms of Head-Words and all the words belonging to the synsets denoted by these head words is shown in Table 11[13].

4.4 Implementation

From the structured lexicon table and the table of relations we build data bases using the built-in dbm engines of unix and the programes are written in Perl. These dbm engines use hashing techniques to enable fast retival of the data by key.

Following three hash tables are built from the structured lexicon.

a) Head-word hash
where Key=stem and Value=head-word

[13] Till 16th April 2010

Table 11. Relational statistics

No.	Relation	Headwords	Words
1	is_a_kind_of	2239	6807
2	is_a_part_of	560	1654
3	janya-janaka	17	193
4	sva-svāmī	36	122
5	ājīvikā	30	106
6	pati-patnī	25	105

b) Synset hash
with Key=head-word and Value=synset

c) Word-info hash
generated by Key=stem and Value=word-index and gender

From the table of relations, corresponding to each relation R, we built a hash table which returns the associates a head-Word W with another head-word W', if W' is related to W by relation R.

Amarakośa-jñāna-jāla is presented as a web application developed with 'apache' web server and 'perl' for CGI script. User submits a query a word and a relation, machine produces all the words related to the given word by the chosen relation. The word here may be either a stem or an inflected word form. In the case of inflected word form, machine consults the morphological analyser to get the stem. Figures in appendix - 1 give sample results of queries for different word-relation combinations. When a cursor is placed on a word a tool tip shows its word-index and gender(as shown in Fig. 2.).

5 Conclusion

The study of Amarakośa from a point of view of exploring the relations was undertaken to reveal the implicit knowledge and make it explicit. The resulting computational tool helps a Sanskrit reader to get a feel for various kinds of relations mentioned in the Amarakośa and thereby its richness as a knowledge source. The hierarchical relations such as is_a_part_of and is_a_kind_of will be of help in information extraction, while the associative relations help a reader to get the cultural knowledge.

Sanskrit has a rich tradition of kośas. Most of them are arranged as a list of words with similar meaning (synonymic) or a list of words indicating various shades of a given word (polesemic). *Nāmamālā, Śabdaratnākara, Śabdacandrikā* are a few among the first type and *Nānārthasaṅgraha, Anekārthadhvanimañjarī, Viśvaprakāśa* are a few examples of the second type. *Amarakośa, Abhidhānaratnamālā* and *Vaijayantīkośa* has both kind of entries.

This implementation may serve as a model to build similar tools for various other kośas mentioned above.

The Amarakośa is now available with various kinds of search facilities as a web service at

`http://sanskrit.uohyd.ernet.in/~anusaaraka/sanskrit/samsaadhanii/amarakosha/home.html`

References

1. Bharati, A., Kulkarni, A., Nair, S.S.: 'Use of Amarakosha and Hindi WordNet in Building a Network of Sanskrit Words. In: International Conference On Natural Language Processing, C-DAC Pune (2008)
2. Colebrooke, H.T.: Kosha or Dictionary of the sungskrita language by Umura singha with an English interpretations and annotations. Serampore (1808)
3. Fellbaum, C.: WordNet An Electronic Lexical Database. MIT Press, Massachusetts (1999)
4. Jha, G.N., Chandrashekar, R., Singh, U.K., Jha, V.N., Pandey, S., Singh, S.K., Mishra, M.K.: Online Multilingual Amarakośa: The Relational Lexical Database. In: 5th Global Wordnet Conference, IIT Mumbai (2010)
5. Kulkarni, M., Pushpak, B.: Verbal roots in the Sanskrit WordNet. In: 2nd International Sanskrit Computational Linguistics Symposium, Brown University (2008)
6. Kulkarni, M., Dangarikar, C., Kulkarni, I., Nanda, A., Bhattacharya, P.: Introducing Sanskrit Wordnet. In: 5th Global Wordnet Conference, IIT Mumbai (2010)
7. Moosath, P.T.C.: Paarameswarii: malayalam commentory of Amarakosha. National Book Stall, Kottayam (1914)
8. Moosath, P.T.C.: Triveni: malayalam commentory of Amarakosha. National Book Stall, Kottayam (1956)
9. Nair, S.S., Swain, P., Kulkarni, A.: Developing network of Sanskrit words across Part-Of-Speech categories. In: CSATS 2009, Rashtriya Sanskrit Vidhyapeeth, Tirupati (2009)
10. Oka, K.G.: The Namalinganushasana: Amarakosa of Amarasimha with the commentary (Amarakoshodghaatana) of Ksheraswamin. Upasana prakashan Delhi, Varanasi (1981)
11. Online Multilingual Amarakosha: `http://sanskrit.jnu.ac.in/amara/index.jsp`
12. Patkar, M.M.: History of Sanskrit Lexicography. Munshiram Manoharlal Publishers Pvt. Ltd., Delhi (1981)
13. Pandit, S.: Namalinganusasana (Amarakosha) With the commentary of (Vyakhyasudha or Ramasrami) of Bhanuji Dikshit. Tukaram javaji proprietor of the "Nirnaya Sagar" Press, Bombay (1915)
14. Tiwari, D.: Devakośaḥ ardhāt amarakośaḥ. Nag prakashak, Delhi (1989)
15. Wood, A.: Knowledge before printing and after – The Indian tradition in changing Kerala. Oxford University Press, Delhi (1985)

A Appendix - 1

आजीविका

धीवर

अर्थः :: धीवरः | वर्गः :: वारिवर्गः | कैवर्त, दाश, धीवर, जालिक

आजीविका

अर्थः :: मत्स्यः | वर्गः :: वारिवर्गः | अण्डज, झष, मत्स्य, मीन, पृथुरोमन्, शकुली, वैसारिण, विसार, अनिमिष

Fig. 3. Example of *ājīvikā*

अवयवी(Holonymy)

अङ्गुली

> अर्थः :: अङ्गुली | वर्गः :: मनुष्यवर्गः | अङ्गुली, करशाखा, कर्णिका
>अर्थः :: हस्तः | वर्गः :: मनुष्यवर्गः | पञ्चशाख, पाणि, शय, हस्त, कर
>अर्थः :: भुजः | वर्गः :: मनुष्यवर्गः | बाहु, भुज, प्रवेष्ट, दोस्
>अर्थः :: देहः | वर्गः :: मनुष्यवर्गः | गात्र, कलेवर, संहनन, शरीर, वपुस्, वर्ष्मन्, विग्रह, काय, मूर्ति, तनु, तनू, देह, करण, उत्सेध, भूतात्मन्, आत्मन्, धामन्, क्षेत्र, अजिर

Fig. 4. Example of *avayavī*

अवयवः(Meronymy)

देह

अर्थः :: देहः | वर्गः :: मनुष्यवर्गः | गात्र, कलेवर, संहनन, शरीर, वपुस्, वर्ष्मन्, विग्रह, काय, मूर्ति, तनु, तनू, देह, करण, उत्सेध, भूतात्मन्, आत्मन्, धामन्, क्षेत्र, अजिर

अवयवाः

अर्थः :: चक्षुरादीन्द्रियम् | वर्गः :: धीवर्गः | हृषीक, इन्द्रिय, विषयी, ख
अर्थः :: पाय्वादीन्द्रियम् | वर्गः :: धीवर्गः | कर्मेन्द्रिय
अर्थः :: चक्षुरादीन्द्रियम् | वर्गः :: धीवर्गः | हृषीक, इन्द्रिय, विषयी, ख
अर्थः :: गर्भवेष्टनचर्मः | वर्गः :: मनुष्यवर्गः | गर्भाशय, जरायु
अर्थः :: शुक्लशोणितसम्पातः | वर्गः :: मनुष्यवर्गः | कलल, उल्ब
अर्थः :: कुक्षिस्थगर्भः | वर्गः :: मनुष्यवर्गः | भ्रूण, गर्भ
अर्थः :: कृष्णवर्णदेहगतचिह्नः | वर्गः :: मनुष्यवर्गः | जड्ल, कालक, पिप्लु

Fig. 5. Example of *avayava*

पराजातिः (Hypernymy)

गङ्गा

>अर्थः :: गङ्गा | वर्गः :: वारिवर्गः | सुरनिम्नगा, गङ्गा, जह्नुतनया, विष्णुपदी, भागीरथी, भीष्मसू, त्रिपथगा, त्रिस्रोतस्
>अर्थः :: नदी | वर्गः :: वारिवर्गः | नदी, सरित्, आपगा, ह्रादिनी, निम्नगा, शैवलिनी, स्रवन्ती, स्रोतस्विनी, तरङ्गिणी, तटिनी, धुनी, द्वीपवती, कूलङ्कषा, निर्झरिणी, रोधोवक्रा, सरस्वती, भोगवती, सिन्धु, वाहिनी
>अर्थः :: तटागादयः | वर्गः :: वारिवर्गः | जलाशय, जलाधार

Fig. 6. Example of hypernymy

अपराजातिः (Hyponymy)

नदी

अर्थः :: नदी | वर्गः :: वारिवर्गः | नदी, सरित्, आपगा, ह्रादिनी, निम्नगा, शैवलिनी, स्रवन्ती, स्रोतस्विनी, तरङ्गिणी, तटिनी, धुनी, द्वीपवती, कूलङ्कषा, निर्झरिणी, रोधोवक्रा, सरस्वती, भोगवती, सिन्धु, वाहिनी

अपराजातिः

अर्थः :: देवगङ्गा | वर्गः :: स्वर्गवर्गः | सुरदीर्घिका, मन्दाकिनी, स्वर्णदी, वियद्गङ्गा
अर्थः :: नरकस्थ नदी | वर्गः :: नरकवर्गः | वैतरणी
अर्थः :: गङ्गा | वर्गः :: वारिवर्गः | सुरनिम्नगा, गङ्गा, जह्नुतनया, विष्णुपदी, भागीरथी, भीष्मसू, त्रिपथगा, त्रिस्रोतस्
अर्थः :: यमुना | वर्गः :: वारिवर्गः | कालिन्दी, शमनस्वसृ, सूर्यतनया, यमुना
अर्थः :: नर्मदा | वर्गः :: वारिवर्गः | मेखलकन्यका, नर्मदा, रेवा, सोमोद्भवा
अर्थः :: कार्तवीर्यावतारित नदी | वर्गः :: वारिवर्गः | बाहुदा, सैतवाहिनी
अर्थः :: गौरीविवाहे कन्यादानोदकाज्जातनदी | वर्गः :: वारिवर्गः | करतोया, सदानीरा

Fig. 7. Example of hyponymy

Gloss in Sanskrit Wordnet

Malhar Kulkarni, Irawati Kulkarni, Chaitali Dangarikar,
and Pushpak Bhattacharyya

IIT Bombay, Mumbai, India
{malhar,irawati,chaitali,pb}@iitb.ac.in

Abstract. Glosses and examples are the essential components of the computational lexical databases like, Wordnet. These two components of the lexical database can be used in building domain ontologies, semantic relations, phrase structure rules etc., and can help automatic or manual word sense disambiguation tasks. The present paper aims to highlight the importance of gloss in the process of WSD based on the experiences from building Sanskrit Wordnet. This paper presents a survey of Sanskrit Synonymy lexica, use of Navya-Nyāya terminology in developing a gloss and the kind of patterns evolved that are useful for the computational purpose of WSD with special reference to Sanskrit.

Keywords: Sanskrit Wordnet, Wordnet, Word Sense Disambiguation, Lexical databases, Sanskrit Lexicography.

1 Introduction

Wordnet is a lexical database. Words belonging to four lexical categories, i.e., noun, adjective, adverb and verb are stored in this database. Words are grouped in such a way that their cognitive synonym will form a set and this set will express a distinct concept. This set of cognitive synonyms is called '*synset*'. Synsets in a wordnet are interlinked by means of conceptual-semantic and lexical relations. The resulting network of meaningfully related words and concepts can be navigated with the browser.

The first wordnet was built in 1985 for English language.[1] Then followed wordnets for European languages: EuroWordNet.[2] Since 2000, wordnets for a

[1] WordNet® as a lexical database for English was developed at Princeton University under the direction of George A. Miller. The word "WordNet" (with capital N) is a registered trademark of Princeton English WordNet. Therefore, names of all other wordnets are written without capital N in it. WordNet's latest version is 3.0. As of 2006, the database contains 155,287 words organized in 117,659 synsets for a total of 206,941 word-sense pairs. In a compressed form, WordNet database is about 12 megabytes in size. It is available for free online use at `http://wordnetweb.princeton.edu/perl/webwn` and can also be downloaded from `http://wordnet.princeton.edu/wordnet/download/`

[2] `http://www.illc.uva.nl/EuroWordNet/`

G.N. Jha (Ed.): Sanskrit Computational Linguistics, LNCS 6465, pp. 190–197, 2010.

number of Indian languages are getting built, led by Hindi Wordnet[3] effort at Indian Institute of Techonlogy Bombay (IITB). All wordnets in the world are connected through Global Wordnet Association (GWA).[4] Similar activity started in India with Hindi Wordnet. It was started in IIT Bombay and is going on for last eight years. As of now, wordnets are getting created for 16 out of 22 official languages of India.[5] IndoWordNet group is formed and wordnets for these 16 Indian languages are getting interlinked via Hindi Wordnet [3].

In this paper, we share our experience of building Sanskrit wordnet. Since it is built following expand model of wordnet creation, Sanskrit lexicographers create synsets based on base synsets of Hindi Wordnet. While creating synsets, lexicographers collect synonyms from traditional Sanskrit kośas and modern Sanskrit dictionaries. Kośas like *Śabdakalpadruma* and *Vācaspatyam* contain Sanskrit glosses. These glosses are very much helpful in creating glosses in Sanskrit. But the etymology based approach of these traditional lexica can be very harmful in this ontology based lexical structure. Since glosses attached to the synsets can become helpful in disambiguating polysemous words, Sanskrit lexicographers take help from already existing wordnets and extract information from them to make Sanskrit glosses more refined and apt. We also show how glosses, given in such a way can be helpful in disambiguating Sanskrit words in the context. Our approach draws a large amount of inputs from the challenegs IITB researchers faced while doing wordnet based word sense disambiguation task for Hindi and Marathi languages. We compare Hindi, Marathi and English glosses to make Sanskrit glosses more appropriate by adding distinct features of the concept in the gloss.

2 Importance of Gloss in Wordnet Structure

Synsets are the basic blocks of Wordnet. The member words of a synset perform the task of disambiguating the co-occurred words (which have multiple meanings). These words are helped in this task to a large extent by the glosses provided of the synsets. These glosses aim to capture the concept for which the words in the synset stand for. Thus they play a very crucial role from the structure of the Wordnet in particular and Word Sense Disambiguation (WSD) in general.

In general we attempt to study the various facets of this aspect of the Wordnet, namely 'The Gloss' on the basis of the experience in building Sanskrit Wordnet with expansion approach [4]. Further, in future, we aim to concentrate on the gloss of the verbs as given in the Sanskrit Wordnet and study the implications from WSD point of view.

---

[3] `http://www.cfilt.iitb.ac.in/wordnet/webhwn/`

[4] List of all wordnets can be found at `http://www.globalwordnet.org/gwa/wordnet_table.htm`

[5] These languages are (1) Hindi, (2) Marathi, (3) Konkani, (4) Sanskrit, (5) Nepali, (6) Kashimiri, (7), Assamese, (8) Tamil, (9) Malyalam, (10) Telugu, (11) Kannad, (12) Manipuri, (13) Bodo, (14) Bangla, (15) Punjabi and (16) Gujarati.

2.1 Importance of Gloss in Wordnet from Computational Point of View

Computationally, the main importance of *glosses* attached to synsets is Word Sense Disambiguation. WSD is the problem of identifying the correct sense of the word given the context in which it appears. Many words have multiple senses. Humans can identify the correct sense of the word from the context. For example, when we encounter a sentence like *shall we go to the bank for fishing*, we are able to recognize that the meaning of **bank** is *river bank* and not *financial institution* from the contextual clue word **fishing**. How can the computer do the same task?

WSD is important for any Natural Language Processing (NLP) application that involves understanding the meaning of words. These include Machine Translation (MT), Information Extraction (IE), Information Retrieval (IR), document summarization, question answering, *etc.* For example, in IR we can increase the precision of the retrieved documents by understanding the sense (meaning) of the words in user queries and retrieving documents pertaining to that sense.

WSD approaches can be divided into two broad categories: *knowledge based* and *machine learning based.* The latter again can be divided into *supervised machine learning* and *unsupervised machine learning.* We will describe the first approach, *viz.*, knowledge based approach which makes heavy use of glosses.

2.2 Knowledge Based Approach to WSD

Knowledge based approaches such as WSD using Selectional Preferences [8], Lesk's algorithm [5], Walker's algorithm [10], WSD using conceptual density [1] and WSD using Random Walk Algorithm [6] make use of dictionary definitions or **glosses**. They are easy to implement requiring only a lookup of a knowledge resource like a Machine Readable Dictionary (MRD). When the MRD is the wordnet, this lookup takes the form of consulting or overlapping **gloss of synsets, synsets** and **recursively for semantically related synsets**. Further, they do not require any corpus- tagged or untagged.

Illustration of the algorithm. Following Lesk (1986), we give an intersection similarity based WSD approach for Hindi WSD using Hindi Wordnet [7].[6] For the word to be disambiguated, we call the collection of words from the CONTEXT of the word in question the *Context Bag* and the related words from the WORDNET as the *Semantic Bag.* Fig. 1 gives the pictorial view of the approach and table 1 gives the algorithm used in this approach.

The word 'bag' here means a set of words. 'Context bag' means the words in the context of the word that is candidate for disambiguation. Sense bag, on the other hand, means words which have various lexical and semantic links with the candidate word, as depicted in the wordnet. This would consist of synonyms, hypernyms, meronyms, words in gloss and so on.

We have used Intersection similarity to measure the overlap. The idea of Intersection similarity is to capture the belief that there will be high degree of

[6] <http://www.cfilt.iitb.ac.in/wordnet/webhwn>

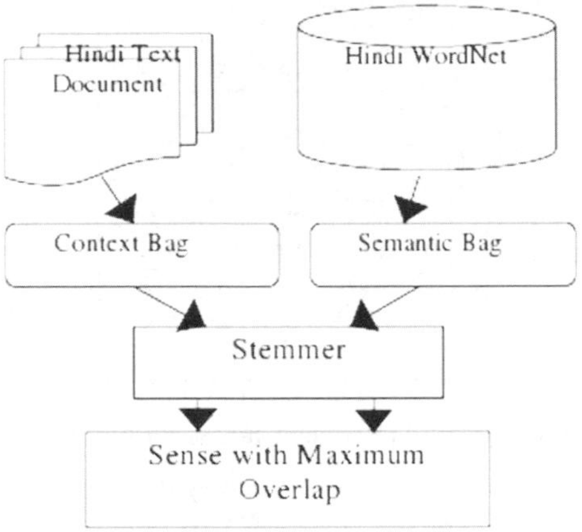

Fig. 1. Overlap of the Context Bag (words in the context) and Dictionary Bag (words of gloss and synsets

Table 1. Lesk like algorithm for WSD

1. For a polysemous word W needing disambiguation, a set of context words in its surrounding *window* is collected. Let this collection be **C**, *the context bag.* The *window* is the current sentence and the *preceeding* and the *following* sentences.
2. For each sense s of W, do the following.
   Let **B** be the bag of words obtained from the
   a. Synonyms in the synset.
   b. Glosses of the synsets.
   c. Example sentences of the synsets.
   d. Hypernyms (recursively up to the roots).
   e. Glosses of Hypernyms.
   f. Example sentences of Hypernyms (recursively up to the leaf).
   g. Glosses of Hyponyms.
   h. Example sentences of Hyponyms.
   i. Meronyms (recursively up to the beginner synset).
   j. Glosses of Meronyms.
   k. Example sentences of Meronyms.
   Measure the overlap **C** and **B** using intersection similarity.[7]

overlap between the words in the context and the *related words* extracted from the WordNet for a sense, and that sense will be a winner sense.

Evaluation. We perform Hindi WSD experiment on the corpora provided by Central Institute of Indian Languages (CIIL) , Mysore [9]. The system has been

[7] The idea of intersection similarity is to capture the belief that there will be high degree of overlap between words in a context and the *related words* extracted from wordnet for a sense, and that sense will be a winner sense [9].

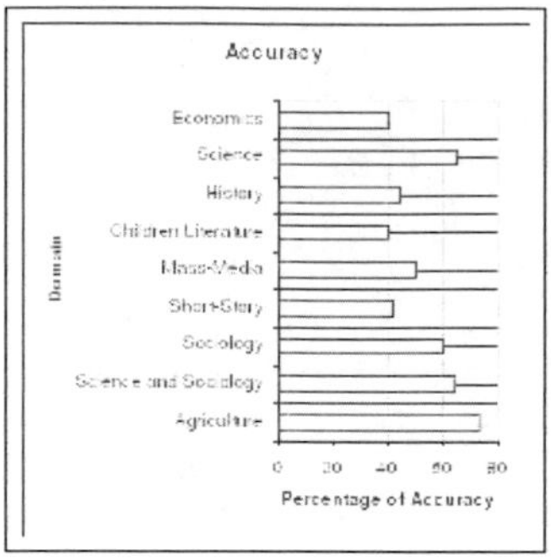

Fig. 2. Histogram showing the WSD accuracy across domains for Hindi Words

Table 2. WSD accuracy across domains for Hindi Words

Domain	Percentage of Accuracy
Agriculture	73.20
Science and Sociology	64.40
Sociology	60.22
Short-Story	42.22
Mass-Media	50.11
Children Literature	40.00
History	44.44
Science	65.00
Economics	40.55

tested on nouns. The domains of the experiment and the accuracy values are mentioned in the table 2. The histogram of figure 2 shows the WSD accuracy across domains.

In this paper an attempt is made to study gloss as designed in Sanskrit Wordnet. Thus we try to see if glosses in Sanskrit Wordnet provide us with surface cues that help get the selectional preferences of different senses of a word.We describe below the glosses of a polysemous noun "*hari*" in Sanskrit. Below we discuss the structure of gloss and application of the algorithm.

Strcture of a Gloss. Verb sense definitions are, in general, infinitive verb phrases with adverbials (often prepositional phrases) and additional restrictions on the semantic class of agents and objects [2]. Table 3 shows how structures can be derived from verb sense definition: Similarly, we can form structures using the *Dhātupāṭha* meanings as the class. For example table 4 shows the structure in the gloss of the verbal root *cup*. In such classification, Navya-Nyāya terminology can be helpful for Sanskrit Wordnet. Therefore, an attempt is made to construct verbal glosses using some structures and combining these structures by either Navya-Nyāya terms or by use of certain cases-endings; like instrumental for instruments involved in performing that action, locative for the situation or the condition in which that action takes place.

Table 3. Structures in verb sense definition (Alshawi, 1987)

launch) (to send (a modern weapon or instrument) into the sky or space by means of scientific explosive apparatus) ((CLASS SEND) (OBJECT ((CLASS INSTRUMENT) (OTHER-CLASSES (WEAPON)) (PROPERTIES (MODERN)))) (ADVERBIAL ((CASE INTO) (FILLER (CLASS SKY))))

Table 4. Structure in the sense definition of the verb *cup gatau*

((CLASS *gati*) ((CLASS *gati*) (Manner *manda*) (AGENT (CLASS ANIMATE)))

The motivation behind systematizing glosses came from lexicographer's experience of understanding Hindi Wordnet (HWN) glosses in the expansion approach. Many times, when HWN glosses were not clearly understood, Sanskrit lexicographers looked for the glosses of the same concept in other wordnets; i.e., English WorNet (EWN) and Marathi Wordnet (MWN).

WSD using Sanskrit Wordnet Glosses. Sanskrit wordnet is still a small database compared to Hindi Wordnet. It contains 5000 synsets. We take following three cases to show how this database can be used in disambiguating polysemous words in following three cases.

1. *Hariḥ vṛkṣaśākhām añcati|*
2. *Bhaktaḥ harīm añcati prasannaḥ mādhavaḥ tasya iṣṭaṃ varaṃ dadāti|*
3. *Hariḥ śaśakaṃ bhakṣayati|*

Here noun *hari* and verb *añcati* are polysemous. In the existing Sanskrit Wordnet, noun *hari* has five senses. These senses are shown in the table 5 and their application in the WSD is shown below.

Table 5. Possible synsets for the word *hari* in the present Sanskrit Wordnet

	Gloss	Synset members
1.	*Vanyapaśuḥ- mārjāra-jātīyaḥ hiṃsraḥ tathā ca balavān paśuḥ*	*siṃhaḥ, mṛgendraḥ,* **hariḥ** *... pañcāsyaḥ ...*
2.	*Devatā-viśeṣaḥ- hindudharmānusārī jagatpālanakartā*	*viṣṇuḥ, nārāyaṇaḥ ... keśavaḥ,* **hariḥ**... *mādhavaḥ ... cakrapāṇiḥ*
3.	*Yaduvaṃśīyasya vasudevasya putraḥ yaḥ viṣṇoḥ avatāraḥ iti manyate*	*kṛṣṇaḥ, devakīnandanaḥ, ...* **hariḥ**, *mādhavaḥ, ... kaṃsāriḥ ...*
4.	*Vanyapaśuḥ- yaḥ vṛkṣāt vṛkṣam pucchena utplutya gacchati yaśca manuṣyāṇām pūrvajatvena khyātaḥ*	*vānaraḥ, kapiḥ, ...* **hariḥ**, *... tarumṛgaḥ*
5.	*Ubhayacharaḥ prāṇī yaḥ varṣārtau jalāśayasya samīpe dṛśyate*	*maṇḍūkaḥ, bhekaḥ, plavagaḥ, ... darduraḥ, ...* **hariḥ**

Sentence 1. Word *hari* in the sentence 1 can be disambiguated using the Lesk like algrithm described in the table 1 in the following way:

***Context Bag* C will consist {hari, vṛkṣa-śākhā, añc}**

- Consider synset 1 as the first candidate for the sentence 1.
  The bag **B** for synset 1 mentioned above will consist {vanya-paśu- mārjāra-jātīya hiṃsra, balavat paśu, siṃha, mṛgendra, **hari**, pañcāsya, ...}.
  Overlap between **C** and **B** will be **1** for synset 1 mentioned above.
- Consider synset 2.
  The bag **B** for synset 2 mentioned above will consist {devatā-viśeṣa, hindu-dharmānusārī jagat-pālana-kartā viṣṇu, nārāyaṇa, keśava, **hari** mādhava, cakrapāṇi ...}
  Overlap between **C** and **B** will be **1** for synset 2 mentioned above.
- Consider synset 3.
  The bag **B** for synset 3 mentioned above will consist {yadu-vaṃśīya, vasudeva, putra, viṣṇu, avatāra,man, kṛṣṇa, devakīnandana, **hari**, kaṃsāri, ...}
  Overlap between **C** and **B** will be **1** for synset 3 mentioned above.
- Consider synset 4.
  The bag **B** for synset 4 mentioned above will consist {vanya-paśu, **vṛkṣa**, puccha, utplu, gam, manuṣya, pūrvajatva, khyāta vānara, kapi, **hariḥ**, tarumṛga...}
  Overlap between **C** and **B** will be **2** for synset 4 mentioned above.
- Consider synset 5.
  The bag **B** for synset 5 mentioned above will consist {ubhayachara, prāṇin, varṣā-ṛtu, jalāśaya, samīpe dṛś, maṇḍūka, bheka, plavaga, dardura, **hari**,...}
  Overlap between **C** and **B** will be **1** for synset 5 mentioned above.

Synset 4 gets the highest overlapping score and therefore it can be considered as the most appropriate sense for the word *hari* in the sentence 1.

Sentence 2. If we perform similar analysis on the sentence 2, we find that *Context Bag* **C** consists {bhakta, hari, añc, prasanna, mādhava, iṣṭa, vara, dā}. We will get overlap score **2** for synsets 2 and 3 as *hari, mādhava* these two candidates overlap. In this case, words *iṣṭa, vara* in the context bag indicate that the sense *devatā* should be selected.

Sentence 3. The *Context Bag* **C** for the sentence 3 consists {hari, śaśaka, bhakṣ}. In this case, information extracted from the relation node of Hypernymy where the sense *siṃha* of the word '*hari*' is linked with the synset of *māṃsāhārī.*

3 Conclusion and Future Work

In this paper, we have shown the importance of glosses in WSD in the context of the noun Hari. Same application can be extended to the verb 'ancati'. In the same way, at least primarily, same application can be extended across POS categories polysemous words in Sanskrit. In this task, the use of Navya-Nyāya terminology and semantic structure of the verb root may play a crucial role. We reserve to explore this aspect in future. Also in future, we propose to examine how *Gloss* in Sanskrit Wordnet becomes useful in the task of WSD across Indian languages, in the context of Indo-Wordnet (Wordnet effort connecting all the available Wordnets of Indian languages). According to us, describing the application of the Sanskrit Wordnet glosses for the specific NLP task of WSD for Sanskrit is the first attempt of its kind. We faced several challenges in building Sanskrit Wordnet through expansion approach. We also did not get much positive support from the traditional lexica in creating glosses. However, we tried to use the traditional śāstraic background in creating glosses of verbs (in particular) and other POS categories to overcome such methodological problems. For one thing, we are sure at the moment that glosses specially designed in the way shown in this paper, certainly help WSD for Sanskrit.

Acknowledgements

Authors thank all the reviewers for their invaluable comments that helped revise the draft of this paper. Authors also thank the editor for cooperation.

References

1. Agirre, E., German, R.: Word sense disambiguation using conceptual density. In: Proceedings of the 16th International Conference on Computational Linguistics, COLING (1996)
2. Alshawi, H.: Processing dictionary definitions with phrasal pattern hierarchies. Computational Lingustics 13, 195–202 (1987)
3. Bhattacharyya, P.: Indowordnet. In: Lexical Resources Engineering Conference, Malta (May 2010)
4. Kulkarni, M., Dangarikar, C., Kulkarni, I., Nanda, A.P.B.: Introducing sanskrit wordnet. In: Bhattacharyya, P., Fellbaum, C., Vossen, P. (eds.) Proceedings on the 5th Global Wordnet Conference (GWC 2010), Narosa,Mumbai, pp. 287–294 (2010)
5. Lesk, M.: Automatic sense disambiguation using machine readable dictionaries: how to tell a pine cone from an ice cream cone. In: Proceedings of the 5th annual international conference on Systems documentation, Toronto, Ontario, Canada, pp. 24–26 (1986)
6. Mihalcea, R.: Unsupervised large-vocabulary word sense disambiguation with graph-based algorithms for sequence data labeling. In: HLT/EMNLP 2005, Human Language Technology Conference and Conference on Empirical Methods in Natural Language Processing, Proceedings of the Conference, Vancouver, British Columbia, Canada, October,6-8 (2005)
7. Narayan, D., Chakrabarty, D., Pande, P., Bhattacharyya, P.: An experience in building the indo wordnet- a wordnet for hindi. In: International Conference on Global WordNet (GWC 2002), Mysore, India (2002)
8. Resnik, P.: Selectional preference and sense disambiguation. In: Tagging Text with Lexical Semantics: Why, What, and How? In: Proceedings of ACL SIGLEX Workshop on Tagging Text with Lexical Semantics: Why, What, and How?, pp. 52–57. ACL (1997)
9. Sinha, M., Reddy, M., Bhattacharyya, P.: An approach towards construction and application of multilingual indo-wordnet. In: 3rd Global Wordnet Conference (GWC 2006), Jeju Island, Korea (2006)
10. Walker, D., Amsler, R.: Analyzing Language in Restricted Domains. In: The Use of Machine Readable Dictionaries in Sublanguage Analysis, pp. 69–83. LEA Press (1986)

Vibhakti Divergence between Sanskrit and Hindi

Preeti Shukla, Devanand Shukl, and Amba Kulkarni

Department of Sanskrit Studies,
University of Hyderabad,
Hyderabad
shukla.preetidev@gmail.com, dev.shukl@gmail.com, apksh@uohyd.ernet.in

Abstract. Translation divergence at various levels between languages arises due to the different conventions followed by different languages for coding the information of grammatical relations. Though Sanskrit and Hindi belong to the same Indo-Aryan family and structurally as well as lexically Hindi inherits a lot from Sanskrit, yet divergences are observed at the level of function words such as *vibhaktis*. Pāṇini in his Aṣṭādhyāyī has assigned a default *vibhakti* to *kārakas* alongwith many scopes for exceptions. He handles these exceptions either by imposing a new *kāraka* role or by assigning a special *vibhakti*. However, these methods are not acceptable in Hindi *in toto*. Based on the nature of deviation, we propose seven cases of divergences in this paper.

Keywords: Pāṇini, *kāraka, vibhakti,* semantics, translation divergence.

1 Introduction

Different languages may follow different conventions for coding the information of grammatical relations. This leads to divergence between languages at various levels in translation in general and in Machine Translation(MT) in particular. Translation divergence patterns have been discussed by Dorr [Dorr,1994], and based on those patterns considerable work is done for Indian Languages, viz., English-Sanskrit-Hindi MT [Goyal and Sinha, 2009], English-Sanskrit MT [Mishra and Mishra, 2008], English-Hindi MT [Dave et al., 2002] to name a few.

Sanskrit and Hindi belong to the same Indo-Aryan family and Hindi inherits a lot from Sanskrit in terms of structure as well as lexicon. As such, when we look at the divergences between Sanskrit and Hindi, most of the cases discussed in Dorr's divergence turn out to be either of rare occurrence, or do not pose much problem as far as accessing the source text using machine tools is concerned. On the contrary divergences are observed at the level of function words such as *vibhaktis*. While both Sanskrit and Hindi are inflectional languages, Sanskrit is synthetic in nature while Hindi is analytic [Dwivedi, 2006]. In Sanskrit both *sup* and *tiṅ* suffixes are termed as *vibhakti pratyayas*.[1] *Sup* suffix is added to a nominal stem and *tiṅ* suffix is added to a verbal root. But in Hindi only the suffix added to the nominal stem is called as a *vibhakti pratyaya*. In this paper

[1] Vibhaktiśca (suptiṅau vibhaktisanjñau staḥ -S.K.) (P-1.4.104).

G.N. Jha (Ed.): Sanskrit Computational Linguistics, LNCS 6465, pp. 198–208, 2010.

we discuss the divergences in the use of nominal suffixes (*vibhakti pratyayas*) between Sanskrit and Hindi.

2 Classification of Divergences

Pāṇini uses *kāraka* - a syntactico-semantic relation as an intermediary step to express the semantic relations through *vibhaktis*. The assignment of *kāraka* to various semantic categories is not one-to-one. Rama Nath Sharma (2002) observes-

> Pāṇini specifies his *kāraka* categories based upon the principle of *sāmānya* 'general', *viśeṣa* 'particular' and *śeṣa* 'residual'. The six categories are identified by general rules formulated based upon linguistic generalizations. Particular rules form exceptions to them. Usage which cannot be accounted for by the above two rule types is governed by rules relegated to the residual category. It is obvious that these exceptions are necessary to capture the peculiarities of usage falling outside the scope of the general rules.

Each *kāraka* in his system has a default *vibhakti*. But as is well-known, there are exceptions and hence there is no one-to-one mapping between the *kārak*a relations to *vibhaktis*. Pāṇini handles these deviations by employing two methods: (a) imposing a *kāraka* role and (b) assigning a special *vibhakti*. For example through the *sūtra ādhāro'dhikaraṇam* (P-1.4.45) a *locus* is mapped to *adhikaraṇa* and then it takes seventh case suffix by default (*saptamyadhikaraṇe ca* P-2.3.36). However the locus of the action related to the verbal roots *śīṅ, sthā and āsa* preceded by the *upasarga adhi* is termed as *karma* (*adhiśīṅsthāsāṃ karma* P-1.4.46) and then by the sūtra *karmaṇi dvitīyā* (P-2.3.2) this karma takes the second case suffix. Thus the deviation of the adhikaraṇa taking the second case suffix instead of seventh is handled by imposing a karma *kāraka* role in place of adhikaraṇa. Similarly, when the linguistic generalisations cannot be captured, he treats the cases as exceptional as in *rucyarthānām prīyamāṇaḥ* (P-1.4.33), where he assigns a *sampradāna kāraka* to the one who desires. In Fig 1, we summarise Pāṇini's way of mapping semantic relations to *vibhaktis* through kārakas.

This classification does not hold good as-it-is for Hindi. If a parallel grammar for Hindi in the Aṣṭādhyāyī style were available, it would have been a simple task to arrive at the divergence cases. In the absence of such a grammar, the grammar rules corresponding to exceptional cases and cases which can not be captured by linguistic generalisations, were all checked for corresponding *vibhaktis* in Hindi. Leaving aside those cases where *vibhakti* is the same in Sanskrit and Hindi and taking up those cases where it diverges, we found that the cases of divergences may be classified into seven types, such as -

1. **Optional Divergence:** *Vibhaktis* optionally found in Sanskrit but absent in Hindi. For instance, in Sanskrit the *karma*[2] of the verb is expressed by the default second case suffix and optionally the fourth case suffix (bālakāḥ

[2] Roughly equivalent to the notion of *object* in western linguistics.

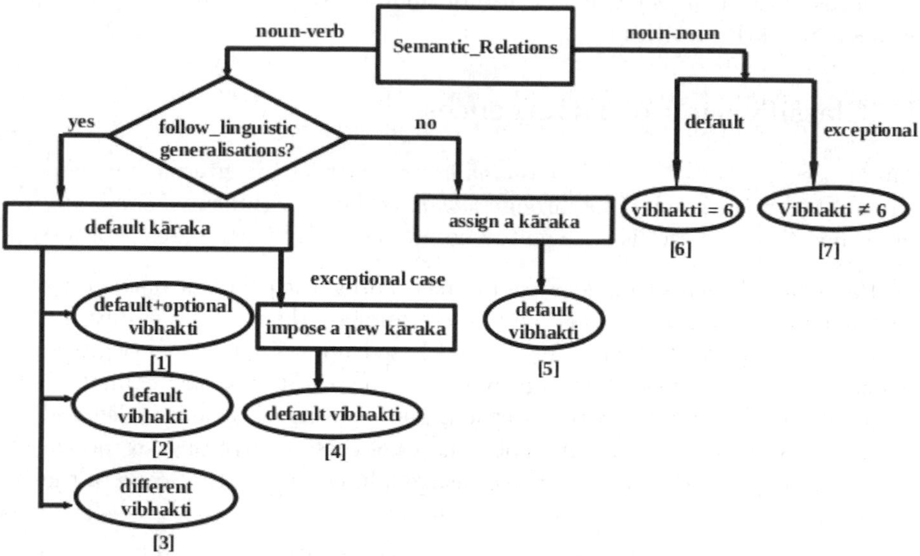

1. Divaḥ karma ca (sādhakatamam, kārake) (P–1.4.43)
2. Karmaṇi dvitīyā (P–2.3.2)
3. Divastadarthasya (karmaṇi, ṣaṣṭhī śeṣe, anabhihite) (P–2.3.58)
4. Adhiśīṅsthāsām karma (ādhāraḥ, kārake) (P–1.4.46)
5. Krudhadruhērṣyānām yam prati kopaḥ (sampradānam, kārake) (P–1.4.37)
6. Ṣaṣṭhī śeṣe (P–2.3.50)
7. Pṛthagvinānānābhistṛtīyānyatarasyām (dvitīyā, pañcamī) (P–2.3.32)

Fig. 1. Semantic–Vibhakti_Mapping

vidyālayāya/*vidyālayaṃ* gacchanti) but in Hindi it takes only the default second case suffix (bālaka *vidyālaya* jāte_haiṁ = 'Boys go to school').

2. **Exceptional Divergence:** Sanskrit has certain exceptional rules which block the default suffixes but Hindi uses only the default suffixes. For example, in Sanskrit the karma of the verb exceptionally takes the sixth case suffix (śakuniḥ *śatasya* dīvyati) but in Hindi it takes the default second case suffix (śakuni *sau_rupae* jītatā_hai = 'Shakuni wins hundred rupees').
3. **Differential Divergence:** Sanskrit and Hindi use different nominal suffixes, for example, in Sanskrit, the person against whom the feeling of hatred *druha* is directed takes the fourth case suffix (durjanāḥ *sajjanāya* druhyanti) while in Hindi it takes the fifth case suffix (durjana *sajjana_se* droha karte_haiṁ = 'The wicked hate the good').
4. **Alternative Divergence:** Sanskrit uses more than one *vibhaktis* but Hindi takes only few among them. For instance, in Sanskrit alternately the sixth as well as the seventh case suffix is used after a word in conjunction with

āyukta (āyuktaḥ *haripūjane/haripūjanasya*) but in Hindi only the seventh case suffix is applicable (hari_kī_ *pūjā_meṁ* līna = 'Deeply absorbed in the worship of Hari').

5. **Non-*kāraka* Divergence:** Divergences at the level of non-*kāraka* nominal suffixes, such as upapada vibhaktis, sambandha vibhaktis, etc. For instance, in Sanskrit the word takes fifth case suffix when governed by the karmapravachanīya *prati* (pradyumnaḥ *kṛṣṇāt* prati asti) while in Hindi it takes the sixth case suffix (pradyumna *kṛṣṇa_ke* pratinidhi hai = 'Pradyumna is the representative of Krishna').
6. **Verbal Divergence:** Divergences due to the special demand of certain verbs. For example, the karma of the verb *ā+ruh* in Sanskrit (vānaraḥ *vṛkṣam* ārohati) takes the seventh case suffix in Hindi (bandara *peṛa_para* caṛhatā_hai = 'Monkey climbs on the tree').
7. **Complex-Predicate Divergence:** This divergence results when a Sanskrit verb is mapped to a complex predicate in Hindi. For instance, the karma of the verb *anu+sṛ* in Sanskrit (*rāmam* anusarati sītā) is expressed by the genitive case in Hindi (*rāma_kā* sītā anusaraṇa karatī_hai = 'Sita follows Rama').

These cases are elaborated in detail.

2.1 Optional Divergence

When a *sūtra* assigns optionally two different vibhaktis in Sanskrit but Hindi allows only one vibhakti, we term the resulting divergence as *Optional Divergence.*

Illustration

***divaḥ karma ca (sādhakatamam and kārake)* (P-1.4.43).** That *kāraka* which is supplemental in the accomplishment of the action of the verbal root *diva* (to play) is termed as *karma*, in addition to *karaṇa*. e.g.,

(1) San: *akṣān/akṣaiḥ* dīvyati
Hin: *pāsoṁ_se* khelatā_hai
gloss: dice_with plays
Eng: '(He) plays with dice.'

In the above example (1), *akṣa* is the instrument of the verbal root *diva* expressed through the third case suffix derived from the general rule *sādhakatamaṃ karaṇam* (P-1.4.42) but due to the optional rule *divaḥ karma ca*, *akṣa* optionally takes the second case suffix. Whereas in Hindi this optional rule does not apply and thus Hindi allows only the third case suffix.

Table 1 lists the *sūtras* which account for an additional optional *vibhakti* in Sanskrit either on account of imposed *kāraka* (Nos.1-3) or on account of special *vibhakti* assignment. In all these cases Hindi allows only the default *kāraka* and hence the default *vibhakti*.

Table 1. Optional Divergence

Sr.No.	Sūtra		Default		Optional	
			kāraka	Vibhakti	*kāraka*	Vibhakti
1	1.4.43	divaḥ karma ca	karaṇa	3	karma	2
2	1.4.44	parikrayaṇe.	karaṇa	3	sampradāna	4
3	1.4.53	hṛkronyatarasyām	kartā	3	karma	2
4	2.3.12	gatyarthakarmaṇi.	karma	2	-	4
5	2.3.22	sañjño'nyatarasyāṃ.	karma	2	-	3
6	2.3.31	enapā dvitīyā	-	6	-	2
7	2.3.59	vibhāṣopasarge	karma	2	-	6
8	2.3.71	kṛtyānāṃ karttari.	kartā	3	-	6

2.2 Exceptional Divergence

When sūtras give exceptional rules for cases by restricting the general rules but these rules are not applicable in Hindi, the divergence is termed as *Exceptional Divergence.*

Illustration

***adhiśīṅsthāsāṃ karma (ādhāraḥ* and *kārake*) (P-1.4.46).** That *kāraka* which is the locus of the verbal roots *śīṅ* (to lie down), *sthā* (to stand), *āsa* (to sit), when preceded by the *upasarga adhi* is termed as karma. e.g.,

(2) San: bālakaḥ *paryaṅkam* adhiśete
Hin: laḍakā *palaṃga_para* sotā_hai
gloss: boy bed_on sleeps
Eng: 'Boy sleeps on the bed'.

In this example, the locus *paryaṅka* of the verbal root *adhi + śīṅ* is termed as karma and thus takes the second case suffix. But this rule does not apply in Hindi. So *paryaṅka* which is the locus and hence *adhikaraṇa* by the rule *ādhāro'dhikaraṇam* (P-1.4.45) takes the seventh case suffix. This applies to other verbs *sthā and āsa* with prefix *adhi* as well.

Table 2 lists the *sūtras* which account for non-default *vibhaktis* in Sanskrit by imposing a new *kāraka* (as in Nos.1-3) or by imposing an altogether different *vibhakti* for the *kāraka.* In all these cases, Hindi however takes the default *kāraka* and hence default *vibhakti.*

2.3 Differential Divergence

All those cases where Pāṇini's rule assigns a certain *vibhakti* either through *kāraka* or by direct case assignment and Hindi uses altogether different *vibhakti* are termed as the case of *Differential Divergence.*

Table 2. Exceptional Divergence

Sr.No.	Sūtra		Default		Exceptional	
			kāraka	Vibhakti	*kāraka*	Vibhakti
1	1.4.46	adhiśīṅsthāsāṃ karma	adhikaraṇa	7	karma	2
2	1.4.47	abhiniviśaśca	adhikaraṇa	7	karma	2
3	1.4.48	upānvadhyāṅvasaḥ	adhikaraṇa	7	karma	2
4	2.3.58	divaḥ tadarthasya	karma	2	-	6
5	2.3.61	preṣyabruvorhaviṣo	karma	2	-	6
6	2.3.64	kṛtvo'rthaprayōge	adhikaraṇa	7	-	6

Illustration

apavargē tṛtīyā (kālādhvanoḥ atyantasaṁyoge) (P-2.3.6). The third case-suffix is employed after the words denoting the duration of time or place (*adhvan*), when accomplishment (*apavarga*) of the desired object is meant to be expressed. e.g.,

(3) San: *ahnā* anuvākaḥ adhītaḥ
Hin: *dinabhara_meṁ* anuvāka (veda kā eka bhāga) paṛha_liyā
gloss: whole day_in anuvaka (veda of one part) read
Eng: 'Anuvaka was perseverely and effectually studied (by him) in one day'.

In example (3), the person not only studied *anuvāka* but completely understood and memorised them. Attainment of knowledge is the fruitful result and hence the word *ahnā* which denotes the duration of time takes the third case-suffix. But in Hindi it takes the seventh case-suffix.

Table 3 lists the sūtras which describe cases where Hindi *vibhakti* differs from the Sanskrit vibhakti.

2.4 Alternative Divergence

In case of *Alternative Divergence* Sanskrit allows more than one case suffixes and Hindi takes only a few of them, and rarely an altogether different case suffix.

Illustration

dūrāntikārthaiḥ ṣaṣṭhyanyatarasyām (pañcamī) (P-2.3.34). When in conjunction with *dūra* (distant,far), and *antika* (near) and their synonyms, the sixth case suffix is optionally employed and alternately the fifth. e.g.,

(4) San: *grāmasya/grāmāt* vanaṃ dūram asti
Hin: *gānva_se* jaṅgala dūra_hai
gloss: village_from forest far is
Eng: 'Forest is far from the village'.

Table 3. Differential Divergence

Sr.No.	Sūtra		Sanskrit		Hindi
			kāraka	Vibhakti	Vibhakti
1	1.4.34	ślāghahnuṇsthā.	sampradāna	4	2,4,6
2	1.4.35	dhāreruttamarṇaḥ	sampradāna	4	6
3	1.4.37	krudhadruherṣyāsūyā.	sampradāna	4	5,7
4	1.4.38	krudhadruhorupasṛṣṭa.	karma	2	5,7
5	1.4.39	rādhīkṣyoryasya.	sampradāna	4	6
6	2.3.6	apavargē tṛtīyā	-	3	7
7	2.3.17	manyakarmaṇyanādare.	karma	2,4	6
8	2.3.23	hetau	-	3	4
9	2.3.37	yasya ca bhāvena.	-	7	6
10	2.3.43	sādhunipuṇābhyām.	-	7	4
11	2.3.67	ktasya ca varttamāne	-	6	3

In example (4), in Sanskrit *grāma* takes optionally the sixth case suffix as well as the fifth when combined with *dūra* as well as *antika* but Hindi takes only the fifth case suffix.

Table 4 lists the sūtras where Sanskrit takes more than one *vibhaktis* but Hindi takes only a few of them.

Table 4. Alternative Divergence

Sr.No.	Sūtra		Sanskrit	Hindi
			Vibhakti	Vibhakti
1	2.3.32	pṛthagvinānānā.	2,3,5	5,6
2	2.3.34	dūrāntikārthaiḥ.	5,6	5
3	2.3.36	saptamyadhikaraṇe ca	2,3,5,7	0
4	2.3.39	svāmīśvarādhipati.	6,7	6
5	2.3.40	āyuktakuśalābhyāṃ.	6,7	7
6	2.3.41	yataśca nirdhāraṇam	6,7	7
7	2.3.44	prasitotsukābhyāṁ.	3,7	2,7
8	2.3.72	tulyārthairatulopamābhyāṁ.	3,6	6
9	2.3.73	caturthī cāśiṣyāyuṣya.	4,6	2,6

2.5 Non-*kāraka* Divergence

Certain words known as *upapada* demand specific vibhaktis called *Upapada Vibhaktis* for nouns with which they combine. These are all typically exceptions to the *ṣaṣṭhī śeṣe* (P-2.3.50). In Hindi, however, in most of the cases sixth case suffix is used.

Illustration

sahayukte apradhāne (tṛtīyā) (P-2.3.19). The word denoting *apradhāna* in conjunction with *saha* and its synonyms takes the third case suffix. e.g.,

(5) San: *rāmeṇa* saha sītā vanaṁ gacchati
Hin: *rāma_ke* sātha sītā vana jātī hai
gloss: rama_with sita forest goes
Eng: 'Sita goes to forest with Rama'.

In example (5), in Sanskrit it is seen that *sītā and vana* are directly related to the verb in the form of *kartā* and karma but *rāma* is not directly related to the verb and hence is called *apradhāna.* The association of *rāma* with *sītā* is denoted by the word *saha* due to which *rāma* takes the third case suffix from the above rule. But in the case of Hindi, *rāma* in conjunction with *saha* takes only the sixth case suffix.

Table 5 lists the *sūtras* where Sanskrit generally takes the second case suffix barring the sixth case suffix but Hindi does not take this exemption into account and takes the sixth case suffix.

Table 5. Non-*kāraka* Divergence

Sr.No.	Sūtra		Sanskrit Vibhakti	Hindi Vibhakti
1	1.4.84	tṛtīyārthe	2	6
2	1.4.85	hīne	2	6
3	1.4.87	upo'dhike ca	2	6
4	1.4.89	lakṣaṇetthambhūtā.	2	6
5	1.4.90	abhirabhāge	2	6
6	1.4.95	atiratikramaṇe ca	2	5,6
7	2.3.4	antarā'ntareṇayukte	2	6
8	2.3.5	kālādhvanoratyantasanyoge	2	7
9	2.3.10	pañcamyapāṅparibhiḥ	5	2
10	2.3.11	pratinidhipratidāne ca yasmāt	5	6
11	2.3.16	namaḥ svastisvāhā.	4	2,4,6
12	2.3.19	sahayukte apradhāne	3	6
13	2.3.26	anyārāditarartedik.	5	5,6

2.6 Hindi Specific Divergences

All the divergences we covered so far were specific to the Aṣṭādhyāyī, and some of them might be attributed to the characteristics of Sanskrit. The default *vibhakti* of *karma* in Sanskrit is dvitīyā. But in Hindi, the karma takes *vibhakti* other than *dvitīyā.* These divergences may be attributed to the nature of Hindi.

On the face of it divergences may be classified into two classes depending upon whether the Sanskrit verb maps to a simple Hindi verb or a complex predicate.

(a) Verbal Divergence
Consider the example,

(6) San: vānaraḥ *vṛkṣam* ārohati
Hin: bandara *peṛa_para* caṛhatā_hai
gloss: monkey tree_on climbs
Eng: 'Monkey climbs on the tree'.

Here, the karma of the verbal root *ā + ruh* is *vṛkṣa* whereas in Hindi the karma *vṛkṣa* takes seventh case suffix. The important question from Hindi grammar point of view then is - whether *vṛkṣa* should be termed as a karma as in Sanskrit or should it be termed as an adhikaraṇa? If we term it as a karma, then the treatment will be close to Sanskrit but one needs to impose a rule accounting for the divergence in the suffix. If we term it as an *adhikaraṇa*, then we deviate from the Sanskrit grammar but can take an advantage of the default vibhakti mapping.

There are many such instances, and these should be treated carefully by Sanskrit scholars who know Hindi well, taking insights from how Pāṇini handled such cases.

(b) Complex Predicate Divergence
If a verb in Sanskrit maps to a complex predicate[3] in Hindi, then the karma of the verb takes the sixth case suffix. e.g.,

(7) San: *rāmam* anusarati sītā
Hin: *rāma_kā* sītā anusaraṇa karatī_hai
gloss: rama_of sita follows
Eng: 'Sita follows Rama'.

Here in example (7), *rāma* which is the karma is expressed with the genitive case in Hindi. This change is a systematic one which may be explained through the rule *kartṛkarmaṇoḥ kṛti* (P-2.3.65). In Hindi *anu + sṛ* is translated as a complex predicate *anusaraṇa_karanā* where *anusaraṇa* is the *kṛdanta* and *karanā* is the main verb.

It is necessary to study these divergences in karma vibhakti in detail. Sanskrit has approximately 2000 verbal roots and many more with the *upasargas*. Mapping the corresponding verb-frames in Sanskrit to the Hindi verb-frames may give some hints towards these divergences. Though it is a voluminous task, any MT system requires this study.

[3] A complex predicate consists of a noun followed by a light verb such as *karanā*, honā, etc. e.g. *vivāha karanā, snāna karanā*, etc.

3 Conclusion

The cases of divergences may be summarised then as:
(A) Divergences originating from Sanskrit Grammar (as shown below)

Sr.No.	Cases	Sanskrit	Hindi
1	Optional	Sanskrit uses optional *vibhakti* in addition to default *vibhakti*.	Hindi allows only the default *vibhakti*.
2	Exceptional	Sanskrit uses different vibhakti than the default (Pāṇini blocks the default *vibhakti* by treating it as an exceptional case).	Hindi uses the default vibhakti.
3	Differential	Pāṇini imposes certain kārakas or vibhaktis which cannot be captured through the semantic generalisations.	Hindi uses different vibhaktis.
4	Alternative	Sanskrit uses more than one alternative vibhakti.	Hindi uses only one of them.
5	Non-*kāraka*	Sanskrit uses more than one alternative *vibhakti*.	Hindi uses the sixth vibhakti.

(B) Divergences due to the idiosyncrasy of Hindi:

1. Special *vibhakti* expectancy of verbs
2. Complex Predicate

These seven cases of divergence may prove useful in resolving the ambiguities at the level of nominal suffixes in Machine Translation and may serve as a platform for looking at vibhakti level divergences between Sanskrit and other Modern Indian Languages.

Acknowledgements. This work is a part of the Sanskrit Consortium project entitled 'Development of Sanskrit computational tools and Sanskrit-Hindi Machine Translation system' sponsored by the Government of India. We gratefully acknowledge Prof. Lakshmi Bai for the discussions at various stages.

References

1. Ramshastri, A.: Saṁskṛta śikṣaṇa saraṇī. Acharya Ramshastri Jñānapīṭha (Hindi) (1998)
2. Cardona, G.: Pāṇini and Pāṇinīyas on śeṣa Relations. Kunjunni Raja Academy of Indological Research, Kochi (2007)
3. Dorr, B.: Classification of Machine Translation Divergence and a Proposed Solution. Computational Linguistics 20(4) (1994)
4. Dave, S., Parikh, J., Bhattacharya, P.: Interlingua Based English-Hindi Machine Translation and Language Divergence. Journal of Machine Translation (JMT) 17 (2002)
5. Dwivedi, K.D.: Bhāṣā vijñāna evam Bhāṣā śāstra. Viśvavidyālaya Prakāśana, Varanasi (2006)

6. Goyal, P., Sinha, R., Mahesh, K.: Translation Divergence in English-Sanskrit-Hindi Language Pairs. In: Kulkarni, A., Huet, G. (eds.) SCLS 2009. LNCS (LNAI), vol. 5406, Springer, Heidelberg (2009)
7. Mishra, V., Mishra, R.B.: Study of Example Based English to Sanskrit Machine Translation. Journal of Research and Development in Comp Sc. And Engg. (37) (January-June 2008)
8. Sharma, R.R., Sharma, M.: Vaiyākaraṇa Siddhānta Kaumudī of Śrī Bhaṭṭoji Dīkṣita (Kāraka Prakaraṇam). Bharatiya Vidya Prakashan, Varanasi-Delhi (Hindi) (1997)
9. Sharma, R.N.: The Aṣṭādhyāyī of Pāṇini, 2nd edn., vol. 1-3. Munshiram Manoharlal, Delhi (2002)
10. Vasu, Ś.C.: The Siddhānta Kaumudī of Bhaṭṭoji Dīkṣita, vol. I. Motilal Banarsidass, Delhi (2003)

Anaphora Resolution Algorithm for Sanskrit

Pravin Pralayankar and Sobha Lalitha Devi

AU-KBC Research Centre, MIT Campus of Anna University,
Chennai, India
sobha@au-kbc.org

Abstract. This paper presents an algorithm, which identifies different types of pronominal and its antecedents in Sanskrit, an Indo-European language. The computational grammar implemented here uses very familiar concepts such as clause, subject, object etc., which are identified with the help of morphological information and concepts such as precede and follow. It is well known that natural languages contain anaphoric expressions, gaps and elliptical constructions of various kinds and that understanding of natural languages involves assignment of interpretations to these elements. Therefore, it is only to be expected that natural language understanding systems must have the necessary mechanism to resolve the same. The method we adopt here for resolving the anaphors is by exploiting the morphological richness of the language. The system is giving encouraging results when tested with a small corpus.

Keywords: Anaphora resolution, Rule based Technique, Finite State Automata, Sanskrit.

1 Introduction

Sanskrit, the language of the Indo-European family has very long writing tradition available in both poetry and prose. It contains many linguistic elements in its rich grammatical and philosophical tradition but Sanskrit syntax has been least focused by the modern linguist. Speyer [15] and Hock [4] have looked on many syntactic aspects of Sanskrit language.

In recent years Sobha et al [11,12,13] has looked on anaphora resolution for Indian languages including Sanskrit in detail from computational perspective. Jha et al [5,6] has discussed about anaphora resolution techniques and the concept of anaphora in different knowledge tradition of Sanskrit including Vyākaraṇa, Nyāya, and Mīmāṃsā. Keeping these in mind, this paper discusses about various types of pronominal constructions in Sanskrit and an algorithm to resolve it.

Anaphora resolution refers to the problem of determining the antecedent of an anaphora in a document. The most common type of anaphora is pronominal anaphora and it can be exhibited by personal, possessive or reflexive pronouns. There are many approaches to solve this problem such as rule based [1,2,3], statistical and machine learning based techniques [7,8,9,10]. This paper describes about a rule based pronominal resolution for Sanskrit and its scope is limited only to the simple prose sentences.

G.N. Jha (Ed.): Sanskrit Computational Linguistics, LNCS 6465, pp. 209–217, 2010.

2 Sanskrit Anaphora

There are two types of anaphoric construction available in Sanskrit; anaphora proper and anaphora-like cases [6] . Sobha et al [13] gives a theoretical description of anaphora proper and in this paper its computational aspects are presented.

2.1 Classification of Sanskrit Anaphora

The different types of anaphora proper are pronominal, reflexives, and reciprocals in a context. Below is the list of Sanskrit anaphora. The reflexive (*sva* and *ātman*) and reciprocal (*parasparam* and *anyonyam*) does not have gender feature so they do not exhibit the gender agreement with their antecedent.

Table 1. Sanskrit anaphora

Anaphors	Sanskrit Anaphor
Pronominal	tat
Demonstrative	idam, etat, adas
Reflexive	sva, ātman
Reciprocal	parasparam, anyonyam

Pronominal Anaphors. The major classification in pronominal is the first, second, and third person pronoun. First and second person are commonly used as deictic, though they are used in discourse. The third person pronoun is the most commonly used anaphora in a language. In Sanskrit *tat* is the third person pronoun which has the following separate forms in different cases for masculine and feminine gender.[1]

In the following set of examples (1-4), we see the distribution of third person pronoun *tat* and its antecedent; a phrase or clause that is referred to by an anaphoric pronoun. Example (1-3) is grammatical because there is a person, number and gender (PNG) agreement between the antecedents and pronominal while (4) is ungrammatical because there is no PNG agreement between them. In example (1a) antecedent *sītā* is in feminine singular form so its anaphor in (1b) *sā* is in feminine singular form. In example (2) antecedent *chātrāḥ* and the anaphor *te* is in masculine plural form while in example (3) antecedent *kālidāsaḥ* and anaphor *saḥ* is in masculine singular form. But in example (4) antecedent *kārtikaḥ* is in masculine singular form but the anaphor *te* is in masculine plural form. This violates the PNG agreement and the sentence is ungrammatical. From the above we find that there should be PNG agreement between pronominal and its antecedent.

[1] Although neuter gender is also available in Sanskrit but we are not giving here the neuter forms because they are same as masculine except in nominative and accusative. In nominative and accusative, the forms are *tat, te, tāni* respectively for singular, dual and plural.

Table 2. Forms of pronominal *tat* in different cases

	Masculine Gender			Feminine Gender		
	Singular	Dual	Plural	Singular	Dual	Plural
Nominative	saḥ	tau	te	sā	te	tāḥ
Accusative	tam	tau	tān	tām	te	tāḥ
Instumental	tena	tābhyām	taiḥ	tayā	tābhyām	tābhi
Dative	tasmai	tābhyām	tebhyaḥ	tasyai	tābhyām	tābhyāḥ
Ablative	tasmāt	tābhyām	tebhyaḥ	tasyā	tābhyām	tābhyāḥ
Genitive	tasya	tayoḥ	tesām	tasyāḥ	tayoḥ	tāsām
Locative	tasmin	tayoḥ	teṣu	tasyām	tayoḥ	tāsu

1.a. [**sīta**$_i$ mama svasā asti]**MC**
sītā.f.nom.sg i.gen.sg sister.f.nom.sg is.prest 3rd sg
b. [**sā**$_i$ pāṭaliputre nivasati]**MC**
she.f.nom.sg pāṭaliputra.loc.sg live.prest.rd pl
'Sita$_i$ is my sister. She$_i$ lives in Pataliputra'.

2.a. [atra **chātrāḥ**$_i$ nivasanti.] **MC**
here student.m.nom.pl live.prest.3rd.pl
b. [**te**$_i$ adhyayanam kurvanti] **MC**
he.m.nom.pl study.m.acc.g do.prest.3rd.pl
'Students$_i$ live here. They$_i$ do study.'

3.a. [**kālidāsaḥ**$_i$ asmākam priya-kaviḥ asti.] **MC**
kālidāsa.m.nom.sg i.gen.pl favourite-poet.m.nom.sg is.prest.3rd sg
b. **saḥ**$_i$ meghadūtasya praṇetā asti.] **MC**
he.m.nom.sg meghadu:ta.gen.sg write is.prest.3rd sg
'Kalidasa$_i$ is our favourite poet. He$_i$ is the writer of Meghaduta.

4.a. *[mama bhrātā **kārtikaḥ**$_i$ asti.] **MC**
i.gen.sg brother.m.nom.sg kārtika.m.nom.sg is.presr.3rd sg
b. [**te**$_i$ pāṭaliputre nivasanti] **MC**
he.m.nom.pl pāṭaliputra.m.loc.sg live.prest.3rd pl
'Kartik$_i$ is my brother. They$_i$ live in Pataliputra.'

The above example shows the distribution of pronominal at the inter-sentential level. In sentence (1a) *sītā* is in nominative form and it is the subject of the sentence and the pronoun is in the next sentence, (1b) takes *sītā* as the antecedent. In sentence (2a) and (3a) *chātrāḥ* and *kālidāsaḥ* having nominative case, are the subject of the sentence and the pronominals *te* and *saḥ* are in the next sentence.

Demonstrative Pronoun. Demonstrative pronouns *idam, etat, adas* are other anaphoric expression in Sanskrit and their uses are often related to the spatial distance. *etat* is used for very close objects, while *idam* is used for distantly situated objects and if the object is beyond the eyes then *adas* is used. In the classical Sanskrit, the use of *idam* and *etat* has hardly any differences. All these have different declinable forms for masculine, feminine and neuter gender. The anaphoric distribution of demonstrative pronoun in Sanskrit can be seen in examples (5-8).

5.a.	[ayam	asmākam	**vidyālayaḥ**$_i$	asti.] **MC**
	this.m.nom.sg	i.m.gen.pl	school.m.nom.sg	is.prest.3rd sg
b.	[**asya**$_i$	bhavanam	bhavyam	asti.] **MC**
	it.m.gen.sg	building	beautiful	is.prest.3rd sg
	'This is our school$_i$. Its$_i$ building is beautiful.'			
6.a.	[mama	**grāmaḥ**$_i$	bihār-prānte	asti.] **MC**
	i.m.gen.sg	village.m.nom.sg	bihār-state.m.loc.sg	is.prest.3rd sg
b.	[**ayām**$_i$	gaṅgāyāḥ	taṭe	asti.] **MC**
	this.m.nom.sg	Ganga.f.gen.sg	bank.m.loc.sg	is.prest.3rd sg
	'My village$_i$ is in Bihar. This$_i$ is situated on the bank of Ganga.'			
7.a.	[gaṅgā$_i$	asmākam	pavitra-nadi	vartate.] **MC**
	Ganga.f.nom.sg	i.m.gen.pl	pious-river.f.nom.sg	is.prest.3rd sg
b.	[**iyam**$_i$	himālayāt	niḥsṛtya	baṅgasāgare
	this.f.nom.sg	himalya.m.abl.sg	originate.ppl	bay of benga.loc.sg
	patati.] **MC**			
	fall.prest.3rd.sg			
	'The Ganga$_i$ is our pious river. Having originated from Himalaya, this$_i$ falls into bay of bengal.			
8.a.	**saṃskṛta-bhāṣā**$_i$	deva-bhāṣā		kathyate.] **MC**
	sanskrit-language.f.nom.sg	god-language.f.nom.sg		say.pst.3rd sg
b.	[**eṣā**$_i$	viśvasya	prācīnatama-bhāṣā.] **MC**	
	this.f.nom.sg	world.m.gen.sg	oldest-language.f.nom.sg	
	'Sanskrit language$_i$ is called language of God. This$_i$ is the oldest language of the world.			

In the sentence (5b) the PNG features of demonstrative pronoun *asya* (5b) matches with *vidyālayaḥ* in (5a), which is the antecedent of *asya* . Similarly the demonstrative pronoun *ayam* in (6b) has antecedent *gr(ma(* in (6a). Here also there is PNG agreement between the pronoun and its antecedent. In sentence (7) *gaṅgā* is the subject of the sentence and is the antecedent of the demonstrative pronoun *iyam* of (7b). Sentence (8a) is in passive voice and the antecedent *sanskrit-bhāṣā* is the direct object position. The pronoun *eṣ(ā* is in sentence (8b) and it has PNG agreement with its antecedent *sanskrit- bhāṣā.*

From the above we arrive at the following algorithm:

1. A pronoun P is coreferential with an NP iff the following conditions hold:
   (a) P and NP have compatible P, N, G features.
   (b) P does not precede NP
   (c) If P is possessive, then NP is the subject of the clause which contains P. If P is non-possessive, then NP is the subject of the immediate clause which does not contain P.

Reflexive Pronoun. *sva* and *ātman* are reflexive pronoun in Sanskrit. In Sanskrit unlike the above pronominal, the antecedent and the reflexives are bound and the antecedent is always the subject of the sentence in which the reflexive occurs. For example, in sentence (9) *rāmaḥ* is the antecedent of the reflexive pronoun *sva* and it is the subject of the sentence. In sentence (10) antecedent *pātravargaiḥ* and reflexive *sveṣu*, which is in re-duplicated form is bound to each other and they are local. Similarly in sentence (11) antecedent *eṣā* and reflexive *ātmānam* are locally bound to each other. But here reflexive is preceding the antecedent due to relatively free word order and focus function.

9.	**rāmaḥ**$_i$	**sva**$_i$-grham	gacchati.	
	ram.m.nom.sg	own-house.loc.sg	go.prest.3rd sg	
	'Ram is going to his own house			
10.	tad	ucyatām	**pātravargaiḥ**$_i$	**sveṣu-sveṣu**$_i$
	then	tell	actor.m.nom.pl	own.m.loc.pl-own.loc.pl
	pāṭheṣu	a-samuḍhaiḥ	bhavitvyam	iti
	leson.loc.pl	not-carefree	become	so
	'So let the actors are told to be careful about thir own lessons.'			
11.	bhartṛ-gatayaā	cintayā	**ātmānam**$_i$	api
	husband-go	contemplation	self	even
	na	**eṣā**$_i$	vibhāvayati.	
	not	she	conscious.prest.eg	
	'Owing to her contemplation regarding her husband, *she* is not conscious of even *herelf.*'			

2. A pronoun P is coreferential with an NP iff the following conditions hold:
   (a) Reflexive and NP have compatible Number features.
   (b) Reflexive does not precede NP
   (c) The NP is the subject of the clause which contains Reflexive.

3 Architecture

3.1 Pre-processing

The position of the noun plays a major role in identifying the antecedent of a pronoun. From the above examples we could see that the antecedent is mostly the subject or the direct object and it can be in the same sentence or in the preceding sentence. Another feature is the position of the noun with respect to the clause, whether it is in the main clause or subordinate clause. Depending on this we can identify whether the noun is the antecedent of the pronominal or not. For this the clause boundaries have to be identified. We use simple linguistic rules to identify the clause boundaries. In the following paragraphs we discuss about how we identify the clause boundary and the subject and object.

Subject/object Identification. We identify subject and direct object using the morphological markings on the nouns. In this we use that all nominative nouns are probable candidate for subject hood, all dative nouns if the verb is a cognitive verb is probable candidate for subject hood and finally possessive nouns with nominative head or dative head are also possible noun for subject hood. Nouns inflected for all other cases are considered as direct object.

Clause Identification. *yat* is the complement marker in Sanskrit. With the help of *yat* we identify the main clause and the subordinate clause. In both the clause there will be a finite verb. For example,

12.

[sītā	avadat]**MC**	[**yat**$_{comp}$	tasyāḥ
sita.f.nom.sg	say.pst.3rd sg	that	she.f.gen.sg
svasā	delhi-nagare	nivasati]**SC**	
sister.f.nom.sg	delhi-city.m.loc.sg	live.prest 3rd sg	

'Sita aid that her sister lives in Delhi city.'

To mark the conditional clause the, conditional marker *yadi* is used. The main clause in this case contains the use of *tarhi,* which is optional. Both the clause will have finite verb. For example,

13.

yadi	rāmaḥ	adya	āgamiṣyati]**COND**
if	he.m.nom.sg	toay	come.fut.3rd sg
[(tarhi)	saḥ	śvaḥ	gamiṣyati]**MC**
then	he.m.nom.sg	tomorrow	go.fut.3rd sg

'If Ram comes today, then he will go tomorrow.'

We use the clause marker for identifying the presence of the clause. The boundaries are identified using the *yadi* as the beginning and *tarhi* as the end for the conditional clause. Similarly we use *yat* as the marker for complement clause and the finite verb before *yat* is considered as the end of the main clause and finite verb after *yat* as the end of subordinate clause. The beginning of the clause is identified using the noun nominative immediately following *yat* as the beginning of the subordinate clause and the noun nominative before the *yat* as the beginning of the main clause. We have similar rules for other types of clauses and for multiple embedding of clauses. Though this does not resolve all the sentences we got a reasonably good result.

Morphological Analyzer. The anaphora resolution system for Sanskrit is based on the hypothesis that any noun in Sanskrit is the combination of root and case suffixes where case suffixes carry information about PNG features. A morphological generator has been developed using Finite State Automata (FSA) and paradigm based approach. The FSA is built using all possible suffixes and the root dictionary is classified based on paradigms [16] . The suffixes, used in FSA are listed with its grammatical features, is utilized, while giving the output of the morphological analyzer to give the syntactic features of the morphemes. The parse for the words from the FSA are validated with morphosyntactic rules.

3.2 Algorithm for Anaphor Resolution

Hence the algorithm we use for identifying the antecedent NPs is as follows:

```
  Consider all the NPs Preceding the pronoun in sentence (S).

  If the clause marker (CM) is present in S

      Then, Identify the Main (M) clause and the Subordinate (Sub) clause
of S

      Identify the Subject and object of the S.

      Consider all the Nouns in the sentence that precede the pronoun.


  Identify the Pronoun (P) / Reflexive (R) and the Nouns (N) preceding
the P/R


  If P is non-possessive then

      Check the PNG of P and N

      If PNG agrees, then check for subject or object marker

      If N is the subject/object of the clause where P does not occur

      Then N is identified as the antecedent


  If P is possessive then

      Check the PNG of P and N

          If PNG agrees, then check for subject or object marker

          If N is the subject of the clause where P occur

          Then N is the antecedent

  If R then

      Check the PNG of P and N

      If PNG agrees, then check for subject or object marker

      If N is the subject of the clause where P occur

      Then N is identified as the antecedent
```

4 Anaphora Resolution System

The input to the anaphora resolution component is the parsed output which identifies the subject, object and the clause boundaries (The parser is not a sophisticated one, we are identifying it using linguistic rules and a small Dictionary). The different types of clauses present in the input sentence are identified (We are using a rule based approach for identifying the clause boundaries). The NPs which precede P are identified. The NPs which matches according to the pronominal algorithm are identified and checked for PNG agreement between the pronoun and reflexives and the NPs. The NP which matches in PNG and the algorithm is identified as the antecedent of the Pronoun.

5 Evaluation

We have tested the system using 200 sentences taken from the book titled *Sanskṛit Sahacara.* It contains simple sentences. The number of pronominal present in 200 sentences was 40. Among the 40 pronominal only 26 were correctly identified. The details about the correct and wrong identification are given in the table 3 below.

Table 3. Evaluation result of Sanskrit pronominal

		Correct	Wrong
No. of Sentences	200		
No. of Pronominals	40	26	14

6 Conclusion

This is an algorithm we have developed to identify pronominal in Sanskrit. The approach outlined here does not exploit the hierarchy, but exploits the nominal morphology. The work is still in progress and we are reporting an on going work. Hence the system is evaluated with a very small data. Pronominal resolution is necessary for any understanding systems such as question answering systems, information extraction and machine translation.

References

1. Byron, D., James, A.: Applying Genetic Algorithms to Pronoun Resolution. In: Proceedings of the Sixteenth National Conference on Artificial Intelligence AAAI (1999)
2. Carbonell, J., Brown, R.: Anaphora Resolution: A Multistrategy Approach. In: Proceedings of the 12th International Conference on Computational Linguistics, pp. 96–101 (1988)
3. Hobbs, J.: Resolving pronoun references. Lingua 44, 311–338 (1978)

4. Hock, H.H.: Studies in Sanskrit Syntax. Motilal Banarasidas, Delhi (1991)
5. Jha, G.N., Sobha, L., Mishra, D., Singh, S.K., Pralayankar, P.: Anaphors in Sanskrit. In: Proceedings of the Second Workshop on Anaphora Resoplution (WAR-II). Nealt Proceedings Series, vol. 2, pp. 11–25 (2008)
6. Jha, G.N., Sobha, L., Mishra, D.: Discourse Anaphor and Resolution Techniques in Sanskrit. In: Proceedings of the 7th Discourse Anaphora and Anaphora Resolution Colloquium, Goa, pp. 135–150 (2009)
7. Kennedy, C., Boguraev, B.: Anaphora for Everyone: Pronominal Anaphora Resolution without a Parser. In: Proceedings of the 16th International Conference on Computational Linguistics (COLING 1996), Denmark, pp. 113–118 (1996)
8. Lappin, S., Mccord, M.: Anaphora Resolution in Slot grammar. Computational Linguistics 16(4), 197–210 (1990)
9. Lappin, S., Leass, H.: An Algorithm for Pronominal Anaphora Resolution. Computational Linguistics 20(4), 535–561 (1994)
10. Mitkov, R.: Factors in Anaphora Resolution: They are not the only Things That Matter. A Case Study Based on Two Different Approaches. In: Proceedings of the ACL 1997/EACL 1997 Workshop on Operational Factors in Practical, Robust Anaphora Resolution, Spain, pp. 14–21 (1997)
11. Sobha, L., Patnaik, B.N.: An Algorithm for Pronoun and Reflexive Resolution in Malayalam. In: International Conference on Computational Linguistics, Speech and Document processing, pp. 63–66 (1998)
12. Sobha, L., Patnaik, B.N.: A Quantitative and a Non-Quantitative Approach to Resolve Reflexives and Pronouns in Malayalam. IJDL 27(1), 33–40 (1999)
13. Sobha, L.: An Anaphora Resolution System for Malayalam and Hindi (An unpublished doctoral thesis) Mahatma Gandhi University, Kottayam (1999)
14. Sobha, L., Pralayankar, P.: Anaphors in Sanskrit. In: International Conference on Referential Entity Resolution. AU-KBC, Chennai (2007)
15. Speyer, J.S.: Sanskrit Synta. E.J. Brill, Leyden (1886)
16. Viswanathan, S., Ramesh Kumar, S., Kumara Shanmugam, B., Arulmozi, S.: A Tamil Morphological analyser. In: Proceedings of International Conference on Natural Language Processing, pp. 31–39 (2003)

Linguistic Investigations into Ellipsis in Classical Sanskrit

Brendan S. Gillon

McGill University
Montreal, Quebec
brendan.gillon@mcgill.ca

Abstract. Ellipsis is a common phenomenon of Classical Sanskrit prose. No inventory of the forms of ellipsis in Classical Sanskrit has been made. This paper presents an inventory, based both on a systematic investigation of one text and on examples based on sundry reading.

Keywords: anaphora, antecedent, ellipsis, gapping.

1 Introduction

Most, if not all, Sanskritists would agree that ellipsis is a common feature of Classical Sanskrit. Yet, what ellipsis is in Classical Sanskrit and which grammatical factors determine under which conditions it is permitted are questions which, to my knowledge, have never been raised in modern studies of Classical Sanskrit. The aim of this paper is to direct the attention of the community of Sanskrit scholars to these questions.

Below, I report on the forms of ellipsis which I have found in two corpora and discuss their bearing on possible answers to the questions raised above. The two corpora are the prose sentences found in V. S. Apte's *The Student's Guide to Sanskrit Composition* and the sentences up to the thirty-eighth verse of the *Svārthānumāna* chapter of Dharmakīrti's *Pramāṇavārttika* (PVS). I also supplement the discussion with cases found in sundry sentences outside of the corpora.

The term *ellipsis* is often used broadly and without much care, with the result that it fails to encompass anything which would be taken by linguists as a unified phenomenon. To arrive at a linguistically useful characterization, let us consider the English expression *Yajñadatta lentils*. This expression is not a sentence. It seems somehow defective and it fails to express a proposition. Yet, if it is preceded by the English sentence *Devadatta cooked rice*, as in (1.1) below, the expression *Yajñadatta lentils* no longer appears defective and it conveys a proposition, namely the one expressed by the sentence *Yajñadatta cooked lentils*. It is surely no coincidence that what the defective expression *Yajñadatta lentils* conveys when it is immediately preceded by the sentence *Devadatta cooked rice* is precisely what is expressed by the sentence which results from inserting the verb *cooked*, found in the preceding sentence, into the defective expression *Yajñadatta*

G.N. Jha (Ed.): Sanskrit Computational Linguistics, LNCS 6465, pp. 218–230, 2010.

lentils. Moreover, as illustrated by the pair of sentences in (1), what the defective expression *Yajñadatta lentils* conveys varies with proposition expressed by the sentence preceding it.

(1.1) Devadatta cooked rice; Yajñadatta lentils.
(1.2) Devadatta grew rice; Yajñadatta lentils.

Generalizing from this example, I say that an expression contains *ellipsis* just in case the expression, by itself, does not form an acceptable constituent, but which, when supplemented by suitable expressions from its preceding or succeeding text, or cotext[1], with little or no modification to it, yields an expression which forms a constituent and the resulting expression expresses what the unsupplemented expression conveys, relative, of course, to the given cotext. I call the *point of ellipsis* the point in an otherwise defective expression where the insertion of an expression taken from the cotext turns the defective expression into one which is not defective and which conveys what the defective expression conveys relative to the cotext. I call the expressions which are supplied the *antecedent.*[2]

The notation I use below is the familiar notation of labelled bracketing. To enhance readability, I omit all pairs of labelled brackets not essential to the discussion. To explain uses of the notation specific to this paper, let me begin with an application of the notation to the first example in (1).

(2) [S Devadatta [V cooked 1] rice]; [S Yajñadatta [V E 1] lentils]

'S' is the label for sentence or clause and 'V' is the label for verb. The antecedent of the point of ellipsis has a numerical index prefixed to its right hand bracket and the point of ellipsis is marked by a pair of square brackets enclosing 'E' and having the same labels as those of its antecedent. Of course, the notation of labelled brackets was devised for languages such as English in which the linear order of expressions play a syntactic role. As is well known, linear order plays a reduced role in Classical Sanskrit prose. The reader, then, should understand the notation applied to expressions of Classical Sanskrit as indicating solely constituency. Thus, in terms of this use of the notation, the placement of 'E' with the constituent immediately containing the point of ellipsis is arbitrary. However, from the cases examined, it appears that the order of the words within the constituent immediately containing the antecedent of the ellipsis and the order of the words in the constituent immediately containing the point of ellipsis are parallel. For that reason, I have placed the 'E' in the same linear position relative to its sister constituents as its antecedent has relative to its sister constituents. Finally, I have supplemented the usual categorial labels for adjective, noun,

[1] It is useful to distinguish between the physical situation, or setting, in which an expression is uttered and the text, or cotext, which either precedes or succeeds it. I shall use these technically defined terms, rather than the more common but very imprecise term *context*.

[2] As is well known, ellipsis and anaphora are very similar phenomena.

adjective phrase and noun phrase with numerical indices corresponding to the standard enumeration in Pāṇinian grammar of the cases. To highlight structural properties of the Sanskrit expressions, I have also taken the liberty of eliminating sandhi.

There is no exceptionless linguistic regularity. Due to limitations of space and time, I have not mentioned well known exceptions to generally recognized linguistic regularities, nor have I considered alternative analyses to the various examples adduced below.

The predominant forms of ellipsis are those of the ellipsis of the head verb of a verb phrase and of a head noun of a noun phrase. In addition, I have found some cases where the point of ellipsis is the point of a complement to a head verb or a head adjective. I begin by reviewing first the forms of ellipsis within a verb phrase, second the forms within a noun phrase and third the forms within an adjective phrase. I end by bringing to the reader's attention a form of ellipsis which I call *denial ellipsis.*

2 Ellipsis within the Verb Phrase

The most commonly encountered pattern of ellipsis within a verb phrase is the one where the verb itself is ellipted[3], similar to the English case given above. (The English case is known in the literature as *gapping.*) Verb head ellipsis may occur when the ellipted verb occurs in one coordinated clause and the antecedent in another. Thus, in the example here, the verb *vidhīyeta* (*can be affirmed*) occurs in the first clause and it serves as the antecedent for the verbless second clause.

> (3) PVS 5.10
> [S vidhau [S viruddhaḥ vā [V vidhīyeta 1]] [S a-viruddhaḥ vā [V E 1]]]
> When there is an affirmation, either what is incompatible can be affirmed or what is not incompatible can be affirmed.

Such ellipsis may also occur when the clauses are asyndetically coordinated, that is, coordinated without a conjunction.

> (4) MBh 1.241.7 (cited in Scharf [8] p. 110)
> [S loke ca [S ekasmin vṛkṣaḥ iti [V prayuñjate 1]] [S dvayoḥ vṛkṣau iti [V E 1]] [S bahuṣu vṛkṣāḥ iti [V E 1]]]
> In ordinary speech, the word 'tree' applies to a single thing, the word 'tree' [in the dual] to two things and the word 'trees' to many things.

Verb head ellipsis is found when the point of ellipsis occurs in a main clause and the antecedent in a subordinate one. In the next example, the point of ellipsis is in the verbless main clause *atra tathā viraktaḥ api* (*now in the same way an unimpassioned person too*). Its antecedent is the verb of the subordinate clause *yathā raktaḥ bravīti* (*as an impassioned person speaks*), which precedes the point of ellipsis.

[3] I follow here the useful coinage of Quirk *et al* [6], who introduced the verb *to ellipt*, corresponding to the noun *ellipsis.*

(5) PVS 9.7
[S atra [S yathā raktaḥ [V bravīti 1]] tathā viraktaḥ api [V E 1]]
Now an unimpassioned person speaks in the same way as an impassioned one does.

Here is another such case example, this one taken from *Śabarabhāṣya*.

(6) ŚBh (cited in Scharf [8] p. 286)
[S na ca [S yathā daṇḍi-śabdaḥ na daṇḍe [V prayuktaḥ 1]] evam go-śabdaḥ na ākṛtau [V E 1]]
And it is not the case that the word 'cow' does not apply to the form in the same way as the word 'stick possessor' does not apply to a stick.

Verb head ellipsis may also occur when the antecedent is in a preceding sentence. For example, *sādhyate* (*is established*) occurs in the first sentence and serves as the antecedent for the point of ellipsis of the verbless main clause of the second sentence.

(7) PVS 5.24
[S tatra kevalam viṣayī [V sādhyate 1]]
[S asyām api [S yadā vyāpaka-dharma-anupalabdhyā vyāpya-abhāvam [V āha 1]] tadā abhāvaḥ api [V E 1] iti]
In this case, only intentional activity [pertaining to an absence] is established. In this case too, an absence also is established when one states the absence of a pervadee on the basis of a non-apprehension of [its] pervader property.

Like ellipsis in other languages which have been well studied, features required at the point of ellipsis of a word and the features of the antecedent expression may differ. Thus, in the next sentence, *uktau*, a past participle being used as a main verb, is in the neuter dual, whereas the point of ellipsis requires the form of masculine singular.

(8) PVS 2.13
[S etena anvaya-vyatirekau yathāsvam pramāṇena niścitau [V uktau 1]]
[S pakṣa-dharmaḥ ca [V E 1]]
Concomitance and contra-concomitance, ascertained by [their] respective means of epistemic cognition, are thereby mentioned. The pakṣa-property is also [thereby stated].

Unusual for Sanskritists whose first language is a Western European language is the fact that the point of ellipsis may precede the antecedent for the point of ellipsis. Although I have not come across examples in the *Svārthānumāna* chapter of the *Pramāṇavārttika* so far, I have found examples in Karṇakagomin's *ṭīkā* on Dharmakīrti's text. In the example adduced below, the first two clauses are verbless and the antecedent for their points of ellipsis is the verb *kathyate*, which occurs in the last clause.

(9) PVSṬ 108.8
[S atra [S [S [NP3 prathamayā kārikayā] [NP1s dharma-kalpanā-bījam] [V E 1]] [S [NP3 dvitīyayā] [NP1s dharma-kalpā] [V E 1]] [S [NP3 tṛtīyayā] [NP1s pratij nā-artha-ekadeśatā-pariharaḥ] ca [V kathyate 1]]] iti samudāya-arthaḥ]
The overall meaning here is this: the seed for the conceptualization of properties is mentioned by the first verse; the conceptualization of properties by the second; and the rejection of [an inference's grounds] forming part of the state of affairs of [its] conclusion by the third.

This configuration, in which the point of ellipsis precedes its antecedent, does not appear to be uncommon, as I have come across two examples outside my two corpora.

(10) ŚBh (cited in Scharf [8] p. 283)
[S tena atra [S ākṛtiḥ guṇa-bhāvena [V E 1]] [S vyaktiḥ pradhāna-bhāvena [V vivakyṣate 1]] iti
Therefore here a form will be accepted as subordinate, the individual as principal.

(11) Candrakīrti on Āryadeva's *Catuḥśataka* 14.10
[S tatra [S yathā ekasya rūpasya ghaṭatvena avasthānam [V E 1]] tathā anyasya api paṭa-ādi-sambandhinaḥ rūpasya kasmāt ghaṭatvena avasthānam na [V iṣyate 1]]
Now, why is a material form which is connected with such things as cloth not accepted as abiding as a pot in the same way as another material form [is accepted] as abiding as a pot?

The question arises whether or not ellipsis within a verb phrase is confined to the head verb. I have found three cases where, besides the verb, other material in the verb phrase is ellipted. In the first case, the verb and the negative adverb *na* are required for the point of ellipsis.

(12) PVS 18.5
[S tathā [AC prasiddhe tat-bhāve hetu-bhāve vā] [S anityatva-abhāve kṛtakatvam [VP [ADV na 1] [V bhavati 2]] [S dahana-abhāve ca dhūmaḥ [VP [ADV E 1] [V E 2]]]]]
In this way, when [either the relation of something] being something or [the relation of something] being the cause [of something] is established, neither does artificiality exist in the absence of non-eternality nor does smoke exist in the absence of fire.

One might wonder whether or not this case of ellipsis might be better described as the ellipsis of the entire verb phrase. There is some evidence to support this analysis. Thus, in the next example, the verb phrase of the first sentence comprises a direct object and the verb, *kāraṇam anumāpayati* (*implies a cause*). Both are required for the point of ellipsis in the following sentence.

(13) PVS 10.6
[S tasmāt nāntarīyakam eva kāryam [NP2 kāraṇam 2] [V anumāpayati 3] tat-pratibandhāt]
[S na anyat [VP [NP2 E 2] [V E 3]] vipakṣe a-darśane api]
Therefore, only an immediate effect implies a cause because of [its] relation to it [sc., the cause]. No other does, even if there is no observation [of it] in the vipakṣa.

However, it does happen that the head verb and only some, and not all, of its complements are ellipted. Thus, the first subordinate clause below contains a verb *cintayati* (*was wondering*) and two complements, second case noun phrase complement *kim* (*what*) and a prepositional phrase complement *mām antareṇa* (lit. *about me*; *about him*). Yet, the succeeding two clauses omit only the verb and the prepositonal phrase and repeat the second case noun phrase.

(14) K 178.23–179.2 (as cited in SG 3.1.13)
[S asyām velāyām [S [NP2 kim nu] khalu [PP mām antareṇa 1] [V cintayati 2]] Vaiśaṁpāyanaḥ] [S [NP2 kim] vā [PP E 1] [V E 2] varākī Patralekhā] [S [NP2 kim] vā [PP E 1] [V E 2] rāja-putra-lokaḥ] iti cintayan eva saḥ nidrām yayau]
At that moment, he went to sleep just while wondering what Vaiśaṁpāyana was wondering about him, what poor Patralekhā was wondering about him and what the prince's retinue was wondering about him.

3 Ellipsis in Noun Phrase

I now turn to ellipsis within the noun phrase. I begin with ellipsis in non-subject noun phrases, since subject noun phrase present complicating factors whose consideration it is best to put off for the moment. As with verb phrases, so with noun phrases, the head can be ellipted. The next sentence, discussed above, contains both the ellipsis of a head verb and the ellipsis of a head noun. Here, the third case noun phrase *prathamayā kārikayā* (*by the first verse*) in the first subordinate clause is reduced to the adjective phrases *dvitīyayā* (*by the second*) and *tṛtīyayā* (*by the third*) in the second and third clauses respectively.

(15) PVSṬ 108.8 (same as (9) above)
[S atra [S [S [NP3 [AP3 prathamayā] [N3 kārikayā 2]] [NP1s dharma-kalpanā-bījam] [V E 1]] [S [NP3 dvitīyayā [N E 2]] [NP1s dharma-kalpā] [V E 1]] [S [NP3 tṛtīyayā [N3 E 2]] [NP1s pratij nā-artha-ekadeśatā-pariharaḥ] ca [V kathyate 1]]] iti samudāya-arthaḥ]
The overall meaning here is this: The seed for the conceptualization of properties is mentioned by the first verse; the conceptualization of properties by the second; and the rejection of [an inference's grounds] forming part of the state of affairs of [its] conclusion by the third.

What is noticeable about the next case is that the antecedent is the single noun of an absolute phrase and the point of ellipsis is the position the noun would occur in in the absolute phrase of the following clause.

(16) PVS 5.12
[S [NP7 [NP6 a-viruddhasya] [N7 vidhau 1]] sahabhāva-virodha-abhāvāt a-pratiṣedhaḥ]
[S [NP7 [NP6 viruddhasya [PRT api]] [N7 E 1]] anupalabdhi-abhāvena virodha-a-pratipattiḥ]
When there is an affirmation of what is not incompatible [with anything else], then there is no denial [of anything else], because there is no incompatibility of [its] co-occurrence [with anything else].
[And] even when there is an affirmation of what is incompatible, without non-apprehension, there is no awareness of the incompatibility.

Ellipsis also occurs in a predicate noun phrase. In the second sentence of the example below, predicated of its subject noun phrase *ākṛti-pratyayaḥ* (*understanding of a form*) is *vyakti-pratyayasya* (*understanding of the particular*), from which *nimittam* (*basis*), the head of the predicate noun phrase in the first sentence, has been ellipted.

(17) ŚBh (as cited in Scharf [8] p. 285)
tasmāt [S [NP1s śabdaḥ] [NP1 [NP6 ākṛti-pratyayasya] [N1 nimittam 1]]]
[S [NP1s ākṛti-pratyayaḥ] [NP1 [NP6 vyakti-pratyayasya] [N1 E 1]]] iti.
Therefore a word is a basis for the understanding of form, the understanding of a form for the understanding of the particular.

3.1 Ellipsis in Subject Noun Phrase

I now turn to ellipsis in the subject noun phrase. As is well known, an independent clause in Classical Sanskrit does not require a subject noun phrase. Some cases are clearly not cases of ellipsis, for no antecedent is required for understanding the subjectless clause. Thus, for example, the following clause has no subject noun phrase. The agent of the verb *āha* is Dignāga, but Dignāga is mentioned no where in the work.

(18) PVS 2.22
[S tathā ca āha [S sarvaḥ eva ayam anumāna-anumeya-vyavahāraḥ buddhi-ārūḍhena dharma-dharmi-bhedena iti]]
And in this way, one [i.e., Dignāga] has said that the entire difference between grounds for inference and what is to be inferred is due to differentiation between property and property-possessor, which falls within the purview of the intellect.

However, there are cases which satisfy all the usual conditions for ellipsis: while there is no noun in the nominative case, there is a subordinate constituent which make sense only relative to a noun in the nominative case; and, in the preceding cotext, there is a noun in the nominative case whose sense is the sense which is relevant the subordinate constituent. Though the following example contains a case of verb head ellipsis, what is relevant here are the two cases of head noun ellipsis. The head noun of the subject noun phrase of the first sentence *bhedaḥ* (*difference*) and the head noun *bhāvānām* (*of things*) of the subordinate noun phrase *bhāvānām abhinnānām* (*of similar things*) are both omitted from the following sentence, whose verb is also to be supplied from the first sentence, as we saw above.

(20) PVSṬ 108.17
[S na [NP3 [N3 jāti-ādinā 1] tāvat] [NP1s [NP6 [N6 bhāvānām 2] [AP6 abhinnānām]] [N1 bhedaḥ 3]] [VP [V kriyate 4]] ...]
[S na api [NP3 [N3 E 1]] [NP1s [NP6 [N6 E 2] [AP6 bhinnānām]] [N1 E 3]] [VP [V E 4]] ...]
No universal makes similar things different, ..., neither does it make dissimilar things different, ...

The sentences in the next two examples are verbless sentences. In each example, the second sentence contains the point of ellipsis associated with which is a subordinate noun phrase whose proper interpretation requires the head noun of the subject of the preceding sentence.

(21) PVS 8.5
[S iti tatra api [NP1s atīta-eka-kālānām [N1 gatiḥ 1]]]
[S na [NP1s [NP6 an-āgatānām] [N E 1]] vyabhicārāt]
So, in this case, there is knowledge of things contemporaneous and antecedent [to the observation of the ground]. There is no knowledge of things future, because of deviation [from them].

(22) PVS 23.18
[S katham tarhi idānīm [NP1s bhinnāt sahakāriṇaḥ [N1 kārya-utpattiḥ 1]] [S yathā [NP1s cakṣur-rūpa-ādeḥ vijñānasya [N1 E 1]]]]
How then now does an effect arise from distinct ancillary causes, as when there is the arising of awareness from [a variety of ancillary causes such as] eye and form?

Remarkably, the antecedent for noun ellipsis may be part of a compound. In the next case, the antecedent is the head *pratiṣedhaḥ* (*ruling out*) of the compound *eka-siddha-pratiṣedhaḥ* (*ruling out what is established for one*).

(23) PVS 11.3
[S dvayoḥ iti eka-siddha-(pratiṣedhaḥ 1)] [S prasiddha-vacanena saṃdigdhayoḥ śeṣavat-asādhāraṇayoḥ sapakṣa-vipakṣayoḥ api [N1 E 1]]
The expression 'for both' rules out anything which is established for [just] one [of the two interlocutors]. The expression 'established' rules out doubtful properties which are deficient and undistributed with respect to the sapakṣa and vipakṣa.

Such cases are not rare. Next is a case where the required antecedent for the second sentence is *pratipattiḥ* (*knowing*), the head of the compound which is the head of the subject noun phrase, again, of a verbless sentence.

(24) PVS 9.8
[S [NP1s [NP5 vacana-mātrāt] a-(pratipattiḥ1)]]
[S na api [NP1s [N1 E 1] [NP5 viśeṣāt]] abhiprāyasya durbodhatvāt vyavahāra-saṃkareṇa sarveṣām vyabhicārāt]
So, there is no knowing [whether or not someone is impassioned] from his merely speaking.
Nor is [there any knowing whether or not someone is impassioned] from a special kind [of speaking], because, [a speaker's] intentions' being difficult to know, [his various forms of] behavior are diverse and so all [forms of his speaking] are deviant [with respect to whether or not he has passion].

The next example contains head noun ellipsis both in the subject noun phrase and in the predicate noun phrase of the second sentence.

(25)ŚBh (as cited in Scharf [8] p. 286)
[S na tu [S [NP1s [N1 go-śabda-avayavaḥ 1] [AP1 kaścit]] [NP1 [NP6 ākṛteḥ] [N1 pratyāyakaḥ 2]]]
[S [NP1s [N E 1] anyaḥ] [NP1 [NP6 vyakteḥ] [N1 E 2]]]]
But it is not the case that some part of the word 'cow' makes one aware of the form, another of the particular.

To see how subtle the question of whether or not the absence of a subject noun phrase is a matter of ellipsis or not, let me explain, using an example from English, an important distinction which can be easily overlooked. There is an important contrast in English between the pronoun *it* and the pronoun *one*, as seen from considering the following minimally contrasting pair of sentences.

(26.1) Devadatta bought <u>a mango</u> and Yajñadatta too bought <u>it</u>.
(26.2) Devadatta bought <u>a mango</u> and Yajñadatta too bought <u>one</u>.

For the first sentence to be true, Devadatta and Yajñadatta must have bought the very same mango. For the second sentence to be true, it is required only that each bought a mango. The point can be made in another way: should the pronoun in each sentence in (3) be replaced by its antecedent, the replacement preserves truth conditions in the second case but not the first.

(27.1) Devadatta bought <u>a mango</u> and Yajñadatta too bought <u>a mango</u>.
(27.2) Devadatta bought <u>a mango</u> and Yajñadatta too bought <u>a mango</u>.

The difference between *it* and *one* is that the semantic value of *it*, when it has an antecedent, must be a unique value of its antecedent, whereas the value of *one* is any of several values, all of which are in the extension of the antecedent.

How does this bear on ellipsis? According to the characterization of ellipsis given at the start, filling a point of ellipsis with its antecedent (modulo gender and number) yields an expression which expresses what the expression containing the ellipsis conveys.

Consider now the next example.

(28) PVS 2.16
[S [NP7 pradeśa-viśeṣe kva-cit 1] na [NP1s [N1 ghaṭaḥ 2]] upalabdhi-lakṣaṇa-prāptasya anupalabdheḥ]
[S [S yadi [NP7 E 1] syāt [NP1s [N1 E 2]]] [NP1s [N1 E 2]] upalabhya-sattvaḥ eva syāt]
There is no pot in a certain particular place, since [it], satisfying the conditions for apprehensibility, is not apprehended [there].
If one were to exist [there], [then] it would be precisely that thing whose presence is to be apprehended.

There are three points of ellipsis in the second sentence, two of them ellipted subject noun phrases and one of them an ellipted verb complement noun phrase. The value of the first ellipted subject noun phrase can not be the value of its antecedent, for the preceding sentence denies the existence of any pot. This means that the value assigned to it is any value which is in the extension of *ghaṭaḥ* (*pot*). This is reflected in the translation of the ellipted subject noun phrase by the English pronoun *one*. In contrast, the value of the second ellipted subject noun phrase must be the very value assigned to the first ellipted subject noun phrase. This is reflected in the translation of that ellipted subject noun phrase by the English pronoun *it*. Finally, the value of the ellipted verb complement noun phrase must be the value of its antecedent noun phrase. This is reflected in the translation of that ellipted noun phrase by the English pronoun *there*.

4 Ellipsis in Adjective Phrase

Ellipsis in adjective phrases appears to be not so common. The only case found in the two corpora is a case where the adjective head, itself a compound, serves as the antecedent for the point of ellipsis in the second sentence.

(30) PVS 3.1
[S bhedaḥ dharma-dharmitayā [A1 buddhi-ākāra-kṛtaḥ 1]]
[S na arthaḥ api [A1 E 1] vikalpa-bhedānām svatantrāṇām an-artha-āśrayatvāt]
Differentiation [of them] as property and property-possessor is brought about through an image of the intellect. It is not that a thing too is so brought about, since conceptual differentiation, being independent [of things], does not have a basis.

The only other case I am aware of occurs in the *Mahābhāṣya*, where the point of ellipsis is the complement to an adjective.

(31) MBh 1.220.22 (cited in Bronkhorst [2] p. 57)
[S [S yathā tarhi ratha-aṅgāni vihṛtāni pratyekam [PP vrajikriyām prati 1] asamarthāni bhavanti] [S tat-samudāyaḥ ca rathaḥ [PP E 1] samarthaḥ] evam . . .]
Just, then, as the parts of a chariot when taken apart are individually incapable of movement yet the whole of them, the chariot, is capable, so . . .

5 Denial Ellipsis

Let me conclude by drawing attention to what might be still another form of ellipsis, though this is not so clear, as I think the examples below will show. What I call *denial ellipsis* is what occurs when an author wishes to deny an antecedently made claim. It has the form of the negative adverb *na* followed by a fifth case noun phrase which gives the grounds for the denial. Its antecedents are either clauses or noun phrases.

In the first case below, the first sentence is a verbless clause expressing a negative existential. The negative existential is within the scope of a seventh case noun phrase *tat-ātmatve* (*having the other for its nature*), best taken as a verbless absolutive. The object of the denial in the second sentence is the preceding verbless clause, under the scope of the verbless absolutive.

(32) PVS 2.21
[S [NP7 tat-ātmatve] [NP1s sādhya-sādhana-bheda-abhāvaḥ] iti cet]
[S na E dharma-bheda-parikalpanāt]
It might be argued that there is no difference between the establishable and the establisher when [the one] has the other for its nature. [This is] not [so], since differentiation among properties is conceptualized.

In the next case, the antecedent for the point of ellipsis is a fifth case noun phrase *prayojana-abhāvāt* (*because of having no point*), which, at the point of ellipsis, must be understood to have propositional content.

(33) PVS 1.12
[S [NP5 prayojana-abhāvāt] [NP1s an-upacāraḥ] iti cet]
[S na E sarva-dharmi-dharma-pratiṣedha-artha-tvāt]
It might be argued that there is no synecdoche [here] since [it] would have no point. [This is] not [so], since [it] has the purpose of prohibiting [from being a ground] a property property-possessors of which are everything [but the pakṣa].

6 Conclusion

Above, I have examined a diverse set of examples of ellipsis in Classical Sanskrit. The preponderance of cases are those in which the antecedent of the point of ellipsis is a head verb or a head noun. There is only one case where the antecedent is a head adjective. The few remaining cases are those in which a complement of a verb or an immediately subordinate constituent of a noun phrase. The forms of ellipsis differ in important respects from the forms of ellipsis for English known as gapping, verb phrase ellipsis and copular complement ellipsis. Also noteworthy are cases of ellipsis where the antecedent is the head of a compound and cases of ellipsis where the antecedent comes after the point of ellipsis in spite of the fact that the antecedent and the point of ellipsis are in coordinated clauses. Both of these patterns are unknown in English.

The foregoing raises three questions. First, are there other forms of ellipsis in Classical Sanskrit? Second, under what conditions may one expression serve as an antecedent to a point of ellipsis? Third, how is the value of a constituent containing a point of ellipsis determined? To answer these questions satisfactorily, further extensive and careful empirical investigation of prose texts must be undertaken. This requires that texts of Classical Sanskrit be suitably annotated so that facts pertaining to ellipsis can be characterized. The annotation must be fine-grained enough so that the phenomena does not slip through undetected, but not so fine-grained that it prejudices the possiblity of finding data which might draw into question working hypotheses regarding the nature of ellipsis. The preceding presents a first stab at such an annotation.

In closing, let me bring to the reader's attention an observation well known to linguists, especially descriptive linguists (Quirk *et al.* [6] ch. 12.), which will be useful to bear in mind while pursuing answers to the last two questions. There is a close parallel between ellipsis, which, as we saw, requires an antecedent, and anaphora, which comprises a proform with an antecedent. Indeed, it is no coincidence that linguists use the term *antecedent* in connection with both phenomena. This is well illustrated by English nominal head ellipsis, where the pronoun *one* and a point of ellipsis are in complementary distribution.

Table 1. ABBREVIATIONS

A	= *Aṣṭādhyāyī* reference: adhyāya.pāda.sūtra
K	= *Kādambarī*, Peterson (ed) [5]. reference: page.line
MBh	= *Mahābhāṣya*, Kielhorn (ed) [4]. reference: volume.page.line
PVS	= *Pramāṇavārttika, svārthānumāna* chapter, Gnoli (ed) [3]. reference: page.line
PVSṬ	= *Pramāṇavārttikasvārthānumānaṭīkā*, Sāṅkṛtyāyana (ed) [7]. reference: page.line (of ṭīkā)
SG	= *Student Guide to Sanskrit Composition*, Apte [1]. reference: chapter.exerciseset.example
ŚBh	= *Śabarabhāṣya*, Scharf [8].

(34.1) The performance by Alex of Bach is better than the *E/one by Bill of Chopin.
(34.2) Alex's performance of Bach is better than Bill's E/*one of Chopin.

Whether the head noun of a noun phrase in which a determiner or a possessive noun phrase immediately precedes the head noun can be ellipted or replaced by the pronoun *one* depends on the choice of determiner or of a possessive noun phrase. This close parallel between ellipsis and anaphora suggests that investigation into either phenomena should pay attention to the results of investigation into the other.

References

1. Apte, V.S.: The student's guide to Sanskrit composition. A treatise on Sanskrit syntax for use of schools and colleges. Lokasamgraha Press, Poona (1960)
2. Bronkhorst, J.: Three problems pertaining to the Mahābhāṣya. Bhandarkar Oriental Research Institute, Pune (1987)
3. Gnoli, R. (ed.): The Pramāṇavārttikam of Dharmakīrti, the first chapter with the autocommentary, text and critical notes. Istituto Italiano per il Medio ed Estremo Oriente, Rome (1960)
4. Kielhorn, F.: Mahābhāṣya. 4th edition revised by R. N. Dandekar, 4th edn. Bhandarkar Oriental Research Institute, Pune (1985)
5. Peterson, P. (ed.): Kādambarī. Government Central Book Depot, Bombay (1885)
6. Quirk, R., Greenbaum, S., Leech, G., Svartik, J.: A Comprehensive Grammar of the English Language. Longman, London (1985)
7. Sāṅkṛtyāyana, R. (ed): Ācārya-Dhamrakīrteḥ Pramāṇavārttikam (svārthānumānaparicchedaḥ svopajñavṛttyā, Karṇakagominviracitayā taṭṭīkayā ca sahitam. Kitab Mahal, Allahabad (1943)
8. Scharf, P.: The denotation of generic terms in ancient Indian philosophy: grammar, Nyāya, and Mīmāṃsā. American Philosophical Society, Philadelphia (1996)

Asiddhatva Principle in Computational Model of Aṣṭādhyāyī

Sridhar Subbanna[1] and Shrinivasa Varakhedi[2]

[1] Rashtriya Sanskrit Vidyapeetha, Tirupati, India
sridharsy@gmail.com
[2] Samskrita Academy, Osmania University, Hyderabad, India
shrivara@gmail.com

Abstract. Pāṇini's Aṣṭādhyāyī can be thought of as an automaton to generate Sanskrit words and sentences. Aṣṭādhyāyī consists of *sūtras* that are organized in a systematic manner. The words are derived from the roots and affixes by the application of these *sūtras* that follow a well defined procedure. Therefore, Aṣṭādhyāyī is best suited for computational modeling. A computational model with conflict resolution techniques was discussed by us (Sridhar et al, 2009)[12]. In continuation with that, this paper presents, an improvised computational model of Aṣṭādhyāyī. This model is further developed based on the principle of *asiddhatva*. A new mathematical technique called 'filter' is introduced to comprehensively envisage all usages of *asiddhatva* in Aṣṭādhyāyī.

1 Introduction

We (Sridhar et al, 2009)[12] have discussed computational structure and conflict resolution techniques of Aṣṭādhyāyī. In the current paper, the computational model has been restructured and developed further incorporating the principle of *asiddhatva*. In the earlier model, *asiddha* for *ṣatva* and *tuk* were not handled. This model moves closer to the traditional view of Aṣṭādhyāyī.[1]

In Aṣṭādhyāyī the *sūtras* are declared as *asiddha* in the following instances:

1. *pūrvatrāsiddham* (8.2.1)
   Sūtras from 8.2.1 to 8.4.68 (*tripādī*) are *asiddha* to *sūtras* from 1.1.1 to 8.1.74 (*sapādasaptādhyāyī*). Also, in *tripādi* the successive *sūtras* are *asiddha* to their previous *sūtras.*
2. *asiddhavadatrābhāt* (6.4.22)
   The *sūtras* from 6.4.22 to 6.4.175 are deemed as *asiddha* to each other.
3. *ṣatvatukorasiddhaḥ* (6.1.86)
   The *sūtras* 6.1.87 to 6.1.111 are *asiddha* to *ṣatva sūtras* (8.3.39 to 8.3.119) and *tuk sūtras* (6.1.70 to 6.1.75)

[1] There were objections from the traditional grammarians on the assumption of linear application of *tripādī sūtras* in the earlier model. The present model doesn't have any such assumption. The environment is visible through filters to *sapādasaptādhyāyī* even after the application of *tripādī sūtras.*

G.N. Jha (Ed.): Sanskrit Computational Linguistics, LNCS 6465, pp. 231–238, 2010.

The concept of *asiddhatva* is used in Aṣṭādhyāyī:

- to prevent the application of *sūtra* on the substitute
- to enable the application of *sūtra* on the substituent[2] and
- to mandate the order of application of *sūtras*[3]

When *sūtra* A is declared as *asiddha* to *sūtra* B, then the transformed state obtained by application of *sūtra* A is invisible to B. This concept of *asiddhatva* finds its application in vidhi *sūtras.*[4]

Vidhi *sūtras* describe context sensitive transformations. It defines the operation to take effect in a particular context. Each *vidhi sūtra* will inspect the environment to find the context for its application. The environment contains all the transformations that happened due to application of *sūtras*. Even though there is only one environment, for *sūtras*, the object of inspection need not be same as the object of application. It will be same when there is no previous application of another *sūtra* in the environment that is *asiddha* to this *sūtra*. Otherwise, the object of inspection and the object of application will be different.[5] The *sūtra* inspects the state just before the application of the first *asiddha sūtra*. But the application will be on the transformed state due to application of the most recent *sūtra*. This can be explained in following example:

1. *śas hi*
   ↓ 6.4.35
2. *śā hi*
   ↓ 6.4.101
3. *śā dhi*

Suppose the environment is *śas hi* then the condition for *sūtras śā hau* (6.4.35) and *hujhalbhyo herdhiḥ* (6.4.101) are satisfied. As only one *sūtra* is applicable at any given point of time, one of them will be selected by the conflict resolver. Assume that 6.4.35 is applied. Then in the new transformed environment - *śā hi* - the condition for 6.4.101 is not satisfied. However, the *sūtra* 6.4.35 is *asiddha* to 6.4.101. Therefore, the *sūtra* 6.4.101 inspects the state 1 - *śas hi* and finds the condition satisfied. But, when this *sūtra* is applied, the object of application will be state 2 - the latest environment resulted after the application of 6.4.35 - *śā hi.*

The object of inspection for any given *sūtra* can be obtained using a technique called Filter. Filter is a mathematical function that gives an inspection environment for a *sūtra*. The pre-image for the function is the current environment. The object of inspection is the state that exists prior to the first application of any *sūtra* that is *asiddha* to the *sūtra* in question.

---

[2] Asiddhavacanam ādeśalakṣaṇapratiṣedhārtham utsargalakṣanabhāvārtham ca - vartika 6.1.86.1.

[3] See [12].

[4] Vidhi *sūtra* is one among the various types of *sūtra* discussed in the paper[12].

[5] This is similar to the story of *matsya yantra bhedana* during the *svayamvara* of Draupadī in Mahābhārata. Here the participant is supposed to look into the reflection in water of a rotating fish and target the real fish.

2 Model of Aṣṭādhyāyī

The principle of *asiddhatva* discussed above is an important device in the operation of Aṣṭādhyāyī. A complete and consistent process to operationalize Aṣṭādhyāyī can be visualized as an engine comprising of devices like *asiddhatva*, *paribhāṣā* etc. This engine generates words from roots and affixes by following a completely mechanical process.

The core of the engine is built on conventions defined in *paribhāṣā sūtras*. These conventions are framed to interpret metadata. The metadata describe mode of operation, context and the object of operation. The *vidhi sūtras* consist of data marked with these metadata. Interpreting the metadata, the engine carries out the operation on the object with the data supplied.

The mechanics of the engine can be summarized as follows: The *vidhi sūtras* keep inspecting the object(environment) supplied. The *sūtra* that finds the condition satisfied, sends a request to the conflict resolver. More than one *sūtra* may send a request in any particular context. Conflict resolver adopts the resolution techniques and unambiguously selects one among them for application. The selected *sūtra* will be applied to transform the environment.

As described earlier the object of inspection needs to take into account the principle of *asiddhatva*. To achieve this filters at various levels are introduced that are mathematically formulated. The filter $f(A, B, \alpha, x)$ returns the object of inspection $\alpha\prime$ for *sūtra* x in the environment α, where A and B are set of *sūtras* such that ∀ *sūtra* a ∈ A, every *sūtra* b ∈ B is *asiddha*. Set C is an ordered set of *sūtras* that are applied in the environment α arranged in the order of their application. The filter has the following algorithm to return the object of inspection:

If x ∈ A and (C ∩ B) ≠ ∅
then return the state before application of *sūtra* g ∈ α, where g is the first element ∈ (C ∩ B)[6]
else return α

Now, the following sets are defined for usage in specific filter definitions:

1. SA = {1.1.1, 1.1.2, ... , 8.1.140} - *sapādasaptādhyāyī sūtras*
2. TP = {8.2.1, 8.2.1, ... , 8.4.68} - *tripādī sūtras*
3. TU = {6.1.70, 6.1.71, ... , 6.1.75} - *tuk sūtras*
4. AV = {6.4.23, 6.4.24, ... , 6.4.175} - *asiddhavat sūtras*
5. SH = {8.3.39, 8.3.40, ... , 8.3.48} ∪ {8.3.55,8.3.56, ... , 8.3.119} - *ṣatva sūtras*
6. EA = {6.1.87, 6.1.88, ... , 6.1.111} - *ekādeśa sūtras* and
7. PP = g(x) = { x+1, x+2, ... , 8.4.68} : x ∈ TP , which is a dynamic set.

Using these set definitions, specific filter functions can be mapped to the *asiddha* declarations.

[6] C ∩ B is also an ordered set arranged by the order of C.

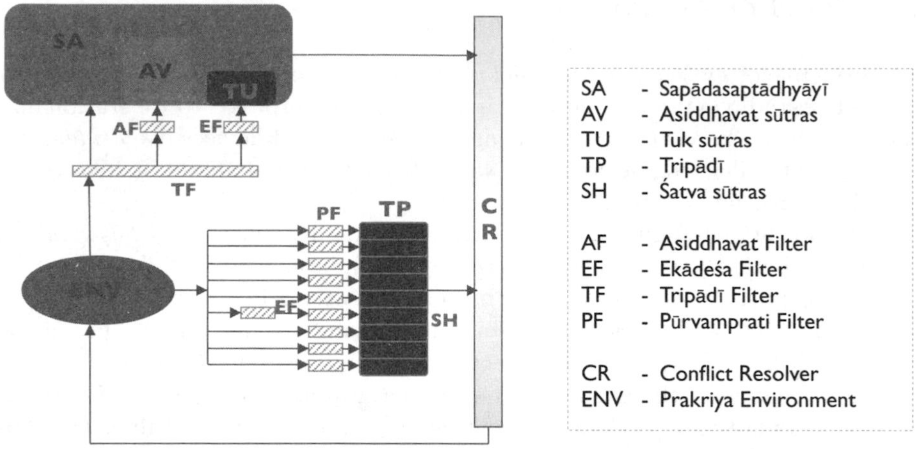

Fig. 1. Computational Structure of *Aśṭādhyāyī* with *Asiddhatva* Principle

Tripādī Filter $TF(\alpha, x) = f(SA, TP, \alpha, x)$ *pūrvatrāsiddham (8.2.1)*
Asiddhavat Filter $AF(\alpha, x) = f(AV, AV, \alpha, x)$ *asiddhavadatrābhāt (6.4.22)*
Ekādeśa Filter $EF(\alpha, x) = f(SH \cup TU, EA, \alpha, x)$ *ṣatvatukorasiddhaḥ (6.1.86)*
Pūrvamprati Filter $PF(\alpha, x) = f(TP, PP, \alpha, x)$ *pūrvatrāsiddham (8.2.1)* is a dynamic filter.

These filters as part of the *Aśṭādhyāyī* engine are depicted in figure 1. In this figure, the environment (ENV) contains all the states due to linear application of all the relavant *sūtras*. Block SA represents the *sapādasaptādhyāyī sūtras*, defined as set SA above. All the *sūtras* in this block will be inspecting ENV through the filter TF, as the *tripādi sūtras* are *asiddha* for the entire *sapādasaptādhyāyī*. Block AV stands for the *asiddhavat* section of *sapādasaptādhyāyī*, defined as set AV earlier. The *sūtras* in this block have the filter AF along with filter TF. Block TU denotes the *tuk* section of *sūtras* of *sapādasaptādhyāyī*, defined as set TU previously. In addition to the filter TF this block has an additional filter EF.

Block TP represents *tripādī*, defined as set TP above. In *tripādī* for every *sūtra* the successive *sūtras* are *asiddha*. Each *sūtra* needs a different form of filter. Hence, this filter has been defined previously as a dynamic filter varying as a function of the given *sūtra*. This dynamism has been represented as a group of multiple filters PF. Block SH denotes the *ṣatva* section of *tripādī*, denoted by set SH earlier. This block has an additional filter EF along with the PF filter. The filter EF is common for Blocks TU and SH as per *sūtra śatvatukorasiddhaḥ* (6.1.86).[7]

The Block CR represents the Conflict Resolution device which receives requests from all the *sūtras* that are applicable in the context and determines a *sūtra* for application by applying the resolution techniques[12].

[7] It is to be noted that set EA is *asiddha* to set SH$\subset$ SA, which is the converse to the general rule that TP is *asiddha* to SA.

3 Examples

The following examples illustrate the current model of Aṣṭādhyāyī and the role of *asiddhatva* in various scenarios:

Example 1. *vanena* (*vana* case 3, number 1)

1. *vana ṭā*
   ↓ 7.1.12
2. *vana ina*
   ↓ 6.1.87
3. *vanena*

This is a simple example where *asiddhatva* has no role. Initially the word *vana* in case 3, number 1 gets the *ṭā* pratyaya from the *sūtra svaujas... (4.1.2)* The *sūtra ṭāṅasiṅasaminātsyāḥ (7.1.12)* finds the condition and sends a request to conflict resolver. The resolver finds no other competing *sūtra* and hence allows this *sūtra* (7.1.12) for application on the environment. It transforms the object of application by changing *ṭā* to *ina* and pushes to the environment stack. Then the *sūtra ādguṇaḥ 6.1.87* goes through similar process and gets applied. The stage *vanena* does not satisfy the condition for any *sūtra* and remains as the final form.

Example 2. *śādhi*

1. *śas hi*
   ↓ 6.4.35
2. *śā hi*
   ↓ 6.4.101
3. *śā dhi*

In this example when the environment is in state 1, two *sūtras śā hau 6.4.35* and *hujhalbhyo herdhiḥ 6.4.101* find the condition and request the resolver. There is no conflict between these two but only one *sūtra* can be applied at a time, hence resolver chooses one among them and applies it. Let us say it picks 6.4.35. After the application of this *sūtra*, the state will be moved to state 2 as shown. The *sūtras* in Block AV will still see state 1, since they are seeing through filter AF. Even at state 2, both the above mentioned *sūtras* find the condition for application as they are seeing through the filter AF. However, the *sūtra* 6.4.35 is not permitted for second application.[8]. Therefore the *sūtra* 6.4.101 will request and is applied on the environment. The object of application will be state 2, even though the object of inspection is state 1. This application transforms the environment to state 3. This state does not satisfy the condition for any *sūtra* and remains as the final form.

[8] *lakṣye lakṣaṇaṃ sakṛdeva pravartate [5]*

Example 3. *jahi*

1. *jan hi*
   ↓ 6.4.36
2. *ja hi*

When the environment in state 1, *hanter jah̄ 6.4.36* is the only *sūtra* that finds the condition and is applied on the environment to transform to state 2. Here the condition is satisfied for the *sūtra ato heh̄ 6.4.105.* Because this *sūtra* is in Block AV, it views the environment through filter AF. Hence the object of inspection for this *sūtra* is state 1 but not state 2. Therefore this *sūtra* is prevented from application. State 2 remains as the final form.

Example 4. *dvā atra*

1. *dvau atra*
   ↓ 6.1.78
2. *dvāv atra*
   ↓ 8.3.19
3. *dvā atra*

The environment is as in state 1, the *sūtra ecoyavāyāvaḥ 6.1.78* finds the condition and is applied. Then environment is transformed to state 2. The *sūtra lopaḥ śakalyasya* 8.3.19 will be applied through similar process. Now the transformed environment is state 3. The question is whether the *sūtra akaḥ savarṇe dīrghaḥ* 6.1.101 will be able to find the condition in this state. The answer is no. The *sūtra* 6.1.101 being in block SA it inspects at the environment through the filter TF. It gets state 2 as the object of inspection, where it does not find condition. Therefore this *sūtra* cannot be applied and the final form is state 3.

Example 5. *kosya*

1. *ko asya*
   ↓ 6.1.109
2. *kosya*

In state 1 the rule *eṅaḥ padāntādati 6.1.109* is applied. Even though in this state, the *sūtra iṇaḥ ṣaḥ* (8.3.39) can find the condition satisfied, but this rule belongs to set SH and inspect ENV through filter EF. So, the *sūtra* 8.3.39 gets state 1 as object of inspection, where it does not find the condition satisfied. Therefore this *sūtra* will not be applied. Hence, state 2 is the final form.

Example 6. *pretya*

1. *pra i ya*
   ↓ 6.1.87
2. *pre ya*
   ↓ 6.1.71
3. *pretya*

In state 1 the rule *aadgunaḥ 6.1.87* is applied. This transforms the environment to state 2. The rule *hrasvasya piti kṛti* tuk (6.1.71) belongs to TU inspects ENV through the filter EF, and gets the object of inspection as state 1. Here, the *sūtra* 6.1.71 finds the condition and applied on the object of application, that is, state 2. Thus, transforming to state 3. It can be seen that in state 2 the *sūtra* do not find the condition satisfied, but this state is not the object of inspection. The final form is state 3.

Example 7. *manorathaḥ*

1. *manasrathaḥ*
   ↓ 8.2.66
2. *manarrathaḥ*
   ↓ 6.1.114
3. *manaurathaḥ*
   ↓ 6.1.87
4. *manorathaḥ*

In state 1 the *sūtra sasajuṣo ruh̄* (8.2.66) is applied transforming to state 2. This is a *rutva vidhi* and is considered as an exception to *asiddha* declaration in the Pāḥinian tradition. Hence *haśi ca* (6.1.114) should find the object of inspection as state 2 not state 1. This is handled in the TF definition. In this context, the *sūtras* 6.1.114 and *rori 8.3.14* find the condition. *sūtra* 6.1.114 is applied instead of 8.3.14 as per resolution technique of *siddha sūtra* precedence. With this application the context is transformed to state 3. Then *sūtra aadgunaḥ* (6.1.87) is applied to get the final form. This has been dealt by Cardona[11]

4 Conclusion

In this paper, the concept of *asiddhatva* has been mapped to the mathematically defined new concept called filter. This definition comprehensively envisages all applications of *asiddhatva* without deviating from the traditional interpretation. Further empirical study to validate the current model will throw more light on apt mathematical representation of *asiddhatva.*

Cardona[11] has discussed many illustrations that show exceptions to *asiddhatva* principle. He has shown that without these exceptions there will be *sūtras* that become purposeless as they would have no scope for application. The feasibility of incorporating these exceptions into the current computational model and filter technique needs a detailed study and further research.

References

1. Giri, S.P., Satyanarayanashastri.: Paṇinīyaḥ Aṣṭādhyāyī. Krishnadas Academy, Varanasi (1984)
2. Bhargava Sastri, B.J.: The Vyākaraṇamahābhāṣya of Patañjali. Chaukhamba Sanskrit Pratishtan, Delhi (1872)

3. Kielhorn, F.: Paribhāṣenduśekhara of Nāgojibhaṭṭa. Parimal Publications, Delhi
4. Mishra, S.: Kāśikā. Chaukamba Samskrita Samsthan, Varanasi (1979)
5. Abhyankar, K.V.: Paribhashasamgraha. Bhandarkar Research Instt, Puna (1967)
6. Kak, S.: The Paninian approach to natural language processing. International Journal of Approximate Reasoning 1, 117–130 (1987)
7. Kiparsky, P.: On the Architecture of Panini's Grammar
8. Bhate, S., Kak, S.: Panini's Grammar and Computer Science
9. Vasu, S.C.: The Aṣṭādhyāyī of Pāṇini. Motilal Banarasidas Publishers, New Delhi (2003)
10. Vasu, S.C.: Siddhānta Kaumudi, Motilal Banarasidas Publishers, New Delhi (2002)
11. Cardona, G.: pūrvatrāsiddham and āsrayātsiddham, MS
12. Sridhar, S., Srinivasa, V.: Computational Structure of Aṣṭādhyāyī and Conflict Resolution Techniques. In: Kulkarni, A., Huet, G. (eds.) Sanskrit Computational Linguistics. LNCS (LNAI), vol. 5406. Springer, Heidelberg (2009)

Modelling Aṣṭādhyāyī: An Approach Based on the Methodology of Ancillary Disciplines (*Vedāṅga*)

Anand Mishra

Department of Classical Indology
Ruprecht Karls University, Heidelberg, Germany
anand.mishra@urz.uni-heidelberg.de

Abstract. This article proposes a general model based on the common methodological approach of the ancillary disciplines (*Vedāṅga*) associated with the Vedas taking examples from *Śikṣā*, *Chandas*, *Vyākaraṇa* and *Prātiśākhya* texts. It develops and elaborates this model further to represent the contents and processes of Aṣṭādhyāyī. Certain key features are added to my earlier modelling of Pāṇinian system of Sanskrit grammar. This includes broader coverage of the Pāṇinian meta-language, mechanism for automatic application of rules and positioning the grammatical system within the procedural complexes of ancillary disciplines.

Keywords: Sanskrit Grammar, Modelling, Aṣṭādhyāyī, Ancillary Disciplines, Vedāṇga, Computer Simulation, Natural Language Processing.

1 The Methodological Approach of the Ancillary Disciplines

The ancillary disciplines (*Vedāṅga*) associated with the Vedas exhibit a certain degree of commonality in their general methodological approach. These disciplines were developed with the principal aim of retaining the given vedic phenomena by envisaging a system for its structured and standardized transmission. The main object of their enquiry are vedic utterances and sacrificial activities which can be empirically observed and analysed.

These disciplines were regarded as empirical sciences (*aparā-vidyā*) in contradistinction to the knowledge of the Absolute (*parā-vidyā*). For example, in MUṆḌAKOPANIṢAD it is stated that "Those who know *brahman* say that there are two sciences which should be known—the spiritual (*parā*) and the empirical (*aparā*). The empirical sciences are (the mastering of) *Ṛgveda*, *Yajurveda*, *Sāmaveda*, *Atharvaveda* (and the ancillary disciplines) *Śikṣā* (phonetics), *Kalpa* (ritual sciences), *Vyākaraṇa* (grammar), *Nirukta* (etymology) and *Jyotiṣa* (astrology). The spiritual knowledge, however, is the one through which that

G.N. Jha (Ed.): Sanskrit Computational Linguistics, LNCS 6465, pp. 239–258, 2010.

Absolute is realized."[1] According to Śaṅkarācārya the word 'Veda' here (as elsewhere in this context) means the collection of vedic utterances.[2] The upaniśadic approach is to know that, which is beyond and behind the given and their effort is not to return back to the given but to cross it. Similarly, the emphasis of the Yoga as well as the buddhist tradition is to get rid of the given. It is *heya*, that which is to be avoided, for which the cause (*hetu*) must be realized, a solution (*hāna*) is to be found and the corresponding prescription (*hānopāya*) must be instructed.[3]

The general procedure employed by the ancillary disciplines can be perceived as consisting of observing a given phenomenon, recording of the observed facts and comprehending the given through these observed facts. The main effort of the authors of the texts belonging to these disciplines is to develop a systematic framework for the description of a given phenomenon. Such a descriptive framework provides

- constituent units associated with a given phenomenon,
- their characterizing features or attributes and
- such statements which interconnect these units and attributes with a view to correlate the description with the original.

In the following, we take up some of the texts of the ancillary disciplines in order to represent them in this framework.

1.1 Prātiśākhya

The given phenomena in this case are the utterances of *saṃhitā-pāṭha*. The constituent *pada*s are units that convey some meaning[4] and these are collected in the *pada-pāṭha*s associated with a particular *saṃhitā-pāṭha*. The main aim of *Prātiśākhya* texts is to synthesize vedic utterances contained in *saṃhitā-pāṭha* out of constituent units, namely, the *pada*s. In this way the constituent *pada*s are correlated with the original *saṃhitā-pāṭha*.

It may be objected that *Prātiśākhya* texts consider *saṃhitā-pāṭha* to be a modification of the *pada-pāṭha* as, for example, the statement *saṃhitā padaprakṛtiḥ* in ṚGVEDA-PRĀTIŚĀKHYA (2.1). Commenting on this, Uvaṭa ([22]) says that *saṃhitā*, whose constituents are *pada*s, is here a modification of the constituting *pada*s. For example, the modifications *ṣatva* or *ṇatva* occur in *saṃhitā* only. Because they are the constituents, therefore, the *pada*s are established original forms.[5] Yāska in his NIRUKTA ([18]) also states that *saṃhitā* is the one

[1] dve vidye veditavye iti ha sma yadbrahmavido vadanti parā caivāparā ca. tatrāparārgvedo yajurvedaḥ sāmavedo'tharvavedaḥ śikṣā kalpo vyākaraṇaṃ nirukto chando jyotiṣamiti. atha parā yayā tadakṣaramadhigamyate (Mu. 1.4-5).

[2] veda-śabdena tu sarvatra śabdarāśirvivakṣitaḥ (Mu. Bhā. 1.5).

[3] For an analysis of this *caturvyūha* approach see [29].

[4] arthaḥ padam (Vā. Prā. 3.2).

[5] padāni prakṛtibhūtāni yasyāḥ saṃhitāyāḥ sā padaprakṛtiḥ saṃhitātra vikāraḥ. tathā hiṣatvaṇatvādayo vikārāḥ saṃhitāyā eva bhavanti. prakṛtibhūtatvācca padānāṃ siddhatvam. (Uv. 2.1).

having *padas* as its constituent and all the branches of the Veda consider it to be so.[6] Commenting upon this, Durgācārya takes up the question in a detailed manner and puts forward two possible cases:

1. That, which is the cause of *padas*, that (*saṃhitā*) is *padaprakṛti*. Why? *Padas* are formed out of *saṃhitā*. Therefore, some consider *saṃhitā* to be the original form (*prakṛti*) and *padas* to be their modifications (*vikāra*).
2. Others, however, understand the statement *padaprakṛtiḥ saṃhitā* to be *saṃhitā*, whose cause are the *padas*. Why? Because *saṃhitā* is gained out of the combinations of *padas* only. Therefore, *padas* are the original form and *saṃhitā* is their modification.[7]

He further raises the question, which option is better: to consider *padas* to be the original form and the *saṃhitā* to be their modification or *vice versa* and decides for the latter giving several justifications based on the earlier usage of *saṃhitā-pāṭha*.[8]

The original phenomena in this case are the given set of vedic utterances, the *saṃhitā-pāṭha*. The set of constituent units consists of the *padas* as well as the basic speech-sounds of vedic utterances (*varṇa*) and syllables (*akṣara*) etc. The rules of *Prātiśākhya* now characterize and combine these constituent units with the aim of correlating them to the original utterances. In this way, they describe the given set and provide a mechanism to retain the original by means of constituents and rules prescribed in a systematic framework. As an example, let us take the following statement:

> *viśva sahabhuvapuṣavasuṣu* (Vā. Prā. 3.101).
> The last vowel /*a*/ of *viśva* becomes long (*dīrgha*) in case *saha*, *bhuva*, *puṣa* or *vasu* follow. For instance: *gandharvastvā viśvāvasuḥ* (Śu. Yaj. 2.3).

This rule operates on constituents like /*a*/, *viśva*, *puṣa* etc. with the help of attributes like *dīrgha*, *svara* etc. and statements like 'add the attribute *dīrgha* to the last *svara* of *viśva* if the unit *vasu* follows' are meant to connect the word *viśva* with the original utterances of the Veda. Without this rule one would have combined *viśva* with *vasu* but would not have changed the concerned vowel long. This would have led the formation of **viśvavasuḥ* which is not to be found in the original utterances.

---

[6] padaprakṛtiḥ saṃhitā. padaprakṛtīni sarvacaraṇānāṃ pārṣadāni. (Ni. 1.17).

[7] padānāṃ yā prakṛtiḥ seyaṃ padaprakṛtiḥ. kiṃ kāraṇam? saṃhitāto hi padāni prakriyante. tasmātsaṃhitaiva prakṛtirvikāraḥ padānītyevameke manyante. *apare punaḥ padaprakṛtiḥ saṃhiteti padāni prakṛtiryasyāḥ seyaṃ padaprakṛtiriti. kiṃ kāraṇam? padānyeva hi saṃhanyamānāni saṃhitā bhavati. tasmāt padānyeva hi prakṛtirvikāraḥ saṃhiteti.* (Durga. 1.17 [28]).

[8] āha. kiṃ punaratra sādhīyaḥ padānāṃ prakṛtitvaṃ saṃhitāyā vikāratvamuta vā vikāratvaṃ padānāṃ prakṛtitvaṃ saṃhitāyā iti? ucyate saṃhitāyāḥ prakṛtitvaṃ jyāyaḥ. āha. kiṃ kāraṇam? ucyate. mantro hyabhivyajyamānaḥ pūrvamṛṣermantradṛśaḥ saṃhitayaivābhivyajyate na padaiḥ. (Durga. 1.17 [28]).

1.2 Śikṣā

The *Śikṣā* texts, e.g. the VYĀSA-ŚIKṢĀ [12] or NĀRADĪYĀ-ŚIKṢĀ [1] undertake a detailed enquiry into the sound units, attach a number of attributes like *svara, vyañjana, sparśa* etc. and also make a number of statements as to the combinations of these units and/or their attributes.[9] For example:

> *ikāraṃ yatra paśyeyurikāreṇaiva saṃyutam |*
> *udāttamanudāttena praśliṣṭaṃ taṃ nibodhata* | | (Nā. Śi. 2.1.6).
> Where a short *udātta* /*i*/ is seen to join a short *anudātta* /*i*/
> the resulting *svarita* should be known as *praśliṣṭa* ([1] p. 109).

Here again a statement is made as to the characterization of a constituent which is the result of some other operation, which in turn is conditioned in terms of certain constituents like /*i*/ and attributes like *udātta* and *anudātta* sounds.

1.3 Chandas

The texts concerned with metrical systems (*chandas*) also function in a similar manner. The given phenomena here are the vedic and secular (*laukika*) Sanskrit expressions. The constituent units in this case are the fundamental sound units (*varṇa*). It is this much smaller set with which the science of prosody works. The attributes are assigned either to a single sound unit or to syllables (*akṣara*) which are certain specific sequences of basic sound units. The statements or rules specify which attribute should be assigned to which particular unit or to which specific sequence. An example from the CHANDAḤŚĀSTRA of Piṅgala [4]:

> *gāyatryā vasavaḥ* | (Cha. Śā. 3.3).
> A verse quarter (*pāda*) containing eight (*vasu*) syllables (*akṣara*)
> is called *gāyatrī.*

The most important unit in case of CHANDAḤŚĀSTRA is an *akṣara* or syllable. This unit has two basic attributes, namely light (*laghu*) and heavy (*guru*). A verse or linguistic expression is seen as a sequence of syllables. Depending upon the number and distribution of syllables in a verse, different verse names or characterizing attributes are assigned to that sequence. The conditions are expressed using one or more of the following considerations:

1. Number of syllables in a sequence
2. Sum of the weights (*mātrā*) of the syllables in a sequence
3. Distribution of syllables in a sequence
4. Number of sub-sequences (*pāda*s)
5. Syllable at which a *pāda* or word starts
6. Syllable after which a pause is to be made

---

[9] For an exhaustive list of the topics covered in the VYĀSA-ŚIKṢĀ see [12] pp. 5-16.

The only operation here is that of attribute assignment. The entire field of operation is considerably simple and straight forward compared, for example, to the operations of AṢṬĀDHYĀYĪ. CHANDAḤŚĀSTRA deals primarily with numbers. It uses various names for numbers and provides groupings based on numerical values. Since, there are two basic attributes *laghu* and *guru* which a syllable can take, it is akin to the binary sequences of 0 and 1. The conditions are such, that the numerical characteristics are to be exploited.

Methodologically, however, the process can again be seen as follows: Given a linguistic expression, it is first analyzed into constituent units. In this case the units are syllables and the process of analysis can also be automatic and rule-based. The units are characterized by using the two basic attributes (*laghu* and *guru*) which sometimes depend on the neighboring units as well. Now the combination of these units in that sequence provides conditions for assigning the verse characterizing attributes to that linguistic expression. In this manner, the expression gets a prosodical identity. This assures the reproduction of that expression according to its standard rendering. This way, the prosodical characteristics of a given expression are retained.

1.4 Vyākaraṇa

The given phenomena which the grammarians like Pāṇini are representing are vedic utterances as well as the correct Sanskrit expressions according to the standard usage prevalent among the cultivated people during their time.[10] Given a collection of such linguistic expressions, AṢṬĀDHYĀYĪ provides a description in terms of constituent units out of which they are composed. It characterizes these units by attaching a number of attributes to them and provides for their appropriate combinations. A process of synthesis is specified which forms the given linguistic expressions from these building blocks. The grammar, thus, connects the basic units with the original expressions. The constituent units and the rules of synthesis together provide a systematic structure to represent the standard usage. In this way, grammar provides a description of the object set and helps in retaining and upholding the given. This is one of the main purposes of its composition (See Sect. 3).

The brief account of the basic methodological approach of some of the *Vedāṅga* texts is to illustrate the possibility to read them from the perspective of the framework mentioned above. This involves recognizing and identifying the components, scrutinizing their character and detecting the relationships between these components. A model which may account for a broader coverage of the processes of ancillary disciplines (*Vedāṅga*) should first and foremost concentrate on these core processes. A particular system can then be modelled as a special instance of this general model. We now turn our attention to propose a general model first.

---

[10] The question as to what exactly is the content of this object set is of no direct relevance to the model presented in this chapter. What is necessary is to acknowledge that some object set is being analyzed and described. For a discussion about the nature of this set see [3] pp. 183-206.

2 A Model for the General Methodology of *Vedāṅga*

2.1 Phenomenon, Description and Its Model

The general effort which is evident in the ancillary texts is towards observing and describing a set of some given phenomena. An instance can be the collection of utterances (*saṃhitā-pāṭha*) in the ŚUKLA-YAJURVEDA or the collection of correct and standard Sanskrit expressions. Yet another example may be even the rituals performed during a vedic sacrifice. Let us say, we represent this set of given phenomena by the symbol Ω. The object set Ω, therefore, is the collection of given phenomena which is to be analysed and described.

$$\Omega : \text{a set of given phenomena} \tag{1}$$

Let us now represent by the symbol Δ the description of the object set Ω. This is what the different ancillary disciplines provide.

$$\Delta : \text{a description of } \Omega \tag{2}$$

We can now specify the relationship between the object set (Ω) and the description (Δ) as follows:

$$\Omega \approx \Delta \tag{3}$$

This description Δ represents the works like the CHANDAḤŚĀSTRA of Piṅgala, the AṢṬĀDHYĀYĪ of Pāṇini or the VĀJASANEYI-PRĀTIŚĀKHYA. A model $\mathcal{M}$ for this description Δ is what we seek. Let the symbol $\mathcal{M}_\Delta$ represent this model.

$$\mathcal{M}_\Delta : \text{a model of } \Delta \tag{4}$$

We specify this relationship as follows:

$$\Delta \simeq \mathcal{M}_\Delta \tag{5}$$

Now let us say that Δ_p represents the descriptions as mentioned in the VĀJASANEYI-PRĀTIŚĀKHYA. Then $\mathcal{M}_{\Delta_p}$ is the corresponding model of this description. Similarly, if Δ_c represents the description which the CHANDAḤŚĀSTRA provide and Δ_a represents the description which the AṢṬĀDHYĀYĪ gives, then the corresponding models would be given by $\mathcal{M}_{\Delta_c}$ and $\mathcal{M}_{\Delta_a}$ respectively.

The hypothesis is that the models $\mathcal{M}_{\Delta_p}$, $\mathcal{M}_{\Delta_c}$ and $\mathcal{M}_{\Delta_a}$ can be generalized to a general model $\mathcal{M}_\Delta$ which corresponds to the core structure and processes common to all the descriptions.[11] In the following, we specify a general model

[11] Owing to the constraints of space, only the above mentioned three descriptions are dealt with. The hypothesis can include descriptions coming from *Śikṣā*, *Nirukta*, *Kalpa* and *Jyotiṣa* texts as well. Apart from the linguistic aspects which these texts contain, the other facets, e.g. the performances of sacrificial rituals as contained in the GṚHYA-SŪTRA also exhibit a similar approach. The structural analysis of rituals is one of the prime foci of the ongoing special research area "The Dynamics of Rituals (*Ritualdynamik*)" at the University of Heidelberg (www.ritualdynamik.de). Ritual experts (Staal [25], Oppitz [19], Michaels [13] [14], Houben [10]) acknowledge the compositional character of rituals. Mishra [17] connects it with the grammatical process of AṢṬĀDHYĀYĪ.

of this methodological approach and then show that the descriptive techniques employed by the AṢṬĀDHYĀYĪ of Pāṇini can be modelled within this general framework.

2.2 Towards a General Model for the Ancillary Disciplines

First let us look into the description Δ. What does Δ look like? For that, we first consider the descriptions as provided by

Δ_p : the VĀJASANEYI-PRĀTIŚĀKHYA
Δ_c : the CHANDAḤŚĀSTRA of Piṅgala and
Δ_a : the AṢṬĀDHYĀYĪ of Pāṇini.

Three basic categories are evident here with which the descriptive technique works. These are:

U : the constituent units of a given phenomenon
A : the nature, characteristics or attributes of these constituent units
S : statements about constituent units, attributes and about the statements themselves.

Constituent Units. If we look into the corpus of these texts, a big collection of constituent units can be identified: speech-sounds or phonemes like *a*, *i*, *u* etc., morphemes like *bhū*, *l(a)(ṭ)*, *(ś)a(p)*, *ti(p)* etc. and words like *sarva*, *ananta* etc.

Attributes. The constituent units possess a number of attributes that reflect their physical, linguistic or structural peculiarities. We use the general term attribute to represent a broader category of designations (*saṃjñā*) or expressions that encode some information — be it the system internal or the system external — and associate it with the appropriate unit. For example, the technical term *guru* is an attribute carrying information that is gained from the distribution of sound units in a string. On the other hand, belongingness to a particular group of literature is an attribute that carries system external information.[12]

Statements. The *Vedāṅga* texts enunciate constituent units together with their attributes. Further, they correlates them. For this purpose, a number of statements are made. Given units and attributes, the statements tell us which attributes are associated with which units or how are two units connected with each other. Many of them tell us the consequences when certain units come closer or are present adjacent to each other. Some statements also inform us about the

[12] These external world information can be of different sorts: phonetic peculiarities, concepts, meanings, pragmatic considerations, some special occasion, opinions, social conventions, a particular group of texts, special geographical regions etc. See, for example *alaṃkhalvoḥ pratiṣedhayoḥ prācāṃ ktvā* (A. 3.4.018), *udīcāṃ māṅo vyatīhāre* (A. 3.4.019), *liṅāśiṣi* (A. 3.4.116), *chandasyubhayathā* (A. 3.4.117) and a number of such rules.

working of the system, the meta-level information for proper interpretation and application of other rules. By providing the combinational possibilities of units, statements constitute a connecting link between these units and the original or given linguistic expression.

The description Δ as provided in texts like AṢṬĀDHYĀYĪ can be modelled on the basis of these three categories.[13] We propose:

Proposition 1 (A General Model $\mathcal{M}_\Delta$ for Δ). *Given a description Δ of some object set Ω. A general model $\mathcal{M}_\Delta$ of this description consists of a triplet (U, A, S), where U is the set of constituent units, A the set of attributes and S the set of statements.*

$$\mathcal{M}_\Delta = (U, A, S) \tag{6}$$

To sum up, our hypothesis is as follows:

1. There is a methodological similarity in the approach of ancillary disciplines (*Vedāṅga*).
2. They start with observing a set of given phenomena (Ω).
3. They provide a description (Δ) of the given phenomena Ω.
4. This description is achieved by providing the constituent units U, assigning them attributes A and making a number of statements S. We represent this through the symbol $\mathcal{M}_\Delta$.
5. $\mathcal{M}_\Delta$ finally connects the constituent units to the original or the given phenomenon.

Symbolically, the methodological approach can be represented as:

$$\Omega \approx \Delta \simeq \mathcal{M}_\Delta = (U, A, S) \tag{7}$$

Before we make some observations on this hypothesis (in Sect. 3.1 and 4), a few examples are needed to elucidate the model and to indicate its feasibility.

2.3 Enunciating the Units, Specifying Attributes and Operations

The authors of *Vedāṅga* texts can be seen as constantly and consistently forming groups and sub-groups of units they enunciate. The purpose of group formation is to club together necessary constituents which undergo some similar operation. This way higher degree of generalization is possible. This is the core philosophy of *lāghava*: to reduce the load by specifying some statement for all the units which share the relevant characteristics and do not handle them individually.

Every element of a group shares a common characteristic: this is the characteristic that *it is a part of that group*. This characteristic can be expressed in terms of an attribute. The name of the subset is the attribute defined and each element of that subset is assigned that attribute. Grouping together related units is one way of assigning an attribute to them.

[13] For the purpose of computer modelling an appropriate data structure is required (see Sect. 3.3).

Example 1. Consider the subset *a(c)* formed in ŚIVASŪTRAs:[14]

a(c) = {*a*, *i*, *u*, *ṛ*, *ḷ*, *e*, *o*, *ai*, *au*}[15]

This is equivalent to the statement that the units {*a*, *i*, *u*, *ṛ*, *ḷ*, *e*, *o*, *ai*, *au*} are assigned the attribute *a(c)*.

a-a(c), *i-a(c)*, *u-a(c)* etc.

Every set of units can be seen as *unit-attribute pair* and conversely *unit-attribute pairs* can be seen as constituting a set. Looking from this perspective, the *pratyāhāra*, for example, are nothing but attributes.

Example 2. Let us now look at the enunciation of the basic speech-sounds in VĀJASANEYI-PRĀTIŚĀKHYA.[16] Here there is a very clear grouping of the listed speech-sounds in different sets like *varṇa*, *svara*, *sandhyakṣara*, *vyañjana*, *kavarga*, *sparśa* etc.

svara = {*a*, *ā*, *ā3*, *i*, *ī*, *ī3*, . . . , *au*, *au3*}
sandhyakṣara = {*e*, *e3*, *ai*, *ai3*, *o*, *o3*, *au*, *au3*}

This again can be seen as assigning the attribute *svara* or *sandhyakṣara* to each element of the corresponding set. In other words, the units like *a* get the attributes like *svara*. Symbolically, *a-svara*, *e*-{*svara*, *sandhyakṣara*}.

Example 3. Consider the rule A. 1.1.001: *vṛddhirādaic* [Assign *vṛddhi* to units having *ā(t)* or *ai(c)*].

This rule performs the operation of attaching an attribute to certain specific units. The condition is that the units to which this attribute can be attached must possess either the attribute *ā(t)* or *ai(c)*. Let us say, we have a unit u_i with either *ā(t)* or *ai(c)*:

u_i-{*ā(t)*, *ai(c)*}

Then the present rule stipulates that the attribute *vṛddhi* is to be assigned to that unit:

u_i-{*ā(t)*, *ai(c)*} $\Rightarrow$ u_i-{*vṛddhi*}[17]

The conditions are expressed here in terms of attributes and the operation performed is assignment of an attribute.

---

14 a i u (ṇ) | ṛ ḷ (k) | e o (ṅ) | ai au (c) | . . . h (l) | (Śivasūtra 1-14).

15 The sub-sets are formed using the *it*-markers and the rule A. 1.1.071: *ādirantyena sahetā* [The first sound and the last *it* sound includes the units occuring in between the sequence].

16 athāto varṇasamāmnāyaṃ vyākhyāsyāmaḥ | *tatra svarāḥ prathamam* | *a iti ā iti ā3 iti . . . ḹ3 iti* | *atha sandhyakṣarāṇi* | *e iti e3 iti ai iti ai3 iti o iti o3 iti au iti au3 iti* | *iti svarāḥ* | . . . *ete pañcaṣaṣṭivarṇā brahmarāśirātmavācaḥ* | (Vā. Prā. 8.1-25).

17 Notation: $A \Rightarrow B$: State B results from state A after the application of a rule.

Example 4. Let us take another example of attribute addition. The groupings in the DHĀTU-PĀṬHA can be seen as assigning the attributes like *bhvādi, curādi* etc. to respective units. The rule *bhūvādayo dhātavaḥ* (A. 1.3.001) assigns the attribute *dhātu* to certain units grouped in a collection. So a unit like *bhū* gets this attribute. The condition here is expressed not in terms of attributes but in terms of certain units like *bhū, ad(a), hu* etc. If there is a unit u_i which is identified as *bhū*, then that unit gets the attribute *dhātu*:

$$u_i\text{-}\{bh\bar{u}\} \Rightarrow u_i\text{-}\{dh\bar{a}tu\}$$

Example 5. The rule *halo'nantarāḥ saṃyogaḥ* (A. 1.1.007) assigns the attribute to two units sharing the attribute *h[a](l)* if they occur consecutively.[18] This condition is based on the phonetic characteristics as well as their positional (or structural) relatedness.

Example 6. The assignment of the attribute *samprasāraṇa* in the rule *igyaṇaḥ samprasāraṇam* (A. 1.1.045) is based not only upon the attribute *i(k)* but also upon the process of vocalization.[19] So, an operation or a statement can also be a condition for another operation or statement.

Example 7. Let us now take an example for combination of units. The rule *guptijkidbhyaḥ san* (A. 3.1.005) stipulates the combination of units *gup(a), tij(a)* and *kit(a)* with the unit *sa(n)*. Now, the units *gup(a), tij(a)* and *kit(a)* get the attribute *dhātu* by the rule *bhūvādayo dhātavaḥ* (A. 1.3.001). The rule *pratyayaḥ* (A. 3.1.001) assigns the attribute *pratyaya* to the unit *sa(n)*. Now, the rule *paraśca* (A. 3.1.002) says that those units, which are having *pratyaya* as attribute must be attached after the units having the attribute *dhātu*. In the light of all this information, it can be said that *sa(n)-pratyaya* is added after *gup(a)-dhātu* etc.

$$u_i\text{-}\{gup(a),\ tij(a),\ kit(a)\} \Rightarrow u_i \prec u_j\text{-}sa(n)^{20}$$
$$u_i\text{-}dh\bar{a}tu \prec u_j\text{-}pratyaya$$

Example 8. The statement *vartamāne laṭ* (A. 3.2.123) facilitates the addition of the unit *l(a)(ṭ)* after a unit having the attribute *dhātu*. The condition for introduction is told in terms of a semantic attribute *vartamāna*. It can be seen as a statement which says that the unit *l(a)(ṭ)* can be introduced after a *dhātu* if the semantic attribute *vartamāna* is present.[21]

These few examples are to illustrate the possibility to read these texts from the perspective of the model mentioned above. This framework needs to be

[18] Anantaraṃ saṃyogaḥ (Vā. Prā. 1.48).
[19] Vocalization is mentioned in the AṢṬĀDHYĀYĪ from *ṣyaṅaḥ samprasāraṇaṃ putrapatyostatpuruṣe* (A. 6.1.013) to *vibhāṣā pareḥ* (A. 6.1.044).
[20] Notation: $x \prec y$ represents 'y follows x'.
[21] The question as to how the semantic attributes like *vartamāna* be made available is an important aspect for computer implementation of the rules. See Sect. 3.4.

tested further for its applicability. One way of testing it is to develop a computer implementation of the same. Next, we mention the main features of the computer implementation of the general model and then its special instance in case of the Aṣṭādhyāyī.

3 Computer Implementation of $\mathcal{M}_{\Delta}$

In Sect. 2.2 we proposed a model $\mathcal{M}_{\Delta}$ for representing the description Δ of some phenomena Ω. The description Δ here denotes the texts like Aṣṭādhyāyī corresponding to the set Ω of linguistic expressions. This model $\mathcal{M}_{\Delta}$ consists of a set of constituent units U, attributes A which characterize them and a number of statements S which tell us which unit gets which attribute and how the units can be modified. All three — units, attributes and statements — can appear as a condition for some operation. Within the context of a particular description, e.g. the Aṣṭādhyāyī Δ_a, there are many conventions which make the description understandable, explicit and precise.

A computer implementation of this model has two concurrent goals: firstly, it serves as a tool to check the suitability of the proposed model for representing the ancillary disciplines; and secondly, it explores the possibility of rendering these texts in a programmed manner, thus underlying their structured composition as well as bringing to surface the structures of the objects of these description.

3.1 Some Pre-suppositions

There are certain assumptions regarding the nature of the grammatical system of Pāṇini that shape the overall architecture of its computer implementation. Without entering into the discussions associated with some of these pre-suppositions (sometimes spanning over more than two millenia) for the sake of presenting a clear picture of the architecture, I think it is appropriate to mention them.[22]

I see Aṣṭādhyāyī within the overall context of the nature of enquiry as contained in the ancillary disciplines (*Vedāṅga*) and the methodological approach adopted by them. There are striking similarities (see Sect. 1). The main objective of these disciplines is to retain and carry the given (vedic phenomena) and use it correctly. In the Paspaśāhnika, while enunciating the purpose of grammar, Vārtikakāra mentions preservation or retention to be the first goal.[23] This is possible only when an exact description of the essential characteristics of the original is composed. It is desired that this description be shorter than the original. Only then can it facilitate a feasible retention of the original. This is

[22] There is, for example, a difference of opinion as to how exactly the Pāṇinian process works: whether it is an automatic device to generate correct Sanskrit sentences, starting from meaning and providing at the end valid linguistic expressions or rather it is a reconstitutive or cyclical process beginning with provisional statements and resulting in perfected (*saṃskṛta*) expressions [9]. For a recent update in some of the important, but still unresolved issues, see [24].

[23] rakṣohāgamalaghvasaṃdehāḥ prayojanam (Pas. Vā. 2).

sought to be achieved by identifying those units which appear recurrently in the original. For reproducing the original, the units need to be combined. For this purpose rules are to be formulated. These rules of synthesis are the connecting threads which connect the description with the original. Securing this connection is of utmost importance. A description that is not properly related with the original is not authoritative. This, for example, is what the *Prātiśākhyas* seek to achieve: to connect the *pada-pāṭha* with the original *saṃhitā-pāṭha*. And this is what a grammarian does by formulating the rules of grammar: connecting the constituents with the original.

A grammarian, however, crystallizes two substantial new steps. Firstly, he opens the object set for expressions which do not necessarily and exclusively belong to vedic expressions.[24] Secondly, he increasingly employs non-real (*kālpanika*) constituent units which do not have any correspondence in the real or given world. Nāgeśabhaṭṭa in his VAIYĀKARAṆA-SIDDHĀNTA-PARAMA-LAGHU-MAÑJŪṢĀ remarks the imaginary nature of the constituent units that are improvised by the teachers and employed only within the grammatical system.[25] For example, units like *l(a)(ṭ)* or *jhi* are not to be found in the world, but are stipulated by the grammarian. This implies, now it is all the more incumbent to provide the connection with the original through the formulation of precise rules. Through his grammar Pāṇini, so to say, converts the imaginary constituent units into real utterances — a kind of *satyakriyā*.[26] Evidently, there is a direct proportionality between the number of imaginary units in a system and the number of rules as well as the complexity of their interdependence. The number of rules to handle the interconnections between abstract units increases with the increase in abstract units.

Why should new imaginary units be introduced at all? For the sake of a compact and concise formulation of the description. Each imaginary unit carries with it the potential to evolve into a much larger number of finished forms. Brevity (*lāghava*), therefore, is not only a feature of the meta-language of Pāṇini, but the main idea behind this style of description, namely, to represent an object set through a description set which is less voluminous than the original. This becomes a necessity as the object set expands immensily once opened up for non-vedic expressions as well.[27]

---

[24] Patañjali, right in the beginning of the PASPAŚĀHNIKA, mentions first the non-vedic expressions and then the vedic ones, which are sought to be instructed. *keṣāṃ śabdānām (anuśāsanam)? laukikānāṃ vaidikānāṃ ca | ... vaidikāḥ khalvapi* | (Pas. Bhā. 2-3).

[25] tatra prativākyaṃ saṃketagrahāsambhavād vākyānvākhyānasya laghūpāyena aśakyatvācca kalpanayā padāni pravibhajya pade prakṛtipratyayabhāgān pravibhajya kalpitābhyām anvayavyatirekābhyāṃ tattadarthavibhāgaṃ śāstramātraviṣayaṃ parikalpayantismācāryāḥ (V. L. [21] p. 7).

[26] Thieme suggested this to be the purpose of grammar, although his idea behind this suggestion was that grammar is a device having magical effect which 'works the truth' of the special nature of Sanskrit, *saṃskṛtasya saṃskṛtatvam* [26]. Houben differs [9], but Scharfe suggests that this direction deserves to be explored further [24] p.87.

[27] rakṣohāgamalaghvasaṃdehāḥ prayojanam | (Pas. Vā. 2). *laghvarthaṃ cādhyeyaṃ vyākaraṇam* | (Pas. Bhā. 20) *lāghavena śabdajñānamasya prayojanam* | (Pradīpa on Pas. Bhā. 20).

This also explains the *generative* potential inherent in the process of synthesis in the Aṣṭādhyāyī. This process of synthesis presupposes an analytical phase. The process of analysis itself is not recorded in the grammatical corpus, but only its results [5]. The process of synthesis from constituent units to the final linguistic expressions is provided in terms of a number of rules allowing certain combinations of units and restricting others. It is this process which the following computer application aims to implement.

3.2 Architecture and Implementation

A brief sketch of the overall architecture of the computer implementation of the entire process is provided in [16] pp. 53-54. Two basic modules are mentioned there: an `Analyzer` and a `Synthesizer`. In the following, I mention a few improvements undertaken since then, especially within the `Synthesizer` module. These improvements are aimed towards incorporating the following features:

1. The modelling for the process of synthesis as implemented in the `Synthesizer` module now reflects the general model for representing the core processes of ancillary disciplines. It thus percieves Aṣṭādhyāyī in the general context of the methodology of *Vedāṅga*. This accounts better for the evolution and development of Aṣṭādhyāyī.
2. While the basic data structure is kept as in [15], a modification in the `ProcessStrips` is implemented to facilitate a better capability for checking conditions for applicability of operations.
3. A key improvement is now towards automatic application of rules. For this several strategies are incorporated (see below).
4. There is now a possibility to ask the user for information in case it is not available within the system.

We now provide a brief account of the main features of the computer implementation of the structure and processes of Aṣṭādhyāyī.

3.3 Basic Data Structure

Language Components and Sound Sets. The core data structure is designed with a view to facilitate the execution of grammatical operations. At any stage of processing, there is string of speech-sounds which gets modified. Certain speech-sounds get added and others get replaced or deleted. This changing string of speech-sounds is represented through a `LanguageComponent` which is a list of sets.

$$\texttt{LanguageComponent} = [\{\}, \{\}, \{\}, \ldots] \tag{8}$$

Each set of this list (called `SoundSet`) represents a speech-sound or a phoneme. It contains attributes which this speech-sound possesses. Moreover, it also contains attributes of units to which it is a part.

$$bh\bar{u} = [\{bh, dh\bar{a}tu, \ldots\}, \{u, a(c), d\bar{i}rgha, dh\bar{a}tu, \ldots\}] \tag{9}$$

Fig. 1. Sound Sets and Language Components

The `LanguageComponent` represents linguistic units at all levels: phonemes, morphemes, words, compounds, sentences etc. On the one hand there is no rigid boundary between two units, allowing them to conjoin at ease, and on the other hand, individual units can be identified on the basis of the characterizing attributes. Another advantage is that it allows, for example, changes at the level of a phoneme, conditioned upon semantic or syntactic attributes. In fact, there is no categorization in terms of phonology, morphology, syntax and semantics; there are only constituent units and attributes. Attributes reflect these characters, e.g. *a(c)* is an attribute reflecting phonetic characteristics, *dhātu* a morphological one, *apādāna* a syntactic one and *vartamāna* has semantic nature. But in the `LanguageComponent` they all occur as an attribute of the corresponding units. The units are, so to say, marked up with the attributes and the appropriate sub-string which corresponds to some attribute can be obtained by performing simple set intersections. For example, consider the following `LanguageComponent` at some stage of derivation:

$$[\{bh, bh\bar{u}, dh\bar{a}tu, \ldots\}, \{u, bh\bar{u}, a(c), d\bar{\imath}rgha, dh\bar{a}tu, \ldots\}] \tag{10}$$

Here, the vowel phoneme can be obtained by intersecting the sound sets with the set $\{a(c)\}$ and the morpheme denoting a verbal root can be gained by noting the `SoundSets` which intersect with the set $\{dh\bar{a}tu\}$.

Basic Operations. An operation brings about some change in the `LanguageComponent`. This can be of two types: either an attribute is added to one or more `SoundSets` or the `LanguageComponent` gets extended at some index. The operations of AṢṬĀDHYĀYĪ can be implemented on the basis of these two fundamental operations.

Representing operational stages. The basic data structures and operations reflect the state of the `LanguageComponent` at a particular stage in the process of synthesis. This is not enough for the application of many rules, where the conditions include information like how the current rule being applied is related to the rule applied in the previous step. In other words, the successful advancement of the process is not just dependent upon a particular stage but sometimes on previous stages as well. The system must, therefore, be aware of not just a particular stage, but the entire process itself. The entire process is now modelled as a chain of `Slices` with each `Slice` representing a given stage in the process of derivation.[28]

[28] This and the following features are new improvements from the previous versions mentioned in [15] and [16].

Slices correspond to those operational statements which bring about some modification in the `LanguageComponent`. It consists of an `InitialLanguageComponent` and a `FinalLanguageComponent`; the latter one is obtained after the application of an operational rule.

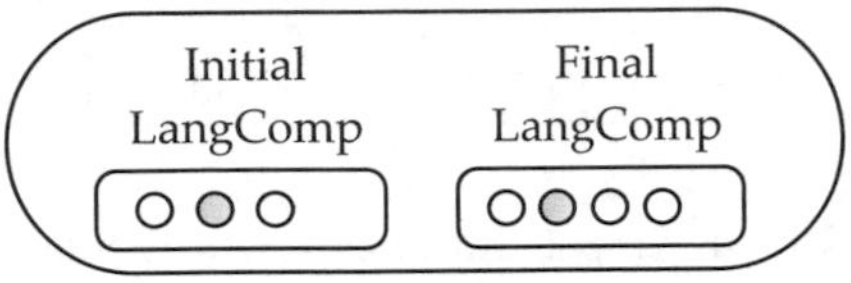

Fig. 2. A Slice

ConditionCodes is a dictionary which stores the necessary information about the specific operation which is executed at a particular stage. The information is gathered by applying the `conditionsCheck` functions. These functions check the applicability of an operation at a given stage. In case the function is applicable, they set the corresponding parameters. `ConditionCodes` associated with each `Slice` make it possible to trace back the operational steps, thus allowing e.g. the application of rules from *tripādī* section.

Ancillary Strip. There is an `AncillaryStrip` now which is the place to assign *static* attributes to a unit which is to be added. Static attributes are those attributes which are inherently associated with a unit and do not change once assigned, e.g. the attribute *a(c)* or *h[a](l)*.[29] From the point of view of data structure, it is similar to the `ProcessStrips`.

Process Strip now is a list of `Slices`. This data structure reflects the entire process of synthesis. During the process of synthesis `ProcessStrip` gets extended, one `Slice` is added at each operational step. We now present a trace of the important procedural steps.

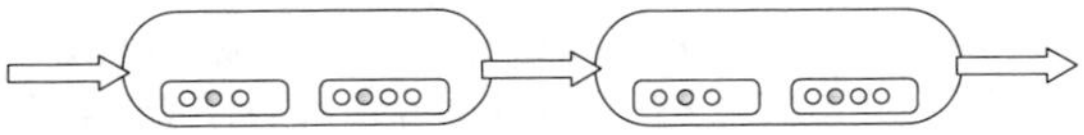

Fig. 3. Ancillary Strip / Process Strip

3.4 Procedural Steps

The process of synthesis is carried on the `ProcessStrip` which carries with it the necessary information about the current and previous stages of development.

[29] For more information, see [16].

Step 1. The first step is to look at the last `Slice` of the given `ProcessStrip`, take the `FinalLanguageComponent` and initialize the next `Slice` by adding it as the `InitialLanguageComponent`.

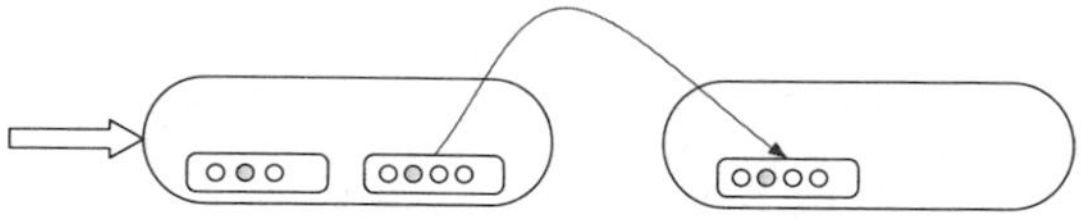

Fig. 4. Step 1: Initializing a new Slice

Step 2. The `ProcessStrip` is sent to `ConditionsCheck` module. At this stage conditions are verified as to which rules can be applied at the current status of the `ProcessStrip`. It results in a set of `ConditionCodes` corresponding to applicable rules. For the sake of checking the conditions, certain *dynamic* attributes (e.g. *saṃyoga*, *ṭi* etc.) are assigned if required.

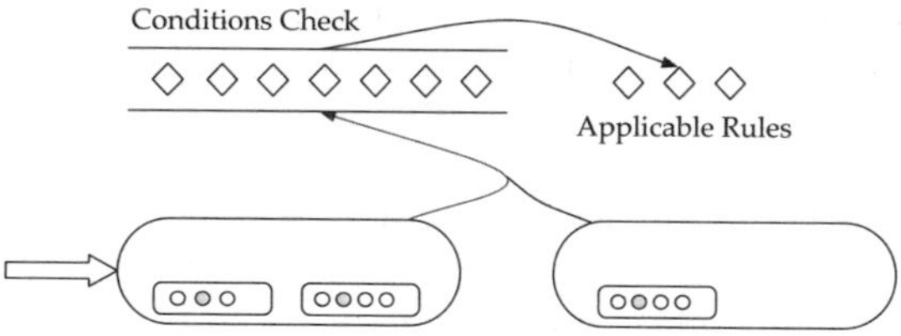

Fig. 5. Step 2: Condition Check

Step 3. The next major step is to select one rule from among the set of applicable rules. This is done in a `Selector` module. The main function of this module takes the set of applicable `ConditionCodes` and makes it a singelton set.

Here, there are different kinds of selection which are to be accounted for.

1. Selection on the basis of semantic conditions / attributes: Even if there is only one rule available for application, there is sometimes a need to select the right unit. For example the rule: *tip tas jhi sip thas tha mip vas mas ta ātām jha thās āthām dhvam iṭ vahi mahiṅ* (A. 3.4.078). If the condition is fulfilled (e.g. the presence of a *lakāra*) still a selection process is required. This is facilitated by rules like *tiṅas trīṇi trīṇi prathama madhyama uttamāḥ* (A. 1.4.101) and *tāni ekavacana dvivacana bahuvacanāni ekaśaḥ* (A. 1.4.102). The attributes *prathama*, *ekavacana* etc. are needed to decide with the help of these rules, which suffix is to be chosen. From an operational perspective, it is not enough to define *prathama*, *ekavacana* etc. It is required to know whether this attribute is acceptable to the `ProcessStrip`. There are two ways to determine this: (i) either from the `analyzedSet` gained from the `Analyzer` module (see [16]) or (ii) from the user.

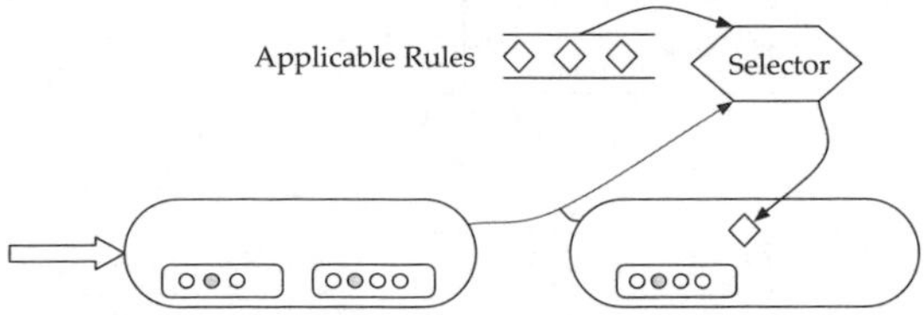

Fig. 6. Step 3: Selecting one rule from the set of applicable rules

2. Selection when two applicable operations have two completely unrelated *sthānins*: Here it is required to select between two operations, which are both applicable at a given stage, but have completely unrelated place of application within the `ProcessStrip`. This can happen e.g. in cases when more than one word is to be processed simultaneously in a linguistic expression. This issue needs more attention in the future development of the system.
3. Conflict Resolution: This is the case when given two applicable rules r_1 and r_2 a definite order, say first r_1 then r_2, must be followed. Otherwise, it leads to wrong results. Here we use primarily the *siddha* principle [7], [8], which involves copying the current `ProcessStrip`, applying both the rules and checking whether one destroys the *nimitta* for the other (see [16]).
4. There is a mechanism to test the mutual rule ordering between two rules which come closer during the process of synthesis. It is effected by noticing the mutual order in their `ConditionCodes`. This may later result in a kind of rule dependency graph of AṢṬĀDHYĀYĪ.

The selected `ConditionCode` becomes now a part of the `Slice`.

Step 4. The entire `ProcessStrip` now enters the `Operations` module where, depending upon the values in `ConditionCode`, an operation is performed. It modifies the `LanguageComponent`, adds it as the `FinalLanguageComponent` to the `Slice` and returns the `ProcessStrip`.

Modification of the `LanguageComponent` is undertaken in the following steps: If a new unit is to be added, then it is passed through a sub-module named `StaticAttributes`. Here a unit is rendered in the form of a `LanguageComponent` and one after another the rules for addition of *static* attributes (like *a(c)*, *h[a](l)* etc.) are assigned. In this module the order of applicability is linear and the attributes which are used as condition for other rules are assigned before that rule. There is no conflict resolution here as it is not needed. Not all the attributes but only the *static* ones are assigned. This is to improve the performance of the system. `ProcessStrip` is extended as long as there is some rule which is available for application. During the applicability check, it is taken care of that the rules do not get applied infinitely. This is achieved in that the condition is so formulated that the application of some rule makes its further application at the same location (*sthānin*) on same conditions (*nimitta*) impossible.

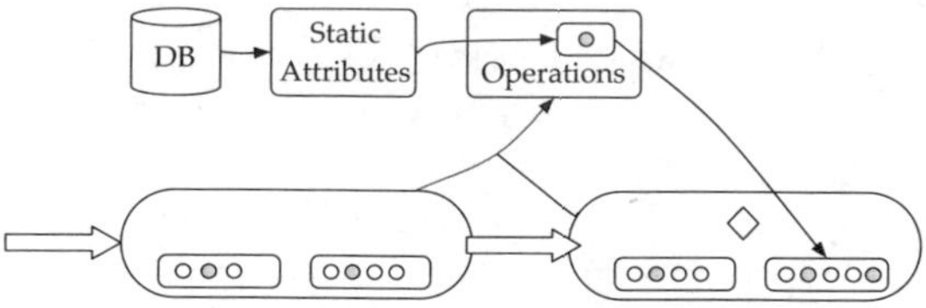

Fig. 7. Step 4: Applying the selected rule

For example, once *l(a)(ṭ)* is introduced after a *dhātu*, it is not introduced again. So the condition check includes checking for the already presence of *l(a)(ṭ)*.

4 Some Observations

The model presented above and the corresponding implementational design is aimed to facilitate the representation of the content, structure and operations of the methodological enquiry of ancillary disciplines (*Vedāṅga*) in general and the AṢṬĀDHYĀYĪ of Pāṇini in particular.

The entire framework is based on a minimal set of categories (*Units*, *Attributes*, *Statements*). The scheme of the general architecture and the core data structures is such as to allow to represent the functionalities of interpretative procedures (e.g. *paribhāṣās*, *siddha*-principle [7]) and additional amendments (e.g. *Vārtika*).

The system is not guided by any fixed external organization, for example, the division of the grammatical corpus in *śiva-sūtra*, *sūtra-pāṭha*, *dhātu-pāṭha*, *gaṇa-pāṭha*, *uṇādi-sūtra* etc. or any reorganization of these under categories of modern linguistics (see e.g. [23] p. 97), but focuses on the core methodology. There is, however, a possibility to make the system aware of these divisions. The approach is to identify the constituent units and assign a number of attributes to them and then on the basis of these represent the external structures.

Table 1. Abbreviations

A.	AṢṬĀDHYĀYĪ
Uv.	UVAṬA-BHĀṢYA on ṚGVEDA-PRĀTIŚĀKHYA
Cha. Śā.	CHANDA-ŚĀSTRA of Piṅgala
Ni.	NIRUKTA
Pas.	PASPAŚĀHNIKA
Pas. Bā.	PASPAŚĀHNIKA BHĀṢYA
Pas. Vā.	PASPAŚĀHNIKA VĀRTIKA
Mu.	MUṆḌAKOPANIṢAD
Mu. Bhā.	MUṆḌAKOPANIṢAD ŚĀṄKARA-BHĀṢYA
V. L.	VAIYĀKARAṆA-SIDDHĀNTA-PARAMA-LAGHU-MAÑJŪṢĀ
Vā. Prā.	VĀJASANEYI-PRĀTIŚĀKHYA
Śu. Yaj.	ŚUKLA-YAJURVEDA

The system allows for the representation of the content, structures and processes of other ancillary disciplines (e.g. *Śikṣā*, *Chandas*, *Nirukta*, *Prātiśākhya* etc.) thus facilitating a more comprehensive and comparative study of these sciences. The possibility of system external (human) input during the process of synthesis is now taken into account.

References

1. Bhise, U.R.: Nāradīyā Śikṣā with the Commentary of Bhaṭṭa Śobhākara. Bhandarkar Oriental Research Institute, Poona (1986)
2. von Böhtlingk, O.: Pāṇini's Grammatik. Olms, Hildesheim (1887)
3. Bronkhorst, J.: Greater Magadha. Studies in the Culture of Early. Brill, Leiden (2007)
4. Dhūpakara, A.Ś.: Śrīpiṅgalanāga-viracitaṃ Chandaḥśāstram. Parimal Publications, Delhi (1985)
5. Deshpande, M.M.: Semantics of Kārakas in Pāṇini: An Exploration of Philosophical and Linguistical Issues. In: Matilal, B.K., Bilimoria, P. (eds.) Sanskrit and Related Studies: Contemporary Researches and Reflections, pp. 33–57. Sri Satguru Publications, Delhi (1990)
6. Gladigow, B.: Sequenzierung von Riten und die Ordnung der Rituale. In: Stausberg, M. (ed.) Zorostrian Rituals in Context. Numen Book Series, Studies in the History of Religions, vol. CII, pp. 57–76. Brill, Leiden (1999)
7. Joshi, S.D., Roodbergen, J.A.F.: On siddha, asiddha and sthānivat. Annals of the Bhandarkar Oriental Research Institute, Bhandarkar Oriental Research Institute, vol. LXVIII, pp. 541–549. Bhandarkar Oriental Research Institute, Poona (1987)
8. Joshi, S.D., Roodbergen, J.A.F.: The Aṣṭādhyāyī of Pāṇini. With Translation and Explanatory Notes, vol. II. Sahitya Akademi, New Delhi (1993)
9. Houben, J.E.M.: 'Meaning statements' in Pāṇini's grammar: on the purpose and context of the Aṣṭādhyāyī. Studien zur Indologie und Iranistik 22, 23–54 (1999-2001)
10. Houben, J.E.M.: Memetics of Vedic Ritual, Morphology of the Agniṣṭoma. In: Griffiths, A., Houben, J.E.M. (eds.) The Vedas: Texts, Language & Ritual, pp. 23–47. Forsten, Groningen (2004)
11. Katre, S.M.: Aṣṭādhyāyī of Pāṇini. Motilal Banarsidass, Delhi (1989)
12. Lüders, H.: Die Vyâsa-Çikshâ besonders in ihrem Verhältnis zum Taittirîya-Prâtiçâkhya. Göttingen (1894)
13. Michaels, A.: 'Le rituel pour le rituel' oder wie sinnlos sind Rituale? In: Caduff, C., Pfaff-Czarnecka, J. (eds.) Rituale heute: Theorien - Kontroversen - Entwürfe, pp. 23–47. Reimer, Berlin (1999)
14. Michaels, A.: The Grammar of Rituals. In: Michaels, A., Mishra, A. (eds.) Grammar and Morphology of Ritual, pp. 15–36. Harrassowitz Verlag, Wiesbaden (2010)
15. Mishra, A.: Simulating the Pāṇinian System of Sanskrit Grammar. In: Huet, G., Kulkarni, A., Scharf, P. (eds.) Sanskrit Computational Linguistics. LNCS (LNAI), vol. 5406, pp. 127–138. Springer, Heidelberg (2009)
16. Mishra, A.: Modelling the Grammatical Circle of the Pāṇinian System. In: Huet, G., Kulkarni, A., Scharf, P. (eds.) Sanskrit Computational Linguistics 2007/2008. LNCS (LNAI), vol. 5402, pp. 40–55. Springer, Heidelberg (2009)
17. Mishra, A.: On the Possibilities of a Pāṇinian Paradigm for a Rule-based Description of Rituals. In: Michaels, A., Mishra, A. (eds.) Grammar and Morphology of Ritual, pp. 91–104. Harrassowitz Verlag, Wiesbaden (2010)

18. Lakshman, S.: The Nighaṇṭu and The Nirukta. Motilal Banarsidass, Delhi (1967)
19. Oppitz, M.: Montageplan von Ritualen. In: Caduff, C., Pfaff-Czarnecka, J. (eds.) Rituale heute: Theorien - Kontroversen - Entwürfe, pp. 73–95. Reimer, Berlin (1999)
20. Sarup, L.: The Nighaṇṭu and the Nirukta. Motilal Banarasidass, Delhi (1967)
21. Shastri, K.: Nāgeśabhaṭṭa-kṛta Vaiyākaraṇa-siddhānta-parama-laghu-mañjūṣā. Kurukshetra University Press, Kurukshetra (1975)
22. Shastri, M.D.: The Ṛgveda-Prātiśākhya with the commentary of Uvaṭa. Vaidika Svadhyaya Mandir, Varanasi (1959)
23. Scharf, P.M.: Modeling Pāṇinian Grammar. In: Huet, G., Kulkarni, A., Scharf, P. (eds.) Sanskrit Computational Linguistics 2007/2008. LNCS (LNAI), vol. 5402, pp. 95–126. Springer, Heidelberg (2009)
24. Scharfe, H.: A new perspective on Pāṇini. In: Indologica Taurinensia, vol. XII, pp. 1–273 (2009)
25. Staal, F.: Rules without meaning: ritual, mantras and the human sciences. Lang, New York (1989)
26. Thieme, P.: Meaning and form of the 'grammar' of Pāṇini. Studien zur Indologie und Iranistik 8/9, 12–22 (1982)
27. Varma, V.K. (ed.): Vājasaneyi-Prātiśākhyam. Chaukhamba Sanskrit Pratishthan, Delhi (1987)
28. Varma, V.K.: Ṛgveda-Prātiśākhyam. Banaras Hindu University Sanskrit Series, Varanasi (1972)
29. Wezler, A.: On the quadruple division of the Yogaśāstra, the caturvyūhatva of the Cikitsāśāstra and the 'four noble truths' of the Buddha. In: Indologica Taurinensia, vol. XII, pp. 289–337 (1984)

Author Index

Printing: Mercedes-Druck, Berlin
Binding: Stein+Lehmann, Berlin